KNOWLEDGE ENGINEERING

Volume I FUNDAMENTALS

KNOWLEDGE ENGINEERING

Volume I   FUNDAMENTALS

Hojjat Adeli, Editor
*Department of Civil Engineering
The Ohio State University*

McGraw-Hill Publishing Company

New York St. Louis San Francisco Auckland Bogotá Caracas
Hamburg Lisbon London Madrid Mexico Milan Montreal New Delhi
Oklahoma City Paris San Juan São Paulo Singapore Sydney Tokyo Toronto

This book was set in Times Roman by the College Composition Unit
in cooperation with Ruttle, Shaw & Wetherill, Inc.
The editors were B. J. Clark and John M. Morriss;
the production supervisor was Leroy A. Young.
The cover was designed by Carla Bauer.
Project supervision was done by The Total Book.
R. R. Donnelley & Sons Company was printer and binder.

KNOWLEDGE ENGINEERING

Vol. I, Fundamentals

Copyright © 1990 by McGraw-Hill, Inc. All rights reserved. Printed in the
United States of America. Except as permitted under the United States Copyright Act
of 1976, no part of this publication may be reproduced or distributed in any form or
by any means, or stored in a data base or retrieval system, without the prior written
permission of the publisher.

3 4 5 6 7 8 9 0 DOC DOC 9 5 4 3 2 1 0

ISBN 0-07-000355-6

Library of Congress Cataloging-in-Publication Data

Knowledge engineering / [edited by] Hojjat Adeli.
   p.   cm.
  Includes index.
  Contents: v. 1. Fundamentals—v. 2. Applications.
  ISBN 0-07-000355-6 (v. 1)—ISBN 0-07-000357-2 (v.2)
  1. Expert systems (Computer science)   I. Adeli, Hojjat, (date).
QA76.76.E95K577   1990
006.3'3—dc20   89-8338

For information about our audio products, write us at:
Newbridge Book Clubs, 3000 Cindel Drive, Delran, NJ 08370

EDITOR'S BIOGRAPHY

Currently professor of civil engineering at The Ohio State University, Hojjat Adeli received his Ph.D. from Stanford University in 1976. He is the editor-in-chief of the international journal, *Microcomputers in Civil Engineering*, and the author or editor of nearly 200 research publications including several books in the fields of knowledge engineering and expert systems, computer-aided design, parallel processing, mathematical optimization and simulation, applied mechanics, and structural engineering. He is also the editor-in-chief of the forthcoming Marcel Dekker book series *New Generation Computing*. The first two volumes of the series, *Supercomputing in Engineering Analysis* and *Parallel Processing in Computational Mechanics* are scheduled for publication in late 1990. He is listed in twelve *Who's Who*'s and biographical listings including *Who's Who in the World, Men of Achievement,* and *International Directory of Distinguished Leadership.*

CONTENTS

	Contributors	ix
	Preface	xiii
1	**Representation of Knowledge** W. Bibel, J. Schneeberger, and E. Elver	1
2	**Rule-Based Expert Systems** F. Golshani	28
3	**Knowledge Acquisition Systems** B. R. Gaines	52
4	**Model-Based Knowledge Acquisition** M. T. Harandi and R. Lange	103
5	**Models of Expertise in Knowledge Engineering** P. E. Slatter	130
6	**AI Planning** M. Drummond and A. Tate	157
7	**Knowledge in the Form of Patterns and Neural Network Computing** Y. H. Pao	200
8	**Machine Learning** Y. Kodratoff	226
9	**Propositional Logic** R. R. Yager	256

10 **Natural Language Processing: Computer Comprehension of Editorial Text**
S. J. Alvarado, M. G. Dyer, and M. Flowers 286

Index 345

CONTRIBUTORS

Sergio J. Alvarado received his Ph.D. in computer science from the University of California, Los Angeles, in 1989. He is currently an assistant professor in computer science and director of the Artificial Intelligence Laboratory at the University of California, Davis.

Wolfgang L. Bibel is a professor of computer science at the University of British Columbia and fellow of the Canadian Institute for Advanced Research. He received his Ph.D. from the University of Munich, West Germany, in 1968. Between 1969 and 1987 he worked at the Technical University of Munich. His nearly 100 publications range over various areas in artificial intelligence such as automated deduction, program synthesis, and knowledge representation. Dr. Bibel is section editor of the *Artificial Intelligence Journal* and the associate editor of the *Journal for Symbolic Computations*. He chaired the International Joint Conference on Artificial Intelligence in 1989.

Mark Drummond received his Ph.D. in artificial intelligence from the University of Edinburgh in 1986. He subsequently worked as a postdoctoral fellow at the University of Edinburgh's AI Applications Institute. He is currently a scientist at the NASA Ames Research Center, Moffett Field, California. His current research interests include the relationship between planning, scheduling, and execution; real-time reactive scheduling in light of execution failures; and the role of dependency information in error management. Of particular interest is the Mars Rover Sample Return Mission.

Michael G. Dyer received his Ph.D. in computer science from Yale University in 1982. He is currently an associate professor of computer science and director of the Artificial Intelligence Laboratory at the University of California, Los Angeles. He is the author of *In Depth Understanding* (MIT Press, 1982), and has authored over 60 articles in books, journals, and conference proceedings, covering numerous areas of cognitive science, including computational linguistics, legal reasoning, modeling of emotions, and artificial neural networks. Dr. Dyer is on the editorial advisory board of the journals, *Expert Systems: Research and Applications, Knowledge-Based Systems,* and *Connection Science.*

Erdal Elver studied informatik at the Technical University of Munich from 1981 to 1986. Since 1986 he has been with Siemens AG in Munich where he is now the project leader of the PRINCESS group and works on the development of expert system shells. His special interests are expert systems and knowledge representation.

Margot Flowers did graduate work in computer science at Yale University and is currently an adjunct assistant professor in the computer science department at the University of California, Los Angeles. She has authored over 20 publications in natural language processing, connectionist modeling, legal reasoning, mechanical device comprehension, and scientific reasoning.

Brian R. Gaines is Killam Memorial Research Professor and director of the Knowledge Science Institute at the University of Calgary. He is also professor of computer science and psychology, and director of the Software Research and Development Group at the University of Calgary. His previous positions include professor of industrial engineering at the University of Toronto, technical director and deputy chair of the Monotype Corporation, and chair of the department of electrical engineering and science at the University of Essex. He received his B.S., M.A., and Ph.D. from Trinity College, Cambridge, and is a chartered engineer, and a fellow of the Institution of Electrical Engineers, the British Computer Society, and the British Psychological Society. He is editor of the *International Journal of Man-Machine Studies* and *Future Computing Systems,* as well as the *Computers and People* and *Knowledge-Based Systems* book series. He has authored over 250 papers and authored or edited 5 books in various areas of computer and human systems. His research interests include the socioeconomic dynamics of science and technology; the nature, acquisition, and transfer of knowledge; software engineering for heterogeneous systems; and expert system applications in manufacturing, sciences, and humanities.

Forouzan Golshani received his Ph.D. from Warwick University, United Kingdom, in 1982. He joined Arizona State University as an assistant professor of computer science in 1984. He has written over 40 publications in the areas of expert systems, parallel architectures, software engineering, and database management.

Mehdi T. Harandi obtained his M.Sc. and Ph.D. from the University of Manchester, England, in 1976 and 1979, respectively. In 1981 he joined the department of computer science at the University of Illinois, Urbana-

Champaign, where he is currently an associate professor of computer science. He is the editor-in-chief of the *International Journal of Expert Systems: Research and Applications*. He was the program chair of the 1987 International Workshop of Software Specification and Design. His current research interests are in the areas of software specification and design, knowledge bases and their application to software development, and distributed expert systems.

Yves Kodratoff is presently director of research at the French National center for Scientific Research. He has done extensive research in the area of machine learning and published a book on the subject recently. He was the program chair of the 1988 European Conference on Artificial Intelligence. His current research interests are program synthesis from specifications and its application to software engineering, and various aspects of machine learning, including the coupling of numeric and symbolic techniques, developing abductive and inductive tools working in the presence of a large knowledge base, and machine discovery and analogy.

Rense Lange received his Ph.D. in psychology from the University of Illinois, Urbana-Champaign in 1981. He is now working toward the completion of a Ph.D. thesis in the area of artificial intelligence at the University of Illinois. He has published in the areas of cognitive processing, problem solving, decision making, and machine learning. His research in expert systems focuses on knowledge representation and knowledge acquisition.

Yoh-Han Pao is professor of electrical engineering and computer science and director of the Center for Automation and Intelligent Systems Research at Case Western Reserve University. He was the chair of Case Western's electrical engineering department during 1969–1977 and director of the National Science Foundation Division of Electrical, Computer, and Systems Engineering during 1978–1980. Professor Pao has received numerous awards including the U.S. Emblem of Merit for Distinguished Civilian Service in 1945 and the Outstanding Educator of America award in 1972 and 1973. He was the technical editor of the *IEEE Journal of Robotics and Industrial Automation*. He is a fellow of the IEEE and the Optical Society of America. His current interests are expert systems, pattern recognition, distributed processing, and neural nets.

Josef Schneeberger studied informatik at the Technical Universities of Munich and Darmstadt. From 1985 to 1988 he was a member and leader (for one year) of the Artificial Intelligence Group of the Technical University of Munich. He is currently on the faculty of the Technical University of Darmstadt.

Philip E. Slatter is a knowledge engineer with Telecomputing plc in Oxford, England, where he has worked on large-scale expert system applications since 1985. Between 1980 and 1985 he was an analyst/programmer with the British

Steel Corporation. His academic qualifications include a B.Sc. in psychology, an M. Phil. for research into cognitive aspects of expert systems, and a Ph.D. in applied cognitive psychology. He is the author of a recent book entitled *Building Expert Systems: Cognitive Emulation* (Ellis Horwood, 1987).

Austin Tate is director of the Artificial Intelligence Applications Institute at the University of Edinburgh. Dr. Tate's research interests relate primarily to knowledge-based planning systems, novel database architectures, and knowledge representations systems. He is a member of the steering group of the United Kingdom Knowledge-Based Systems Research Club and the coordinator for its Planning Special Interest Group. His is on the editorial board of several journals and technical book series.

Ronald R. Yager is currently director of the Machine Intelligence Institute and professor of information systems at Iona College. He received his Ph.D. from the Polytechnic Institute of New York in 1968. During 1983–1984 he served at the U.S. National Science Foundation as program director in the Information Sciences program. He is the editor-in-chief of the *International Journal of Intelligent Systems*. He also serves on the editorial board of a number of other journals. He has published close to 200 articles. His current research interests include multicriteria decision making, uncertainty management in knowledge-based systems, fuzzy set theory, neural modeling, and visual databases.

PREFACE

The first volume of *Knowledge Engineering* presents state-of-the-art reviews and tutorials on fundamental aspects of knowledge engineering. The second volume complements the first by presenting applications of applied artificial intelligence (AI). The field of applied AI and knowledge engineering is very young. Students usually must refer to numerous sources to learn the fundamentals of the subject. The two volumes attempt to present summaries of the various subjects in a single document and are oriented toward practical applications. They are suitable as primary reference books in introductory courses on applied AI and knowledge engineering.

Leading and internationally recognized researchers have contributed to these volumes. We hope this effort becomes a continuing book series with future volumes concentrating on other aspects of knowledge engineering and new applications of AI.

Hojjat Adeli
Editor

KNOWLEDGE ENGINEERING

Volume I   FUNDAMENTALS

CHAPTER 1

REPRESENTATION OF KNOWLEDGE

W. BIBEL
J. SCHNEEBERGER
E. ELVER

1 INTRODUCTION

Engineers of all sorts are increasingly interested in including "knowledge" among the materials from which they construct their artifacts. Whereas traditional materials such as metal, wood, or plastics are tangible, knowledge is an extremely evasive stuff. Thus the need to "materialize" or *represent* knowledge arises in the first place. This chapter introduces some of the techniques that have been developed to allow for the *representation of knowledge,* especially within a machine, and discusses the major problems in such an endeavor.

As a research area in its own right, knowledge representation evolved within the field of artificial intelligence (AI), where it continues to play a central role. This should not come as a surprise since everyone would probably agree that intelligence has a lot to do with how knowledge is being handled in the human mind and, hence, in any kind of intelligent device. From this AI perspective the area started some thirty years ago.

From a more general perspective, knowledge representation has a much longer tradition rooted mainly in philosophy, logic, and psychology. In fact, both notions "knowledge" and "representation" touch on deep philosophical issues that cannot totally be ignored even from an engineer's point of view. What is knowledge? Although our presentation naturally emphasizes the more technical aspects of knowledge representation, we will nevertheless give a brief account of some of these philosophical issues in the subsequent section.

Even in its more technical aspects, knowledge representation has turned out to be a difficult area that does not seem to lend itself to easy solutions of its main problems in the near future. The reader must therefore not expect the presentation of a few simple recipes to be observed in building the next expert system, robot device, or what have you. While researchers in this area begin to agree on what the fundamental problems are, their proposed solutions do not converge yet. For instance, there is still a vast amount to be understood before the knowledge represented in one system can be shared with another system, something that people apparently can easily achieve.

Another major difficulty in presenting this area consists of the fact that it spans a wide spectrum of research topics. In consequence, to write a sort of tutorial about this field in a single chapter is obviously not an easy task. We have chosen to do this in the following way.

First, we have restricted the main discussion to a relatively small subarea of the entire field. In particular, we took knowledge representation in its literal sense to denote ways of just representing knowledge via some formalism or another. Usually much more is covered by this notion since it is impossible to strictly separate the issues of representing from those of processing knowledge. One influences the other in an intrinsic way. To remind the reader once again of this drastic restriction, in Section 5 we offer a few key words and pointers to the important and large remaining part of the field not covered here in detail. In addition, we encourage the reader to consult the literature cited or perhaps to even start reading a more general textbook on AI such as Charniak and McDermott (1985), to mention a recent one.

The core of this chapter consists of Sections 3 and 4. There we first survey some ten formalisms that have been used extensively in the literature and in knowledge-based systems. We use logic as a sort of canonical formalism that provides meaning to all others as well. We then point out the strong points of most of these formalisms relative to each other on the basis of rational arguments rather than personal taste. While the authors' bias toward logical formalisms will not be concealed, we hope to have collected convincing arguments for this preference.

Those readers with no taste at all for the philosophically flavored issues involved might well skip the subsequent section entirely.

2 KNOWLEDGE AND ITS REPRESENTATION

2.1 Basic Concepts

Like many other general concepts, the notion of *knowledge* is ill defined and thus has a different meaning for different people. Primarily it is a cognitive notion: what we "know" is anything that is at our disposal in the conscious mind. To some extent, knowledge can be communicated among people through some sort of medium such as spoken language. For this purpose knowledge is somehow represented in such a medium. That is, we establish a certain relationship between the represented knowledge and the cognitive phenomenon that constitutes this knowledge.

Consider the sentence "The gripper picks up the log" uttered by someone watching a scene of logging. The observer has in mind the knowledge of this scene (depicted as the middle box in Fig. 1-1) of which the sentence is a represented form (the left box in Fig. 1-1). The interpretation of the represented knowledge is the cognitive knowledge in the mind. The figure suggests that this interpretation is a function ι mapping the sentence to the cognitive knowledge.

The cognitive knowledge in this case arises from the observation of a scene in the external or real world that is depicted in the right-hand box of Fig. 1-1. We presume that cognitive knowledge is represented somehow in the brain which allows us to think of it as a represented form of the external-world scene. Hence we have the right arrow from left to right. By transitivity this yields an interpretation of the sentence in terms of the scene in the external world. Keep in mind that such an interpretation is not possible except by way of interaction with an intermediate mind, which plays the central role, as clearly demonstrated by the figure.

Although it may appear as an unusual view, the scene in the external world may well be considered as a representation of the cognitive knowledge as well. Hence we have the arrow from right to left as well. We will not dwell on the resulting one-to-one correspondence further which obviously begs many deep scientific and philosophical questions. Also we mention, as an aside, that for human beings in a sense there might be no "real" world (and thus no represented form of knowledge either) since their experiences take place exclusively in the realm of consciousness. Yet operating under the assumption of its existence clearly is a very successful "working hypothesis".

It is in such represented form only, as in the quoted sentence above, that knowledge becomes more generally accessible and in this way adopts additionally a figurative, noncognitive meaning. Although a lot of confusion may arise if one ignores the cognitive origin of the notion, the main interest from our point of view lies in this figurative meaning. When we refer to knowledge in this noncognitive sense, we also prefer to think of the interpretation ι as a function into the external world, or rather into some other representation of it; that is, ι then abbreviates the composition of the two functions illustrated in Fig. 1-1.

There are many *ways of representing* knowledge (as there are many *kinds* of knowledge). Writing text like this one provides one method of representation. We are particularly interested in ways that use a sufficiently precise no-

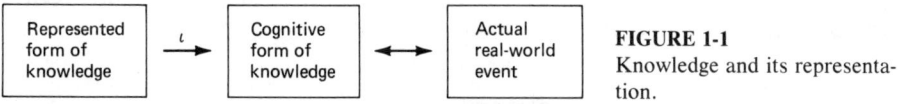

FIGURE 1-1 Knowledge and its representation.

tation that it can be used in, or by, a computer program. A systematic representational method satisfying this requirement is referred to as a *scheme*. A number o such schemes are discussed in the next section. Here we must mention (the language part of) logical calculuses as one type of example. A particular expression in such a scheme is called a *configuration,* such as a particular logical formula in a logical calculus. In other words, a scheme is a set of configurations.

To comply with the aforementioned requirement, we consider it intrinsic to the notion of a scheme that a computer may verify whether a particular arrangement of marks is a well-formed configuration, i.e., a definite notion of *well-formedness* is provided along with the configuration of a scheme. This requirement narrows the concept of a scheme to strictly formal ways of representation and, at least for the time being, eliminates many other ways which humans have of conveying meaning such as drawings, poems, conversational English, musical performances, and so forth. This is to say that we do not exclude altogether the possibility of formalizing also these ways in some approximation.

A second important requirement on a scheme is to provide for an associated *semantic theory* that assigns a meaning to any of its possible configurations. We want to be able to understand the represented knowledge after all. This amounts to capturing in a precise way the interpretation function ι which will be understood as a mapping into the external world, as described above.

The configurations of a scheme need to be constructed in some *medium,* for instance, as marks or symbols on a piece of paper. There are all sorts of such media. And in fact it is rather difficult to give a precise characterization of this notion and its properties, but it is less of a problem for the more practical purposes.

Further, the represented knowledge typically is not just sitting there, but first has to be imprinted into the chosen medium and may then play a number of different roles. The represented knowledge may be accessed, read, changed, linked, or what have you. In other words, a number of *operations* have to be taken into consideration that are to be applied to represented knowledge.

To summarize, the goal of a suitable representation of knowledge requires the consideration of the kind of knowledge to be represented, of the medium available for the representation, of the scheme to be used for the representation, of the meanings that the symbols used in a configuration are supposed to convey, and of the operations that are to be applied in the manipulation of the represented knowledge. This leaves us with a fairly large degree of freedom since quite a variety of choices are to be made here. Moreover, the kind of knowledge as well as the necessary operations may vary quite a bit under different circumstances. It is because of this variety that there exists no such thing as *the* knowledge representation formalism in artificial intelligence or *intellectics*.[1] Rather, there are quite a few competing ones.

---

[1] Our preferred name for the union of artificial intelligence and cognitive science.

2.2 Desiderata in Knowledge Representation

As mentioned in the previous section, there are a number of competing schemes in knowledge representation and hot debates on which of those should be considered the "right" one. As one might expect, at present we are *not* in a position to end these debates through some convincing argument. To get at least some sort of a measure for the evaluation and comparison of the different schemes to be discussed, we state here some of the major goals to be pursued in knowledge representation. Since people provide the best model for how to deal with represented knowledge, it is a good idea to look carefully at some of the features they realize.

Let us first consider the purpose of representing knowledge such as the earlier one expressed in the sentence about the gripper picking up the log. We would like an intelligent device to interpret this sentence in such a way so that the device could answer questions about the current location of the log and availability of the gripper for further tasks and possibly notice some danger for a person standing close by and issue a warning.

What these questions demonstrate is that a lot of additional information goes along with such a sentence. A major goal in knowledge representation is to get a machine to associate such information with a sentence, like the one we consider, when it is representing the sentence internally, an extremely hard problem indeed (known as the *qualification problem*).

But even representing the sentence itself begs a number of questions. Should the sentence be represented just as it is or in some different, possibly less readable form? The behavior of human beings indicates that our brains use the second alternative since we tend to remember the gist of a conversation long after we have forgotten the exact wording. A more technical argument lies in the observation that the sentence alone is *referentially ambiguous* if it is looked at later. "Which gripper, which log are you talking about?" one would ask. So unique names have to be associated with such objects. There are further ambiguities in natural language as well as in pictures and other external sources of information that have to be resolved while representing knowledge internally.

One of the most striking features of human knowledge processing is the fact that the communication of knowledge may change behavior even though the knowledge must not necessarily contain anything like advice for how to behave. If I learned that the road I was about to take was closed, then I would immediately change my plans even though what I had learned was a fact rather than anything of the nature of a command. Apparently people can transform such knowledge into procedural behavior. Thus they are able to accumulate knowledge in an additive way and yet use it at the same time for procedural purposes.

This is not the only way that people acquire knowledge. I learned how to ride a bike, for instance, by some mechanism that we know relatively little about. It is a fact that current expert systems technology has accomplished neither of these two possible ways of acquiring knowledge to a satisfactory degree, probably one of the major reasons for its relatively moderate commercial success thus far. Given their apparent relevance in human knowledge processing, I would put both *additivity* and *learning* of procedural behavior high on the list of desiderata for systems that represent and process knowledge. The scheme used for the representation should support either of these properties if possible.

One reason to feed knowledge into a computer is to aid human experts in their work. This aid will be the more effective as it becomes easier for the expert to quickly grasp the knowledge presented by the system. *Cognitive* or *epistemological adequacy* of the representation might be the right term for denoting this desideratum. This also leads us to the fundamental question of knowledge representation of whether there are *primitive representational structures,* richer than just objects, relationships, and so forth, that pervade human knowledge representation.

With anything related to computers, we have to be concerned about *efficiency* in the processes to be carried out by the machine. Otherwise we could spend most of our time waiting for responses from the computer. This again puts a constraint on the representation since an operation is performed at a different speed in one representation than in another.

These last two desiderata are actually constraints on representations in two different media, assuming current computer technology. While cognitive adequacy addresses the representation on the level of the human/machine interface, efficiency plays a role on the level of the machine (i.e., hardware) only. This demonstrates the importance of the operations for our considerations which are very different ones (human interaction vs. operations on hardware) for these two levels. In fact, this also demonstrates the usefulness of talking of different *levels* of representation and of the representational medium. So far we have mentioned two such levels which could be subdivided into further ones according to the kind of user or level of implementation.

3 FORMS OF REPRESENTATIONS

In this section we briefly describe ten of the major forms of representation in use.

3.1 Natural Language Representation

Natural language provides a form of representation that has become the standard medium for communication among people. Sometimes natural language is complemented by drawings, figures, gestures, and the like. None of these qualify under the notion of a scheme as noted in Section 2.1. Nevertheless it is helpful to take into account these forms of representation as a guide for the development of more formal ways of representation.

3.2 Fregean Representation

Predicate logic emerged from attempts by G. Frege, and others before and after him, to abstract a scheme from natural language and thereby retain most of its expressiveness. Let us first consider the simplest kind of sentences.

The car is red

may be translated to (the language of) predicate logic as

Is_red (car)

In this and similar *singular* statements, a property, i.e., being red, is ascribed to an object, the car. The first is captured as a *predicate,* that is, *Is_red,* the second as a *constant,* or *car.* Any such statement is called a *formula* in predicate logic. The formulas are thus for predicate logic what we called configurations for the more general schemes discussed in Section 2. At present we are dealing with the simplest type of formulas that do not contain any logical operators and are called *atoms.*

Predicate logic is quite liberal in what exactly is to be the predicate. In the present example, we might as well think of two things involved in the statement, the car and the color red, and a property captured by the verb *is* that links the two. Under this view, the formal statement would read

Is (car , red)

It is merely a *convention* to standardize the sequence in which the predicate and the objects are listed. The particular convention followed here is in no way an intrinsic feature of predicate logic. Hence writing

car Is red

would be absolutely legitimate in predicate logic. However, if three objects became involved in such a statement, then the suggestive symmetry in the last expression would be present no more. Hence the convention turns out to cover all cases uniformly well, so we stick to it most of the time. Also the use of parentheses is merely a convention. We could as well dispense with the parentheses and write

Is car red

or some further variation.

Logic differs from natural language in that the symbols used in a formula need to be interpreted in order to acquire a certain meaning; without such an *interpretation* they are meaningless symbols. For instance, the symbol *car* from a logical point of view may be interpreted as any object, say a tree. Logically, a car can be distinguished from a tree only by stating their respective properties in additional formulas such as this more structured one:

All cars have wheels

It translates into

>   for_all x (Car (x) → Has_wheels (x))

literally, "for all objects x, if x is a car, then x has wheels". The logical *operator* (or *connective*) "for_all" is called a *quantifier*. While the previous example expresses a *fact,* this one states a (logical) *rule.* Rules typically contain variables like the x in this formula. Any formula that does not contain variables, such as facts, is called *ground* or a *description.* As with the fact before, a number of conventions apply for rules (or even more general formulas) as well, too many to be discussed in further detail here. Let us just state that the same sentence could be expressed in the following and other ways as well:

>   for_all x in *cars* : Has_wheels (x)
>   for_all x (Car (x) → Has (x , wheels))
>   *Car* (x) → *Has* (x, wheels)
>   *Has* (x, wheels) ← *Car* (x)
>   has (X , wheels) :− car (X)

Which is preferred is a matter of taste (to a large degree determined by our habits). Often there is actually a more rational reason for preferences depending on what we want to do with this knowledge or, more precisely, which operations we are interested in performing on these formulas. If our quantifiers are all the same in any of our statements and never are affected by the operations applied, then we should omit them (as in the third variant of the last set). If we want to have a flexible way of talking about parts of a car other than its wheels, then we should use the two-place predicate where *wheels* can be replaced by other constants or variables.

Often we cannot exactly predict the operations that may become relevant later. Logic is actually flexible enough to cope with this. It is perfectly all right to make a change in the representation in the middle of building the knowledge base, provided the relationship between the old and the new representation is specified by a definition like

>   for_all x (Has_wheels (x) ↔ Has (x, wheels))

This leads us to observe a very important aspect of logic. Namely, the choice of the predicates, functions (such as the *left_front_wheel_of* my car), and constants is not dictated by logic; rather, it is the knowledge engineer who makes the appropriate decisions.

All the sentences considered so far are formalized in what we might more precisely call *first-order predicate logic.* There is also a *higher-order predicate logic* where statements of the following kinds are possible:

>   for_all X (X (my_car) → X (your_car))
>   *COLOR* (*Green*) ∧ *Green* (*your_car*)

These might be read as "All properties of my car are properties of yours as well" and "Green is a color that your car has", respectively. Quantifying in this way over predicates leads us to the *second* level (or order) of higher-order logic, a process that can be iterated so that arbitrary levels are possible. Also, as demonstrated with the second example, it is possible to consider higher-order predicates like *COLOR* that apply to predicates like *Green* as if they were objects.

When the emphasis is not exclusively on the representation by way of general predicates (that may be interpreted in an arbitrary way) but the treatment also accounts for the special predicates with built-in properties like equality, then the simpler name *first-order logic,* or even just *logic,* is preferred. This distinction is not always strictly kept in the literature. There is no way of giving a brief survey of logic simply because it is a field of its own, filling whole sections of libraries. Numerous textbooks are available ranging from easily read ones such as Pospesel (1976) to more advanced texts such as Manna and Waldinger (1985) and Gallier (1986). Some of the specific features of logic are mentioned in later discussions.

We cannot conclude this section before noting one of the most important properties of logic. All that matters from a logical point of view is whether a statement (i.e., formula) is true (or false). We are particularly interested in whether a formula is *logically true* (or *valid*), i.e., true under any possible interpretation (similarly for false). Many tasks may be naturally stated in a way that all that is required is to test for this property. One of the most attractive features of logic is therefore the fact that it provides not only a (formal) language with well-defined *semantics* but also a *calculus* in which *deductions* can be calculated. A deduction is a "logically correct chain of inferences" while *inferencing* in general refers to coming to believe new facts on the basis of other information.

If the fact just mentioned about logic were meant to be emphasized, people would even talk of the predicate *calculus* as another name for logic. On the basis of this calculus, mechanical devices (or *theorem provers*) have been developed which allow us to carry out deductions, i.e., check for the kind of validity just described, in an automatic way. One application is discussed in the next section.

3.3 Horn-Clause Logic

A particularly interesting part of first-order logic is Horn-clause logic. To informally describe this, first we have to look at a few well-known properties of logic.

With the logical operators such as $\wedge$, `for_all`, and so forth, we can construct a formula that is nested in an arbitrarily complex way. Any such formula is equivalent to formulas with a much simpler structure. It is, for in stance, equivalent to a formula for_all $\bar{x}$ F.† Here $\bar{x}$ is short for a sequence of zero or more variables. The main point is that F does *not* contain any quantifiers. From the point of view of the formula's connectives, F therefore may be regarded as a *propositional formula*. Any such propositional formula can be further simplified structurally into a conjunction of *clauses*, i.e. subformulas which in turn are disjunctions of *literals*, i.e., (possibly negated) atoms. Thus any formula can be represented as a set (a conjunction) of clauses.

Horn clauses are clauses in this sense that are restricted to contain at most one atom; i.e., all other literals in it (if there are any) are negated atoms.

$$\text{Has}(x, \text{wheels}) \lor \neg \text{Car}(x)$$

is such a Horn clause, as is

$$\text{Is}(\text{car}, \text{red})$$

In fact, the former is equivalent to

$$\text{Has}(x, \text{wheels}) \leftarrow \text{Car}(x)$$

In this latter form, *Horn-clause logic* is used in the programming language Prolog where such a clause is written

$$\text{has}(X, \text{wheels}) :- \text{car}(X)$$

A Prolog *program* consists of any number of Horn clauses along with a *query*, i.e., a Horn clause with *no* atom, written in the form

$$?- \text{is}(\text{car}, X)$$

To run a Prolog program in effect means to invoke a theorem prover of the kind mentioned at the end of the previous section and to check the validity of the query against the knowledge represented in (the *body* of) the program. The term *theorem proving* is a bit misleading and can be understood in the historical context only.

Only a very small fraction of the potential complexity in the structure of formulas in full first-order logic is actually needed in practice. Indeed, many problems can comfortably be represented in Horn-clause logic. This explains why Prolog is a programming language with satisfactory expressiveness. In more sophisticated applications, the limitations just described are indeed felt, hence the attempts to develop a programming language such as LOP (Bayerl et al., 1988) with the full expressiveness of first-order logic.

3.4 Functional Representations

As an aside in Section 3.2, we mentioned functions as an ingredient of logic. They are an important feature through which an essential part of the logical

---

† This particular equivalence holds when we look at formulas from a *refutational* point of view, which means that our interest is in invalidity rather than validity.

language, i.e., its *term sublanguage,* is built. Just as in mathematics, with functions we can build terms by starting from constants and variables.

left_front_wheel_of (*my_car*)

(5 + x) × 12

× (+(5, x), 12)

(× (+ 5 x) 12)

are such terms, and again the last three are just different versions of the same term.

The term sublanguage, enriched by further constructs not usually incorporated in it within a logic context, is actually so powerful in its expressiveness that it became the basis of the most popular programming language in artificial intelligence, which is LISP. For instance, the last term of those just shown is in a form used in this language. While a LISP program is indeed a function that looks exactly like a functional term in this particular form, a few distinctions should be made in regards to the logical subterm language.

First, LISP is not only a programming language but also an interactive programming *environment.* Apart from that, LISP *evaluates* its terms while logic in its pure form treats them as purely syntactic structures. But this feature of evaluation can be, and has been, added to logic in various ways. One way is to add a special *evaluate* predicate, as done by Weyhrauch (1980) under the notion of *procedural attachment.* Another way is to treat terms within an equational calculus integrated into logic (see Bibel, 1987, section V.3).

Further, in LISP we can *define* functions (with the built-in function *defun*). For instance,

(*defun square* (*num*) (× *num num*))

defines *square* as the function that computes the square of two numbers. The equivalent in logic is

square = λ*num* (× *num num*)

(note the exact correspondences) whereby we use the λ operator as a logical operator (similar to a quantifier). In this case the operator might be read "the set of all pairs" with *num* as the first and × *num num* as the second argument. In other words, this expression (or term) represents a set of pairs, i.e., a function of what, of course, we want *square* to do. What we suggest here is to include this operator in first-order logic. There is nothing wrong with doing so, although usually in the literature reference would, in the context of LISP, be made to a somewhat different logical formalism, called the λ calculus.

From this correspondence between first-order logic and LISP we see that there is a hierarchy of languages, i.e., LISP, Prolog, LOP, with increasing expressiveness. We should say, "with increasing *representationally direct* expressiveness", since logic as well can be embedded in LISP. To see why, note that the definition of formulas in first-order logic is actually a purely functional construction (which is why some authors also talk of a functional formalism in the context of logic). However, while functions typically have values such as numbers or objects, logical formulas are the only functional constructs that exclusively admit truth values which are truly in a special category worthy of a special treatment.

Most major AI programs in the past were written in LISP. With Prolog now catching up in performance and providing a better programming *environment*, this trend is changing and will probably continue to change in favor of Prolog (and later predictably in favor of something like LOP). LISP certainly still has its merits, though, and the reader is encouraged to learn more about it from Charniak et al. (1980).

3.5 Associative Networks

A rather popular kind of representation is known as *associative* (or *semantic*) *networks* (or *nets*). Figure 1-2 shows an example. They are called associative networks because they associate different things, here drawn as ovals, with each other. For instance, the object mycar is associated with the object red by a *labeled directed arc*. The label denotes the predicate that relates the two objects. In logic this particular part of the net would be represented as *Color (mycar , red)*. In other words, each arc represents a fact that logically is a ground literal with a binary predicate. With this explanation you should now understand what the net actually expresses and how the same thing would be represented as a logical formula.

Once one has grasped the idea of representing information as an associative network, it becomes actually very easy to "read" them. Their illustrativeness is one reason for their popularity. Another is their *object-oriented* way of representation. All the properties of an object are associated by the outgoing and incoming arcs at its node. Originally (Quillian, 1967) this was used to represent the meaning of words in natural language understanding. To each word can be attached all its associations, thus defining its "semantics", hence the notion "semantic" networks although they are not more semantic than are logical formulas.

Technically, associative nets are represented in the computer by an indexing scheme that uses pointers (i.e., memory addresses) from one node to the other and vice versa. This makes it very efficient to access any of or all the properties of a given object.

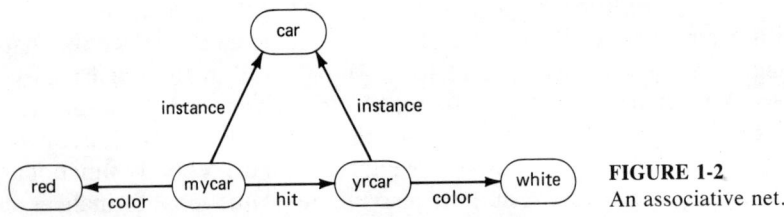

FIGURE 1-2
An associative net.

The way we have described nets so far would allow the representation of facts, but no logical connectives. Work has been done to add these to the associative net formalism (Schubert, 1976), making the two notations equivalent in expressive power. This addition, however, is no longer as illustrative as when the notation is restricted to facts only.

3.6 Frames

Situations in everyday life are often characterized by a stereotypical structure. For instance, if you think of the situation of a car accident, then you immediately expect that a number of *typical* things happened. A car hit another one, a person was hurt, police were called, and so forth. One might speak of the *frame* of an accident in this sense. The observation that our thinking to some degree is guided by such prefabricated frames or *prototypes* led M. Minsky (1985) to introduce this concept as an appropriate technique for representing knowledge. His attempts were also aimed at capturing in this way something like a primitive representational structure, mentioned in Section 2.2. Consider again the information represented by Fig. 1-2.

An insurance agent would be interested in the cars' values rather than their colors, so let us exchange these two properties and thus consider the following logical formula:

$$Hit(mycar, yrcar) \land Instance(mycar, Car) \land Instance(yrcar, Car) \land Value(mycar, 2000) \land Value(yrcar, 5000) \land 0 < 2000 < 40{,}000 \land 0 < 5000 < 40{,}000$$

Although this is a nicely grouped piece of knowledge, in a large knowledge base the pieces would be mingled among many others, so the group could not be easily recognized as belonging together. To achieve this grouping, a *frame* is created and given a name, say, *accident*. In this case it has four *slots*, which might be named

- hitting_car
- hit_car
- value_of_hitting_car
- value_of_hit_car

The idea is that whenever we refer to *accident*, exactly these four items are taken into consideration, in particular filled with values. For such values, a number of *facettes* are associated with each slot, of which one is meant to store the actual value. In this example, the value facette of the slot "hitting_car" would take *mycar* as its content. Other facettes in a slot may refer to the class (e.g., *Car*) of which the value is an instance, or may specify a range of values such as $0 < x < 40{,}000$ in the value slots of our example, or may contain default values in case no value is explicitly specified. Facettes can serve several additional functions such as providing information about inheritance (see the next section), activating programs, and so forth. With all (or at least a sufficient number of) such values entered into the facettes, we get a particular instance of the general concept *accident* represented as a frame.

The idea of frames is to group pieces of knowledge that belong to a certain situation (such as accidents) or that characterize a certain concept (such as a class of objects with a number of stereotypical properties). Frames thus provide an object-oriented form of representation, as did the associative networks. This supports the process of acquiring knowledge into an expert system, since the system may prompt the knowledge engineer to fill the various slots once a frame is activated. The grouping may also contribute to the efficiency in processing the knowledge since it allows us to focus on relatively small chunks of knowledge, possibly linked to other such chunks, as discussed in the next section.

There are ways to realize the frame idea within logic. One uses the λ operator already mentioned in Section 3.4. In our present example the definition of *accident* would simply be

$$\lambda w, x, y, z \, Hit(w, x) \wedge Instance(w, Car) \wedge Instance(x, Car) \wedge Value(w, y) \wedge Value(x, z) \wedge 0 < y < 40{,}000 \wedge 0 < z < 40{,}000$$

The particular accident of our example would be named *accident* (*mycar* , *yrcar* , *2000* , *4000*). For a more detailed comparison of logic and frames, see Hayes (1985).

3.7 Inheritance

One of the arcs in Fig. 1-2 expresses the fact that my car is an instance of the class of all cars. In Section 3.2 we considered the piece of knowledge of cars having wheels. From both statements we can conclude that my car has wheels. As an instance of the class of cars, it *inherits* this property from the class.

Much of human knowledge is structured in such a hierarchical way. To mention another, often used example, consider a particular elephant named Clyde. He is an instance of the class of mammals, call it MAMMAL. Another subclass of mammals is humans (HUMAN) in which we consider a person named Fred as an instance. This configuration is shown in Fig. 1-3 as an inheritance network. Here we let circles picture instances (objects such as

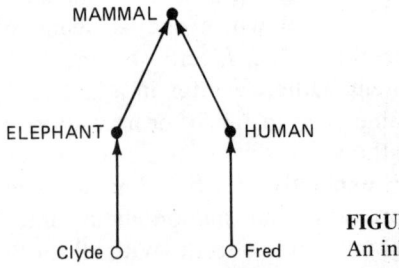

FIGURE 1-3
An inheritance network.

Clyde) and solid circles picture classes (such as ELEPHANT). The arrows represent the relationships *instance_of* or *subclass_of*, whichever applies. For both these relationships the names *AKO* (for "a kind of") or *ISA* (for "is a"—whence *ISA hierarchies*) are also in use. Note that such a network could also be looked at as an associative one, as discussed in Section 3.5.

As illustrated already with the *has_wheels* property, the idea of inheritance networks is that properties assigned to classes higher up in the hierarchy are inherited by those lower than them. This idea can be combined with the representation of concepts in the form of frames. To see how, imagine each of the nodes in the figure being associated with a frame for the concept whose name is shown in the picture. One of the slots in each of these frames might be named *#_of_legs*. In the knowledge acquisition phase, the only slots among these, which had to be filled with appropriate values, would be those in the MAMMAL and HUMAN frame. All others could then be determined by inheritance in the following way.

Say we want to know how many legs Clyde has. The slot *#_of_legs* in the frame Clyde is examined for its value, but none is found. Therefore the same slot is examined in the frame one level up in the hierarchy, ELEPHANT, again with no result. However, one further level up the value 4 is found and used as the answer (possibly also filled in the slots examined for future queries of that sort).

Unfortunately, this process works only for *single* inheritance systems without complications. These are networks of the sort shown in Fig. 1-3 with each node having at most a single outgoing arc. For realistic cases this is too strict a limitation. The standard example for multiple inheritance nets is the following one presented as a number of statements.

- Elephants are gray.
- Royal elephants are elephants.
- Royal elephants are not gray.
- Clyde is a royal elephant.

As we see, the first rule in this scenario has exceptions, as is often the case in reality. Now the point is how a system could arrive at a reasonable answer to the question of Clyde's color. Clyde is a royal elephant, so his color should be different from gray. But being a royal elephant implies being an elephant, by way of the second rule, and they are gray. Hence Clyde's color should be gray according to this line of reasoning, while naturally one would expect to get the first answer.

A number of proposals have been made to resolve such conflicts in accordance with our intentions. One uses the notion of *inferential distance ordering* (Touretzky, 1986). The idea is to prefer the first answer on the basis of the fact that the chain of reasoning involved, roughly speaking, passes through a common node (namely, ROYAL_ELEPHANTS) which therefore is considered "closer" to Clyde than the node ELEPHANT and hence is preferred for inheritance.

This kind of problem is closely related to that of nonmonotonic reasoning mentioned again in Section 5.2. Here we conclude by emphasizing that inheritance systems use the hierarchical ordering to *implicitly* control inferences in one way or another, such as the one just described, but exclusively based on the information represented within the hierarchical order. That this kind of representation of knowledge has a straightforward logical equivalent should be obvious from the observation that the arcs represent either a predicate *instance* or a rule (in the case of subclasses).

3.8 Procedural Representation

Some of the most successful expert systems have been using a formalism known as a *rule-based* (or *production*) *system* for representing knowledge. It consists of

- A working memory
- A set of *production rules*
- A rule interpreter

The working memory consists of a set of data structures representing the current state of the system. The production rules are of the following form:

   if ⟨*conditions*⟩ then ⟨*action*⟩

The conditions are stated in some form that may be regarded as a logical formula, mostly a boolean combination (or even just a conjunction) of literals. The action is something that changes the state of the working memory. If no conclusion with the logical rules (as discussed in Section 3.3) is likely, then we simply speak of rules here as well. A robotics example of such a rule is

   if *block is clear* then *pick-up block*

The rule interpreter *matches* the conditions to the entries in the working memory. If this matching process is successful, then the rule "fires," i.e., the action specified in the right side of the rule is executed. Matching is a special form of *unification* used in theorem proving (Section 3.2) whereby variables are instantiated in the conditions only, not in the working memory. For instance, if the working memory contained the information *block_031 is clear,* and if *block* in the above rule were defined as a variable, then the matching process would succeed. In detail, the substitution of the constant *block_031* for the variable *block* in the rule would make the condition identical with the entry in the working memory. Hence the action *pick-up block_031* would be executed.

Various regimes are used to control the behavior of the interpreter. The two major ones are *forward chaining* and *backward chaining*. In the forward-chaining mode, the process starts with the entries in the working memory and fires rules until a desired goal is achieved. A particularly efficient algorithm, called RETE, has been developed to achieve this control regime. For a comprehensive treatment of this and other rule-based systems in general, see Brownston (1985).

In the backward-chaining mode, the process starts with the desired goal and traces the rules back until their conditions match with the working memory. Prolog (discussed in Secton 3.3) uses this control regime. But note the difference in the kind of rules used here and in Prolog. Many problems (such as planning robot actions) are dynamic and therefore seem difficult to model in a declarative, logical language (such as Prolog). However, not only have a number of such rule-based expert systems successfully been implemented in Prolog, but also there is work demonstrating how actions may be naturally modeled in a logical setting (Bibel, 1986a).

3.9 Neural Nets

In the early days of the computer age, first attempts to design machines in a way similar to the human brain's structure and operations were made. With the technology available at that time, the chances of success were low because any realistic simulation would need not just a few but at least thousands of neuronlike processors, a requirement that can only now be met with the advanced technology of VLSI and computer-aided chip design. With the advent of this technology, a new and vigorous wave of interest in neural-net-like computer structures is under way. While biological fidelity of such computer models would be of major concern for understanding the brain's mechanisms, a derivative class of models, called *connectionist networks,* aims rather at the computational power achieved by exploiting parallelism.

Such structures consist of massive numbers of processing elements, called *units*. They interact by using weighted connections. Each unit has a *state* or *activity level* that is determined by the input received from other units in the network. Among various refinements of this general model is the one that takes the combined effects of the rest of the network (the *total input*) on the jth unit as a scalar quantity, say x_j, to be considered as its state. x_j is nonzero only if it exceeds a certain threshold value θ_j. In addition, one usually assumes that this total input is a *linear* function of the states of all connected units providing input and of the weights w_{ji} of the connection from the ith to the jth unit. Thus

$$x_j = -\theta_j + \sum_i y_i w_{ji}$$

All the *long-term knowledge* in a connectionist network is encoded, i.e., represented, by the locations of connections or by their weights. So processing knowledge in this model always means adding or removing connections or changing weights. The *short-term knowledge* is normally encoded by the states of the units, but sometimes by the weights or threshold values as well.

One of the main issues in connectionism is how to represent conceptual knowledge such as "all cars have wheels". The approaches here range from *localist* ones, in which each concept is represented by a single unit, to *distributed* theories, in which a concept corresponds to a pattern of activity over a large part of the network. There is no way to cover this interesting, but still somewhat uncertain, ground here in more detail. The interested reader should consult the relevant literature, such as Rumelhart et al. (1986), Hinton (1989), and Feldman and Ballard (1982) for details. But let us state a few observations at this point.

A fresh representational approach has often inspired scientists to find better solutions to known problems. In this respect, connectionistic ways of representation provide one more opportunity to uncover solutions to interesting artificial intelligence problems. Among these problems are search and learning. Connectionist approaches to modeling search and learning exhibit great potential because they would be less noise-sensitive than the apparently rigorous representations modeled in some logical formalism. For instance, learning to drive a bike could perhaps be modeled in a connectionist way more easily than via a logical formalism. A strong learning capacity would actually seem crucial for connectionist models to lead to successful applications since it is unlikely that all the states of the units, the connections, and their weights could be set "by hand" if there are millions of them.

But at present it seems likely that connectionist procedures could be programmed as well as conventional ones. Both compute partial recursive functions after all. In any case, it seems premature to expect quick and easy successes in areas that have frustrated artificial intelligence researchers quite a bit.

3.10 Others

There is no way to cover all the grounds of knowledge representation in a single chapter. Here we mention some of the omissions.

First, there are other extensions of first-order logic that play an important role in representing knowledge, especially modal logics. Here modalities are modeled by way of additional operators such as □ (e.g., to be read "necessarily") and ◇ ("possibly").

A particularly important issue for knowledge engineering is the representation of measures of evidence, certainty, or uncertainty. There are a number of ways to incorporate such features in logical formalisms. Some just add extra parameters to the predicates that carry certainty values. Others add new logical operators (such as fuzzy quantifiers like "most" or "a few") to the formalism.

A vast area in itself is low- and higher-level representation of visual information. In vision, the discussion is at least as much focused on *what* is actually to be represented as it is on *how*. To understand why, note that the flow of the raw visual information coming in from a scene consists of a vast amount of barely structured information from which there is a long way to go until, say, a tree is recognized as such. In a more comprehensive treatment one would also cover what is called *analogical representations* that model shapes and other structures in a more direct way than realized by Fregean representations.

Because of these and a number of further omissions, the interested reader is encouraged to consult the literature, especially the text by Brachman and Levesque (1985b).

4 COMPARATIVE EVALUATION

Neither computer science nor artificial intelligence (or any other area for that matter) has come up with a solution for the Babylonian problem of how to cope with the variety of languages. An important part of the knowledge representation problem is basically a language problem as should have become clear from the survey of the previous section. Obviously, the presence of different languages impedes communication. On the other hand, it is a common experience that rephrasing a complex structure in different ways often is quite enlightening.

One goal in knowledge representation is to rationalize the virtues and disadvantages of different formalisms. At this point in time we have not reached a situation where for a given problem everyone would agree on the most suitable formalism to be used for its representation. In the current section we nevertheless try to provide a number of arguments that might serve as a pointer into the direction in which to proceed with further research. Deliberately the section is structured by issues rather than by pairwise comparisons.

4.1 Linear vs. Two-Dimensional Formalisms

Logic in its usual setting is a one-dimensional (or linear) language while semantic nets (and in a sense also frames) are two-dimensional languages. From the point of view of cognitive adequacy (see Section 2.2) of a representation, this is an important difference since humans seem to be more capable of grasping a two- (or three-) dimensional "description" than a one-dimensional one. Just think of how many words it would take to describe the content of a cartoon that can be understood with one glimpse.

From the viewpoint of the processing characteristics, however, this difference is completely irrelevant in present machines, since the two-dimensional information is projected down to a linear representation in the computer anyway. Note that we restrict our discussion to "present machines", which hints at the multiprocessor machines, discussed in Section 3.9, with inherently nonlinear topologies that are possibly capable of analogical representations (as probably realized in the brain).

To the extent that the transformation from the two-dimensional to the one-dimensional representation (and vice versa) is possible at high speed (as is the case for semantic nets), it does not make any significant computational difference which formalism we prefer. So the cognitive aspect may be taken as decisive, except if issues discussed further below become relevant.

Having emphasized the importance of nonlinear representations such as semantic nets with regard to user convenience, from now on we restrict the discussion to the one-dimensional forms that by way of transformation may be regarded as canonical.

4.2 Object-Oriented Descriptions

Objects are identified by their names and characterized by their properties and relations with other objects. For instance, consider the description "Mary is an American living in Paris". In logic one could represent it by

$$Nation\ (\ mary\ ,\ amer\ )\ \wedge\ Live\ (\ mary\ ,\ paris\ )$$

If the focus is on the object, one would prefer a structure like

$$mary\ :\ Nation\ (\ amer\ )\ \wedge\ Live\ (\ paris\ )$$

If there are many more properties stated in the knowledge base for Mary, the convenience of the second object-oriented form of representation becomes obvious. But there is also a computational aspect, since the second form is shorter and suggests a direct access from an object to each of its properties, without searching.

The second form is very much like the frame representation from Section 3.6 where the predicates are listed top-down along with their arguments (objects and values) listed left to right, resulting in a two-dimensional table named by the object in question. But at the same time it is also a variant of the semantic net representation where the arguments are arranged around the object in question and linked to it with arcs labeled by the predicates.

So, except for the dimensional aspect discussed in the previous subsection, we have identified a second difference between logical and the object-oriented formalisms. In essence, it amounts to extracting the common factor, i.e., the object in question (*mary* in the two formulas above). This has an organizational effect of grouping the respective properties.

While this clearly points out a disadvantage of the *standard* form of logic during a phase of interaction with a system where the focus is on a certain object, it by no means affects the nature of logic itself. On one hand, the concept of logic is not restricted to a unique syntactic version of language; i.e., the second version of the above formula may well be regarded as a syntactically correct formula (in a "nonstandard" language) of logic. On the other hand, the form of representation must be regarded as separate and independent from the computational efficiency of its implementation. This is because the first formula above may well be (and has occasionally been) implemented in such a way that the efficiency suggested by the second formula is achieved even for the first one (e.g., by internally using a *dag* as data structure). Conversely a weak implementation of the second formula may well be worse than a straightforward one for the first. In other words, while there is a computational advantage in carrying out certain tasks in a way suggested by the second formula, this advantage may be achieved independent of the actual representation on the user level.

The same difference just discussed also distinguishes semantic nets and frames. For instance, if we additionally characterize the object *paris,* then the frame notation requires an extra mention of this object (as the name of a new frame) while in a semantic net the object *paris* is still represented by a single node only, connected with further ones. Again, while this might be worth consideration in view of the user interface, the difference may disappear in an appropriate implementation that provides for the elimination of this redundancy.

Other useful variants deviate from the standard language of logic in a way similar to the one described before. We mention only one more example, taken from Michalski et al. (1980, chapter 4). Instead of representing "The car's color is black or blue" as

Color (car) = black or Color (car) = blue

Michalski prefers to use

Color (car) = black or blue

Again, the common part "*Color (car) =* " is extracted like a common factor.

4.3 Unary Functions as Description Language

Apart from the difference identified in the previous subsection, the language of semantic nets and frames differs from that of logic by its restriction to literals that pair argument and value of a unary function. Instead of using the predicate *Nation* above, full advantage is taken of the functional dependence between *mary* and *amer*. So in logical terms the literal *Nation (Mary , amer)* is actually interpreted as *Nation (mary) = amer,* where *Nation* plays the role of a function rather than that of a predicate.

First note that, on the basis of what was said in Section 4.1 about infix notation, *Nation (mary) = amer* still may be regarded as a literal in logic. The functionality made explicit in this way can enhance the system's performance and thus offers a computational advantage. But what we said about the independence of form and implementation in Section 4.2 applies here in the same way. The knowledge acquisition component might itself extract this functionality from a literal like *Nation (mary , amer)* in a given context defined by the state of the knowledge base. Under this aspect, it appears as an unnecessary burden on the user to always provide this extra information about the functionality. Once it is available, however, its explicit representation clearly offers an advantage for the user.

As a further example, assume we want to express that a_1, a_2, and a_3 form a triangle *Tria* (a_1, a_2, a_3) in logical terms. In the functional notation inherent in the semantic net or frame formalisms, one has to take into account the formed object as a whole, say *t,* and then write, for instance,

$$form(t) = triangle \wedge part_1(t) = a_1 \wedge part_2(t) = a_2 \wedge part_3(t) = a_3$$

A comparison with the three-place predicate shows that the functional notation may well become awkward. This inconvenience is not removed if we generalize by allowing binary predicates, as some semantic net formalisms do.

We also note that unary properties (which are the same as sorts) must be categorized in order to fit into the functional scheme; e.g., we cannot simply say *Amer* (*mary*) but need a category *Nation* of which *Amer* is an instance, perhaps a negligible inconvenience only.

A number of advantages have been claimed for semantic nets or frames in comparison with logic beyond the issues raised here, but none withstood a thorough analysis in terms of the computational relevance or expressive power. For instance, the fixed number of arguments of a predicate was meant to cause an inflexibility not present in nets and frames. By defining *Tria* (a_1, a_2, a_3) as *Tria* ($a_1 \cdot (a_2 \cdot (a_3 \cdot nil))$), where " $\cdot$ " is the usual *cons* operation (a binary function), apparently this simple solution provides predicates in effect with a variable number of arguments. To mention a last example of alleged advantage, the identification of classes as explicitly addressable objects again is not an exclusive feature of nets and frames; namely, the logical formula $\lambda x_1 x_2 x_3$ *Tria* (x_1, x_2, x_3) does exactly that, naming the class of triangles.

4.4 Assertional Representation

We have noticed some of the advantages of language constructs exploited in the formalisms of semantic nets and frames. While they differ from the standard form of logic, they nevertheless may be regarded as logic constructs. This possibility of embedding becomes relevant as we proceed to express assertions rather than mere descriptions. Although the distinction between descriptions and assertions is an intuitive one, we might define descriptions as formulas in logic that are ground after possible λ reductions. That is an operation which, for instance, reduces [$\lambda x_1 x_2 x_3$ *Tria* (x_1, x_2, x_3)] (a_1, a_2, a_3) to *Tria* (a_1, a_2, a_3) and is, in fact, part of what takes place when LISP evaluates a function. In other words, descriptions are formulas without variables or quantifiers except those in λ constructs.

While semantic nets and frames are particularly specialized in describing structured objects, "logical languages like that of first-order predicate calculus are far superior at meeting the needs for belief representations, since they can provide for a kind of noncommittal expression not possible with frame languages" (Brachman et al., 1985a). The example with the triangle mentioned in the previous subsection has already hinted at this strength of (even the standard form of) logic. Disjunctive knowledge and null values, both topics in databases, point in the same direction, not mentioning complex assertions crucially involving quantifiers as they occur in mathematical statements.

4.5 Semantics for Multistyle Representations

In this entire section we have identified a number of issues that have been a focus of discussions on the representation of knowledge. Each could be subsumed under the concern for either cognitive adequacy or efficiency (see Section 2.2). In terms of cognitive adequacy, these concerns are well justified. It is extremely important to provide people with all kinds of means for representing different sorts of knowledge. In fact, even for the same sort of knowledge, different people prefer different representations. The conclusion from these observations must be that knowledge engineers have to create systems with *multistyle representations* rather than dictate what kind of representation a particular individual has to choose in a certain context.

Yet different individuals (and systems) need to communicate with each other. So they must be able to understand what the others' constructs mean in their own terms. Here is where logic can provide one of its fundamental contributions. Recall that for most of the constructs discussed we were able to show an equivalent logical expression. In this sense, logic is the prime candidate for playing the role of a *canonical language* (Bibel, 1984). It has a well-defined and extensively studied semantics which can be imported to such language constructs as used in associative nets, frames, or what have you, as well.

The advanced deductive mechanisms available for logic form another fundamental contribution to knowledge representation, as pointed out in Section 3.2. However, these mechanisms have actually been regarded as a major problem with logic because of the computational complexity inherent in these mechanisms in their most general form. Now there are two sides of this argument to be considered. One is the complexity "in the small". Recall the concerns for quick access to the properties of an object described in Section 4.2. It is inconceivable that any such concern could not as well be accommodated within a logical framework on the level of implementation. The examples given in this section have supported this claim.

The other side of the concern for a feasible computational complexity "in the large" is more serious. All current reasoning mechanisms in general (mind you, not in practice) are actually infeasible. This, however, is not so much of a problem with some particular formalism such as logic, but it is with any formalism expressive enough to handle the kinds of problems envisaged. We only can continue our efforts to identify a computationally feasible part of logic. The range of such efforts includes the enhancement of current theorem provers (Stickel, 1988; Bayerl et al., 1988) and studies in the logical power actually needed (Levesque, 1989).

5 FURTHER ISSUES AND APPLICATIONS

To balance the biased review of the field of knowledge representation provided so far, we raise a number of further issues and point out work that has been done or is still going on.

5.1 Some Kinds of Knowledge

The issue of knowledge representation is tightly coupled with what we want to do with the represented knowledge (to be addressed in the next section). This in turn depends on the kind of knowledge that is actually to be represented. So far the focus has been on descriptional and assertional knowledge. There are many kinds to be taken into account, even kinds that in the natural meaning of the term would not even be subsumed by it.

Much of what we think we "know" is actually just *beliefs*. From a representational point of view, the difference does not matter that much, if at all. Only in the way that beliefs are used in the reasoning processes could the distinction become relevant. Further, there are *judgments, desires,* and *intentions,* concepts that look less far-fetched, even from an engineer's point of view, if one considers robots to become integrated in a natural environment. As with beliefs, these might, at least in a first approximation, simply be regarded as assertions operated upon in a certain way.

More diffuse is what is usually termed *commonsense knowledge* which is vital for humans to get around properly in a natural environment. This notion refers to the vast amount of experience everyone collects in the span of life, but which a robot, or any system for that matter, with current technology would need programmed into it explicitly. In particular, commonsense knowledge includes naive physics knowledge, such as that you would spill your coffee if you failed to hold your cup in an upright position. Commonsense knowledge is always in the background when people draw conclusions in any kind of situation.

Now, the particular knowledge representation problem is, How could we, for a given situation described in one of the discussed formalisms, get exactly that part of commonsense knowledge associated with this situation which is indeed relevant to it? To understand the difficulty of the problem, think of any situation and of the infinite degree of detail necessary to describe it in a way that no doubt about even the slightest point would be possible. The power of human reasoning is just that humans can dispense with most of these details and yet understand what is going on. How do we do this? How can we represent knowledge in machines so that they achieve a similar behavior?

Always present in commonsense knowledge and elsewhere is the need to represent knowledge that is subject to changes in time. What is time from a knowledge point of view? These issues are particularly relevant for programming and, of course, robotics applications, touched on already in Section 3.8.

5.2 Reasoning

Processing knowledge can mean a number of things. In the simplest case, knowledge is just stored, to be retrieved at some later time. The real challenges come if we expect that the system under consideration uses its knowledge to infer other knowledge that is not explicitly stored but can be deduced. Reasoning, to put it in different words, lies at the heart of intelligent behavior.

As we said earlier, this chapter restricts itself to presenting material on the strictly representational side of knowledge representation, thus ignoring reasoning issues (the distinction is rather artificial, though). But to emphasize its importance, let us at least mention a few key words from the reasoning area.

There are the logical deductions mentioned in Section 3.2 and elsewhere before. Theorem provers were mentioned as the devices that do this kind of reasoning, some at an impressive rate of performance. Of similar importance for knowledge engineering is *abduction* that, for a given phenomenon and a number of rules, tries to establish a number of facts that imply this phenomenon, hence *explain* it in this way. Abduction is different from deduction because deduction has the facts provided in the knowledge base while abduction has to establish them.

Induction is the capability to infer a general statement from a number of particular observations. Rules of induction are the inverse of logically valid rules, but they are not equivalent. For instance, if we know that a property holds for all objects under consideration, then we can logically conclude that it holds for two particular objects. But the inverse is not a logically valid conclusion; rather, it can be used as an inductive inference.

Evidential or *probabilistic reasoning* attempts to account for the fact that often the information available is just too sparse to be used in the way that deductive or abductive reasoning is done. On the basis of some evidence, one might still go on to decide on further actions. Humans are almost always in this kind of situation. Figuring out whether there are certain rules that guide this kind of reasoning, and which these are, is the topic of investigations in this subject (Pearl, 1988).

Finally, let us mention *commonsense reasoning* (also called *nonmonotonic reasoning*) which under circumstances just described allows us to "jump to conclusions" by taking into account the "right" circumstantial knowledge. This emphasizes once again the inseparable relationship between representing and processing knowledge.

The reader is offered more information about these topics in later chapters and in further work by Bibel (1987, 1986b, 1988).

5.3 Applications

The material presented in this chapter has applications in almost any kind of technology that incorporates information processing. This book in particular provides an insight into many of them. We therefore confine ourselves to just mentioning a system that has been developed along the lines given in Section 4.5 on multistyle representations. This system is called PRINCESS and is being considered for marketing by Siemens, München (under the name DOMINO-EXPERT).

The system is a toolbox that allows for the representation of knowledge in various forms discussed in Section 3. It realizes the idea set forth there—that logic provides the common basis in which these different forms are actually interpreted. The system supports the acquisition of knowledge to be represented in one of these forms and provides the mechanisms for reasoning with such represented knowledge. It is meant to provide the tools for building expert systems in various application areas, and it has been successfully tested for this purpose. The interested reader might wish to consult Elver et al. (1989) for more details.

6 SUMMARY

In this chapter we briefly reviewed some of the basic notions associated with the concept of knowledge. With this background a number of different formalisms for representing knowledge have been explained and related to logic as a canonical formalism. Also their virtues have been pointed out in relation to the purposes of such a formalism. The treatment has mentioned the many topics in knowledge representation which, for lack of space, had to be left out.

ACKNOWLEDGMENTS

This work was partially supported by Siemens AG, München.

REFERENCES

Bayerl, S., R. Letz, and J. Schumann (1988). SETHEO—A fast theorem prover for the first order logic. Technical report, Technische Universität München, Institut f. Informatik.
Bibel, W. (1984). Knowledge representation from a deductive point of view. In: V. M. Ponomaryov (Ed.), *Proceedings of the IIFAC Symposium on Artificial Intelligence*, pp. 37–48, Pergamon Press, Leningrad, U.S.S.R.
Bibel, W. (1986a). A deductive solution for plan generation. *New Generation Computing* 4:115–132.
Bibel, W. (1986b). Methods of automated reasoning. In: W. Bibel and Ph. Jorrand (Eds.), *Fundamentals of Artificial Intelligence—An Advanced Course*, pp. 173–222, Springer, LNCS *232*, Berlin.
Bibel, W. (1987). *Automated Theorem Proving*, 2d ed. Vieweg Verlag, Braunschweig.
Bibel, W. (1988). Advanced topics in automated deduction. In: R. Nossum (Ed.), *Advanced Topics in Artificial Intelligence*, pp. 41–59, Springer, LNCS 345, Berlin.
Brachman, R. J., V. P. Gilbert, and H. J. Levesque (1985a). An essential hybrid reasoning system: Knowledge and symbol level accounts of KRYPTON. In: *Proceedings IJCAI-85*, Kaufmann, Los Altos, CA.
Brachman, R. J., and H. J. Levesque (Eds.) (1985b). *Readings in Knowledge Representation*. Morgan Kaufmann, Los Altos, CA.
Brownston, L., R. Farrell, E. Kent, and N. Martin (1985). *Programming Expert Systems in OPS5: An Introduction to Rule-Based Programming*. Addison-Wesley, Reading, MA.
Charniak, E., and D. McDermott (1985). *Introduction to Artificial Intelligence*. Addison-Wesley, Reading, MA.

Charniak, E., C. K. Riesbeck, and D. McDermott (1980). *Artificial Intelligence Programming.* Lawrence Erlbaum, Hillsdale, NJ.

Elver, E., W. Bibel, and J. Schneeberger (1989). DOMINO-EXPERT [in German]. In: *Expertensysteme II* (Nebendahl, ed.) Siemens, München.

Feldman, J. A., and D. H. Ballard (1982). Connectionist models and their properties. *Cognitive Science* 6:205–254.

Gallier, J. H. (1986). *Logic for Computer Science: Foundations of Automated Theorem Proving.* Harper & Row, New York.

Hayes, P. J. (1985). The logic of frames. In: Brachman and Levesque (1985b), chap. 14, pp. 287–296.

Hinton, G. E. (1989). Connectionist learning procedures. *Artificial Intelligence.*

Levesque, H. J. (1989). Logic and the complexity of reasoning. Technical Report KRR-TR-89-2, University of Toronto, Canada.

Manna, Z., and R. Waldinger (1985). *The Logical Basis for Computer Programming,* vol. 1. Addison-Wesley, Reading, MA.

Michalski, R. S., H. G. Carbonell, and T. M. Mitchell (Eds.) (1980). *Machine Learning.* Tioga, Palo Alto, CA.

Minsky, M. (1985). A framework for representing knowledge. In: Brachman (1985b), chap. 12, pp. 245–262.

Pearl, J. (1988). *Probabilistic Reasoning in Intelligent Systems.* Morgan Kaufmann, Los Altos, CA.

Pospesel, H. (1976). *Introduction to Logic—Predicate Logic.* Prentice Hall, Englewood Cliffs, NJ.

Quillian, M. R. (1967). Word concepts: A theory and simulation of some basic semantic capabilities. *Behavioral Science* 12:410–430.

Rumelhart, D. E., J. L. McClelland, and the PDP Research Group (Eds.) (1986). *Parallel Distributed Processing: Explorations in the Microstructure of Cognition,* vol. 1, *Foundations.* M.I.T. Press, Cambridge, MA.

Schubert, L. K. (1970). Extending the expressive power of semantic networks. *Artificial Intelligence* 7(2):163–198.

Stickel, M. E. (1988). A PROLOG technology theorem prover: Implementation by an extended PROLOG compiler. *Journal of Automated Reasoning.*

Touretzky, D. S. (1986). *The Mathematics of Inheritance Systems.* Research Notes in Artificial Intelligence. Pitman and Morgan Kaufmann, London.

Weyhrauch, R. W. (1980). Prolegomena to theory of mechanized formal reasoning. *Artificial Intelligence* 13(1, 2):133–170.

CHAPTER 2

RULE-BASED EXPERT SYSTEMS

FOROUZAN GOLSHANI

1 INTRODUCTION

Rule-based systems, or production systems, represent the underlying technology for many of today's artificial intelligence (AI) systems. Emil Post (Post, 1936), whose primary goal was to study computability in formal systems of logic, originated the work on production systems. The Post production system allowed for rewriting of expressions where a rewrite corresponds to forming a conclusion from a set of premises. Later it was shown that production systems were equivalent in power to Turing machines.[1] Although current production systems employed in AI are considerably different from Post's original notions, the fundamental idea, i.e., the notion of premise-conclusion pairs, remains the same.

Rules have been used extensively in AI because rules are simple to work with and because each rule can be considered independent of the others. The latter fact allows for incremental construction of AI programs. Many believe that rules closely represent the way human beings deal with day-to-day problems. In fact, production rules were used with moderate success in the modeling of humans' simple cognitive skills (Newell and Simon, 1972). The rule-based model, which enabled researchers to explain certain results in experimental psychology, has been used to produce many other problem solvers.

Most of the commercially successful expert system shells, such as OPS5, ART, and KEE, support the development of rule-based systems. Among other advantages, these systems allow the knowledge engineer to concentrate on creating well-constrained and concise rules in an incremental manner, without

---

[1] The concept of Turing machines was introduced by A. Turing as a simple model of a computing device. Despite their simplicity, Turing machines are considered to have the capabilities of today's general-purpose computers.

worrying about other aspects of software development. We evaluate rule-based systems with respect to this later.

2 KNOWLEDGE REPRESENTATION BY RULES

Production rules, or IF-THEN rules, are statements of the form

$$\text{LHS} \rightarrow \text{RHS}$$

where LHS (for left-hand side) determines the conditions or situations that must be satisfied for the rule to be applicable and RHS (for right-hand side) is the action(s) that must be taken once the rule is applied. The terms *premise* and *conclusion* are frequently used for LHS and RHS, respectively. Each side of the rule may be in the form of a conjunction. Here is an example:

$$A1, A2, \ldots, An \rightarrow B1, B2, \ldots, Bm$$

The above rule means that whenever A1, A2, ..., An hold, actions B1, B2,..., Bm must take place. Sometimes the above rule is written in a more English-like manner by using the construct IF...THEN...:

$$\text{IF } A1, A2, \ldots, An \text{ THEN } B1, B2, \ldots, Bm$$

For example, the following are examples of rules that express certain regulations in a university environment:

Rule 1: IF John holds a scholarship
         THEN John is a full-time student
Rule 2: IF John is a full-time student
         AND he is a graduate student
         THEN John's academic load is 9 credit hours

3 OBJECT-ATTRIBUTE-VALUE TRIPLETS

An easy way to represent facts is by object-attribute-value triplets, abbreviated OAV triplets. This representation scheme is commonly used in rule-based systems, e.g., MYCIN which is an expert system developed at Stanford University for diagnosing infectious blood diseases. The underlying idea is that *objects* have *attributes,* and each attribute has a *value.* An object is any entity (physical or conceptual) such as an employee, a course, or a department. General properties and characteristics of an object are described by attributes. For example, a course has these attributes: time of offering, location, instructor, etc. As its name suggests, the third member of the triplet gives a value for the specified attribute of the given object. For example, the triplet

(CS-200   location   ERC-357)

states that the location of course CS-200 is ERC-357. Similarly, the triplet

(CS-200   instructor   smith)

represents the factual information "Smith is the instructor of CS-200."

Object-attribute-value triplets can be used as the building blocks for other representational schemes such as semantic nets and frames (see Chap. 1).

Let us now see how we can structure rules by using OAV triplets. Here are some sample rules:

    Rule 3:   IF (John   award   scholarship)
                  THEN (John   status   full-time-student)
    Rule 4:   IF (John   status   full-time-student)
                  AND (John   standing   graduate-student)
                  THEN (John   academic-load   9)

Rules are rarely written for only one specific object. Generally, rules are produced in such a way that they are applicable to a collection of objects. Whenever we need to apply the rule, we *instantiate* that rule to the object (or situation) of interest. By using a variable X, we can state this regulation: Any student who does not have a scholarship and has an academic load of 9 hours must pay $2500 in tuition fees.

    Rule 5:   IF (X   award   none)
                  AND (X   academic-load   9)
                THEN (X   tuition-fee   2500)

As another example, suppose we want to state that any student who is registered for a class and its prerequisite is in violation of regulations. The following rule represents this statement:

    Rule 6:   IF (STUDENT   registered   COURSE-1)
                  AND (STUDENT   registered   COURSE-2)
                AND (COURSE-1   prerequisite   COURSE-2)
              THEN (STUDENT   registration-status   illegal)

In the above, we have made use of three variables, STUDENT, COURSE-1, and COURSE-2. The variable STUDENT, e.g., may be instantiated to a student that is of interest, say, John.

4 AND/OR GRAPHS

The behavior of production systems can be represented by an AND/OR graph (or tree). The nodes of the graph are facts. The arcs represent the condition-action pairs such that there is an arc from a node A to a node B if there is a rule in which A appears on the left-hand side and B appears on the right-hand side. When the left-hand side of a rule consists of several conditions A1, A2,...An, then the corresponding arcs are joined to emphasize the conjunction. As an example, consider the following collection of rules:

    A, B → E
    C → E
    D → F
    E, F, G → H

The tree that represents these rules is presented in Fig. 2-1.

Note that whenever arcs are joined by placing a line over them, the condition part consists of a conjunction. For example, arcs going from E, F, and G to H are joined to represent the fact that E, F, and G must be satisfied in order for the rule to be applicable. However, arcs that are not joined represent choice. For instance, E is the conclusion for two rules—one requiring A and B and the other requiring C. Application of either rule will yield E.

Graphs and trees have been used in AI not only for representation of problems but also for implementation of search methodologies. In the case of rule bases, graphical representation of inference trees helps in understanding the search process. In addition, inference trees may be helpful in validating the rule base.

5 ORGANIZATION OF RULE-BASED SYSTEMS

5.1 History

In the earlier stages, the AI community aimed for developing universal models for general problem solving. Once it was learned that with such generality comes an inability to deal efficiently with specific problems, the attention shifted to creating programs that were more specialized and, consequently, less general. One of the most important steps in that direction was the separation of the representation scheme for the domain knowledge from the techniques by which solutions to various questions could be found. This step was fundamental in keeping the mass of knowledge separate from the reasoning mechanisms. Later this idea evolved into developing systems that consisted of an "inference engine" and a "knowledge base." Essentially, the inference engine is an empty shell that must be filled by the domain knowledge (stored in the knowledge base) in order to perform expert tasks.

5.2 Components of Rule-Based Systems

The structure of rule-based expert systems reflects this separation clearly. Figure 2-2 presents the standard basic architecture for expert systems. Added to

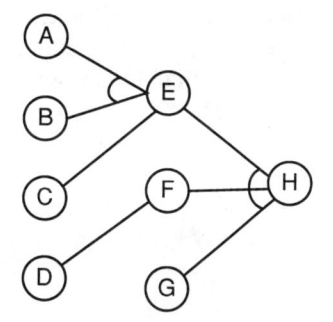

FIGURE 2-1
An AND/OR tree representing a collection of rules.

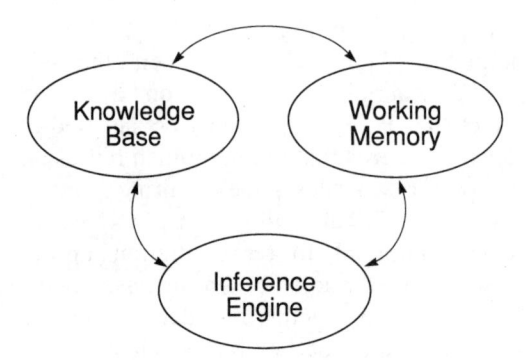

FIGURE 2-2
Main components of a rule-based system.

this bare-bones structure are other components such as the scheduler and the metarule module, which are discussed later.

The knowledge base consists of two parts, known as the "rule part" and the "fact part," in which rules and facts are stored, respectively. The working memory is essentially an extension of the fact base. The difference between the fact residing in the knowledge base and the fact kept in the working memory is that only permanent facts are stored in the knowledge base. The working memory is used for intermediate results. That is, when attempting to solve each individual problem, the system stores in the working memory any new facts that are established, goals to be satisfied, and any intermediary results. That is, prior to the beginning of each consultation, working memory is empty, but its contents change as the search for the solution continues. To make things simpler, consider a database of facts, also called the *context,* with two components: permanent data from the knowledge base and changeable facts, i.e., those related to the particular case at hand. We will see that working memory can play different roles depending on the type of reasoning (e.g., backward or forward).

In explaining how a production system works, for now we consider the simplest case. Other modes of operation will be discussed once different ways of controlling inferences have been studied. The basic algorithm in all cases is a pattern-matching (or pattern-filtering) algorithm. A production system, as presented in Fig. 2-2, operates in cycles. In each cycle, the condition portions (left-hand side) of rules are compared with the current state of facts. When all the conditions of a rule are satisfied, then that rule becomes eligible for execution. Once a rule is executed, the conclusion part (right-hand side) is added to the working memory. The additions may make other rules eligible, and thus another cycle begins in which other rules may be executed. The term *inference engine* is commonly used for the part of the expert system that deals with the selection and execution of rules.

Usually, by using variables that range over classes of objects, rules are written in a general form that makes them applicable to many objects or situations. An example presented previously is rule 5:

```
IF (X    award    none)
   AND (X    academic-load    9)
   THEN (X    tuition-fee    2500)
```

The variable in the above rule may be instantiated to an individual, such as Mary, in order to match the contents of working memory. In that case, we have an instance of the given rule, i.e.,

```
IF (mary    award    none)
   AND (mary    academic-load    9)
   THEN (mary    tuition-fee    2500)
```

5.3   Control Strategies for Reasoning

A number of factors affect the process of reasoning. As part of the process of building an expert system, the knowledge engineer has to make a decision for each one of these issues. For example, the inference engine must know where to start the process, in which direction to pursue the search for a solution, and how to make a choice when several rules are eligible for execution at once. Proper selection of control strategies can greatly affect the efficiency of the system.

BACKWARD AND FORWARD REASONING. One of the useful characteristics of production rules is that they allow for both data-driven and goal-driven computations. Consider the graphical view of a rule base once again. The problem of finding a solution can be seen, essentially, as that of finding a path from the known facts to the desired conclusion on the graph. In searching for such a path, one may start from the initial known facts and go *forward* until the expected conclusion is reached or, alternatively, one may start from the goal and go *backward* until the path is completed. We refer to these two methods as forward reasoning (or forward chaining) and backward reasoning (or backward chaining), respectively. Forward reasoning resembles the data-driven method of computation, and backward reasoning is similar in nature to the goal-directed computations.

Forward chaining can be best described as a "recognize-act" cycle. In each cycle, by matching the contents of the working memory to the left-hand sides of the rules, the expert system determines which rules have become eligible for execution. The interpreter chooses one of these rules, and then it adds the conclusion part of that rule (i.e., the right-hand side) to the working memory. A new cycle can now begin. Let us look at an example.

Example 1. Suppose our knowledge base contains rules 3 and 4 presented previously. Also suppose the following two facts are in the working memory:

```
Fact 1:    (john    award    scholarship)
Fact 2:    (john    standing    graduate-student)
```

There are two relevant recognize-act cycles:

1. Rule 3 can be executed because fact 1 matches its left-hand side. After rule 3 has fired, another fact

    Fact 3:   (john   status   full-time-student)

    will be added to the working memory.
2. Now rule 4 can fire because its left-hand side matches facts 2 and 3. As a result, the fact

    Fact 4:   (john   academic-load   9)

    is added to the working memory.

Since there are no other rules to be executed, the process stops. Thus, by forward chaining, we have concluded that John's academic load is 9 credit hours.

Note that with forward chaining systems, in each consultation, the role of working memory is to contain facts that have been established so far. We will see that when backward reasoning is used, working memory will play an additional role. Note, however, that in both cases the knowledge base and the working memory have completely different purposes.

In *backward chaining,* which is also called *goal-directed reasoning,* we start with the desired conclusion and, by examining the rules, attempt to find a chain of inferences that can establish a link between the known facts and the goal. A better way of illustrating this process is as follows: We assume that our goal is true, and then we try to find evidence to support the assumption. If all the required pieces of evidence are found, we then conclude that the goal is indeed true.

Computationally, backward chaining may be described by a "recognize-reduce" cycle. In many ways, this is similar to a recognize-act cycle, but works in the opposite direction. The basic notion behind the recognize-reduce cycle is that rules are viewed as laws by which we can reduce a goal to a number of subgoals. That is, a rule of the type

$$A1, A2, \ldots, An \rightarrow B$$

may be read, "To show that B holds, show that A1, A2,..., An hold," which means that to solve goal B, we must first solve subgoals A1, A2,..., An.

In each round of the recognize-reduce cycle, we compare the goal (or one of the subgoals) to the right-hand side of rules. When a match is found, then the conditions at the left-hand side of the rule become our new subgoals. The fact base is then searched to see which of these subgoals can be found. The subgoals that are not found will be added to the working memory. One of the subgoals will then be considered in the next cycle.

Example 2. Once again, suppose that we have rules 3 and 4 and facts 1 and 2. This time we attempt to reach the goal

(john   academic-load   9)

by backward chaining. We put the goal in the working memory. The following recognize-reduce cycles will take place:

1. The only subgoal in the working memory awaiting solution is the goal itself. We therefore start with that goal and take it out of the working memory. The goal matches the right-hand side of rule 4. So the conditions at the left-hand side of rule 4 must be satisfied. The triplet

    (john   standing   graduate-student)

    is fact 2 and thus is immediately satisfied. The other condition,

    (john   status   full-time-student)

    will be added to the working memory as a new subgoal.

2. In the next cycle, our only subgoal matches the right-hand side of rule 3. Rule 3 reduces the goal to the subgoal

    (john   award   scholarship)

    which can be found in the fact base. Since all subgoals have been satisfied, we conclude that the original goal

    (john   academic-load   9)

is true.

In contrast to forward chaining where the working memory was used for storing new facts, with backward reasoning all unresolved subgoals are noted in the working memory.

The choice on the direction of search depends on several factors. Some important issues are:

1. *Number of initial facts vs. number of situations acceptable as conclusions:* Obviously the search direction should be toward the larger set. Automated theorem proving (see "Application of Automated Reasoning," by L. Wos, in *Knowledge Engineering: Applications,* vol. II) is a good example of the type of problems that are suitable for backward chaining. It has exactly one goal, i.e., the theorem to be proved, but we had numerous options to begin with. Therefore the reasoning must start from the goal. In fact, we will see that Prolog programs are processed in this fashion. But consider the reduction, i.e., transformation, of mathematical expressions to a different form. We have to start with exactly one situation, but we may have many possibilities for the final form. Thus forward chaining is preferable.
2. *Average number of choices in each direction:* The number of rules that become eligible in each cycle is another important factor. Reasoning should be directed so that the number of possibilities in each cycle is minimized. Let us consider the problem of theorem proving. Forward chaining is not suitable for such problems, since at each step we have numerous choices.

Bear in mind that most of these choices are useless since they would not lead to a solution.
3. *Need for interactive processing:* Many systems allow the user to take part in the process of reasoning by, say, providing additional information. Implementation of such a capability is relatively simpler in backward chaining since when the system cannot establish a certain subgoal, it may consult the user for more information.

 One minor issue must be pointed out. If the fact base is constructed interactively by backward reasoning, then questions that the user must answer may be in the wrong sequence. For example, in MYCIN, which is a rule-based system with backward chaining, the user may be questioned first about the blood type and then about the sex of the patient. This is contrary to the general rule that one of the first things considered in diagnosis of diseases is the sex of the patient.
4. *Need for justification of results:* Backward reasoning facilitates justifications of the suggestions made by the system. When the user asks why a certain conclusion was reached, the system can simply backtrack on the chain that is has found and justify the conclusion by showing that all the evidence necessary for supporting the conclusion has been found.

DEPTH-FIRST VS. BREADTH-FIRST CONTROL. Along with the decision on the direction of search, we must select a methodology for making choices when more than one rule is a candidate for firing. This is a necessity with both forward and backward chaining. Frequently in a step of either the recognize-act cycle or the recognize-reduce cycle we have two or more eligible rules. In such situations, the system must have an effective mechanism to choose between the two. There are numerous strategies for controlling the operation of an inference engine in resolving conflicts. Depth-first and breadth-first strategies are among the more systematic methods. Although these search techniques can be used regardless of the direction of reasoning, it is much easier to discuss depth-first and breadth-first techniques with backward chaining. In a separate section, we present a number of strategies that are primarily used with forward chaining. Consider again the tree that represented our symbolic rule base. To search this tree in depth-first fashion, we pursue each single branch to the end before considering another branch. In backward chaining with a depth-first strategy, the system looks for any opportunity to create a subgoal.

 Example 3. Consider the following collection of rules:

 A, B, C, → D
 A1, A2, A3 → A
 B1, B2 → B
 C1, C2, C3 → C
 A11, A12 → A1
 C11 → C1

The corresponding tree for this rule base is shown in Fig. 2-3.

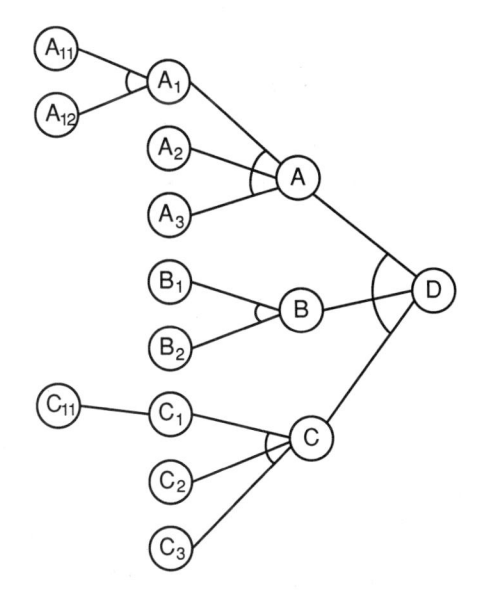

FIGURE 2-3
Tree representation of the rules given in Example 3.

In a depth-first search for establishing goal D, we examine the subgoals in the order shown in Fig. 2-4. Numbers written next to the nodes indicate the order in which the nodes were examined. Note that at each level there are several choices, but we systematically pick one.

With breadth-first search, whenever a goal (or a subgoal) matches the left-hand side of a rule, first we deal with all the premises of that rule before we consider another subgoal. The inference tree presented previously may be searched by the breadth-first method as shown in Fig. 2-5.

Using the analogy of excavation in pursuit of a treasure, in depth-first search we dig deeper and deeper in a spot until either the treasure is found or the limit of the depth is reached. Then we examine and dig another spot. By contrast, when doing a breadth-first search, we sweep across all possible areas before going to a deeper level.

Computationally, the algorithms for these two types of search are very similar. The difference comes from the fact that one employs a stack and the other a queue.

1 Depth-first search
   a Push the goal onto an empty stack.
   b While not "done," check "top" of the stack.
      (1) If the stack is empty, then "done."
      (2) If "top" can be found in the context (i.e., the fact base), then pop the stack; else if "top" is on the right-hand side of a rule, then pop the stack and push the left-hand side of the rule into the stack; else "done."
   c If the stack is empty, then announce "success."

38 KNOWLEDGE ENGINEERING: FUNDAMENTALS

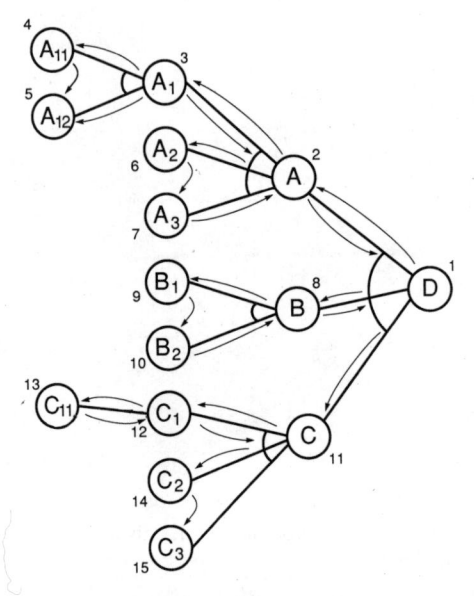

FIGURE 2-4
Depth-first search in an inference tree.

2 Breadth-first search
  a Put goal into an empty queue.
  b While not "done," check "front" of the queue.
   (1) If the queue is empty, then "done."
   (2) If "front" is found in the fact base, then remove "front" from the queue; else if "front" is on the right-hand side of any rule, then re-

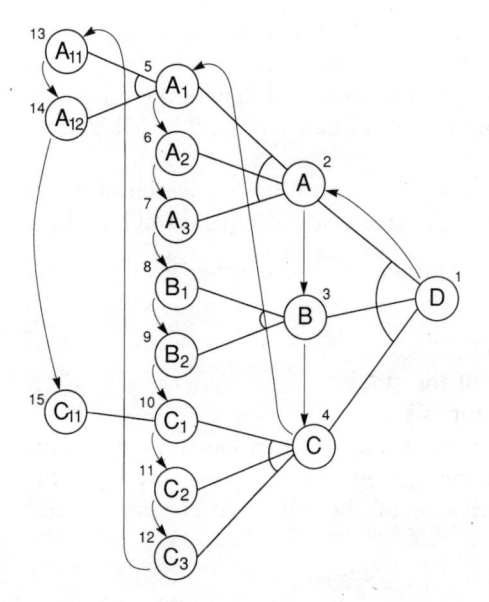

FIGURE 2-5
Breadth-first search in an inference tree.

move "front" from the queue and add the left-hand side of the rule to the (end of the) queue; else "done."
c If the queue is empty, then announce "success."

Note that we have covered only two types of search that are easily applicable to rule bases. There are many other methods for searching on trees (or, more generally, directed graphs) that are less applicable to the type of knowledge base. See, for example, Rich (1983) for a discussion of various search methodologies.

CONFLICT RESOLUTION IN FORWARD CHAINING. With forward chaining, after each step of the recognize-act cycle, one or more rules become candidates for execution. These rules are collected in the *conflict set*. When there is more than one rule in the conflict set, we need to have a strategy (or a combination of strategies) for deciding which rule to fire next. This is called the *conflict resolution mechanism*. Most conflict resolution strategies are simple. Below we discuss a few. See Winston (1984) and Barr (1981) for other possibilities. Often a combination of these strategies is used to ensure elimination of all possible conflicts.

1. *Elimination of executed rules:* Under this strategy, we discard from the conflict set the instantiations of the rule just executed. This ensures that a rule is not allowed to fire more than once on the same set of facts. As a consequence, undesirable loops can be avoided. This strategy is called *refractoriness* in Jackson (1986).
2. *Textual position:* With this method, we choose from the conflict set the rule that appears earliest in the list of all rules existing in the knowledge base. This strategy requires the knowledge engineer to place the more important rules higher on the list. While this method works for small collections of rules, the control mechanism becomes increasingly obscure as the knowledge base grows.
3. *Ordering of rules:* This method requires assignment of explicit priority values to the rules. The rule that has the highest priority number will be chosen from the conflict set for execution.
4. *Specificity:* When the conditions (left-hand side) of one eligible rule form a superset of the conditions of another triggered rule, then by specificity we pick the rule with the larger left-hand side. The logic behind this method is that the rule with more conditions is more specialized to the situation at hand.

    Example. Suppose the following two rules have been triggered and are now in the conflict set. By using specificity, rule 2 will be selected because its conditions subsume the conditions of rule 1.

    Rule 1:   A, B, C → F
    Rule 2:   A, B, C, D → G

5. *Recency:* Under this strategy, priority will be given to the rule that has been added to the conflict set most recently. The idea is that we concentrate

on finding more details about the current subproblem. In some ways, it resembles the depth-first method discussed earlier. Note that we can take the recency strategy to work in the opposite way. That is, we can give priority to the least recent rules, i.e., rules that have been in the conflict set the longest. In this case, we will have some resemblance to the breadth-first method. The recency method requires time tags for the elements of the working memory and subsequently for the rules themselves. This is an overhead that will affect the efficiency of the system.

Other strategies include the following:

- *Data ordering,* which gives priority to certain conditions, and rules that have highest priority conditions are selected first.
- *Size ordering,* in which the rule with the longest list of conditions is selected. (This is different from specificity.)
- *Random choices,* where rules are arbitrarily selected from the conflict set. Random choices are made when none of the other techniques can resolve the conflict.

One of the commercially successful expert system shells, OPS5, uses two different combinations of strategies for conflict resolution. They are called LEX and MEA, and the knowledge engineer may select the one that is more suitable to the particular area of concern. Both LEX and MEA make selections on the basis of first recency, then specificity, and finally arbitrary choices. MEA, however, has a provision that allows the inference engine to focus more on the processing of the current subject or task.

SUMMARY. In constructing a rule-based system, several key decisions, which influence the operation of the system significantly, have to be made. The knowledge engineer must choose a direction for reasoning. Forward reasoning is simple and easily understandable but may involve some useless computations. Backward chaining avoids unnecessary searches but, when reasoning is interactive, may create counterintuitive situations for the user.

To resolve conflicts when two or more rules are candidates for execution, the knowledge engineer should select strategies that effectively make the appropriate choice. Among the more systematic strategies, we saw the depth-first and breadth-first methods. In backward chaining with depth-first search, we pursue each branch to the end before moving onto another branch. With the breadth-first method, however, we deal first with the subgoals of each level before going to the next level. The depth-first strategy may be implemented by using a queue, whereas a stack may be used for implementing the breadth-first method.

A number of other strategies for resolving conflicts were studied. Among them are methods that pick rules on the basis of time of candidacy, textual position, priority, and applicability to the task at hand. The combination of strategies must be selected carefully, so that the possibility of random choices is minimized.

6 VALIDATION OF RULE BASES

Earlier literature in the area of expert systems took a simplistic view of the validation of knowledge bases. Traditionally, it was assumed that if the knowledge base provided reasonable answers and if the expert agreed with its contents, then the system was usable. Such simple validation steps amount to the following. The expert critically examines and verifies each and every rule and evaluates various control strategies employed by the knowledge engineer. Essentially, the expert compares her or his own methods for finding solutions with those built into the system. In this process, a number of cases are presented to both the system and the expert, and their solutions are compared. The more cases studied, the more accurate and better refined the system.

In the evaluation of expert systems in various stages of their evolution, the following basic questions should be considered:

- Does the system provide right answers for the right reasons?
- Are the contents of the knowledge base consistent with the experts' knowledge?
- Since expert systems are often expected to be used by those who have no formal training on computers, are the user-end facilities and capabilities adequate?
- Is there any need for expansion or modification of the representation scheme?

Note that expert systems do not operate on exact knowledge and as such are not expected to provide exact answers. The important issue is whether the answers are "correct" and consistent with the experts' opinions.

Greater reliance on expert systems for solving more complex problems mandates a greater emphasis on the validation process. The traditional approach works fairly well for small knowledge bases. However, for complex problems, which may require a knowledge base containing tens of thousands of rules, we will need a lot more than the confidence of an expert (or a group of experts) before we can use the system for anything serious.

In addition to the ad hoc procedures, we can carry out a number of tests on rule bases to detect some of the simpler bugs. For example, syntactically, we can use a simple procedure that detects all those rules whose condition part can never be satisfied and therefore can never be executed. Such a procedure will look for a contradiction within the condition parts of rules. Here is an example:

> If acid is spilt, and lye is available,
> and no basic material is available,
> then use lye.

The real intention here was to say, "and no other basic material is available." However, since the word "other" was omitted and since lye is considered a basic material, the conditions of the above rule will never be satisfied (Hayes-Roth et al., 1983).

Another type of improvement can be achieved by detecting the rules that are dead-end and facts (or propositions) that cannot be reached. A rule is considered *dead-end* when one of its conditions does not match any facts in the knowledge base. A fact is *unreachable* if it cannot be matched with the right-hand side of any rule. Unconnected nodes or branches in the graphical representations of rule bases are manifestations of these two cases. A simple procedure that traverses the reasoning tree can detect this type of bug (Stachowitz and Comb, 1987).

Often redundancies occur in large rule bases. One case of redundancy is duplication, i.e., when two rules have exactly the same LHS and RHS, possibly with different variable names. Here is an example:

>Rule 7:   IF (STUDENT   award   none)
>   AND (STUDENT   academic-load   9)
>   THEN (STUDENT   tuition-fee   2500)

Note that this is exactly the same as rule 5 presented earlier but with a different variable. The new version uses STUDENT instead of X.

Another case of redundancy is subsumption. A rule R1 subsumes another rule R2 if

1. Rules R1 and R2 have the same RHS but the LHS of R1 is a subset of the LHS of R2.
Example:

>Rule R1.i: A, B, C → F
>Rule R2.i: A, B, C, D → F

2. Rules R1 and R2 have the same LHS's but the RHS of R2 is a subset of the RHS of R1.
Example:

>Rule R1.ii: A, B, C → F, G
>Rule R2.ii: A, B, C → F

3. There is a mixture of the two above.
Example:

>Rule R1.iii   A, B, C → F, G
>Rule R2.iii   A, B, C, D → F

Removal of redundant rules not only decreases the size of the rule base but also contributes to having a more understandable inference process.

Another useful test detects cycles within rules. Cycles may occur directly or indirectly. Direct cycles involve only one rule and are much easier to detect. Indirect cycles involve several rules and are less visible at first. The knowledge engineer must be aware of cycles within the rule base since in some cases (particularly when the control strategies have not been chosen appropriately) cycles may cause infinite loops. Here is an example of indirect cycles:

> Rule C1: A, B → C
> Rule C2: C → D
> Rule C3: D, E → A

We should note that the type of cycles discussed here is different from the recursive rules in, say, Prolog. Logic-based languages such as Prolog owe most of their expressive power to their ability in handling recursion. In cases where the language has provisions for recursion, distinguishing between useless cycles and intentional recursive calls should be done by the knowledge engineer.

Last, inconsistencies must be detected. Some inconsistencies are trivial:

> Rule I1: A, B → C
> Rule I2: A, B → not(C)

Others may be indirect and may involve several rules, for example:

> Rule I3: A → B
> Rule I4: A → C
> Rule I5: B → not(C)

The removal of inconsistencies is considered the most essential step in verifying the knowledge base.

Even though with current technology we can never be absolutely certain that our knowledge base is free of bugs, tests such as those just mentioned briefly can improve the knowledge base to a great extent. Note that knowledge bases are constructed in an evolutionary manner which resists formal verification. In addition, because of the ad hoc (e.g., trial and error) development process, the knowledge engineer may not remember all the earlier insertions. Therefore, the above tests, even though simple, are crucial for large applications.

7 INEXACT REASONING WITH RULES

7.1 Motivations

There are several different reasons why one might need the ability for reasoning under uncertainty. One reason is the unavailability of an exact value for a certain proposition. For example, in the chemical spill program, we may know that acid was spilled but have no knowledge about the amount of the spill. Another reason is the uncertainty about the rules and the inferences that can be drawn by using these rules. A good example comes from the medical field where we often have such statements as, "if symptoms S1, S2, and S3 are observed, then *it is likely that* the patient has disease D." Qualifiers such as *it is likely* signify that the rule is less than certain, and so if it is used in a deduction, the degree of uncertainty must be reflected in the conclusions reached. The third type of inexactness stems from the lack of consensus between experts. Consider the domain of automated legal reasoning where there are institutional provisions for disagreements between different attorneys and judges. To manage this type of inexactness, the system not only must be capable of dealing with the weight of each opinion, but also must find an overall value. Here we briefly examine, at the simplest level, how rule-based systems can handle uncertainty values.

7.2 Reasoning under Uncertainty

In many circumstances, a fact may be believed to be true, but not with complete certainty. Similarly, rules may be less than certain. In such cases, a value representing the level of belief is associated with the fact or the rule. These values are called *certainty factors* or *degrees of belief*. To deal with uncertainty, we need to manage three concepts:

1. Degree of certainty (or uncertainty) of facts, for both those given and those concluded
2. Degree of certainty for compound premises of rules
3. Degree of certainty for rules which themselves may be less than definite

Most systems use a number either in the interval [0, 1] or in the interval [−1, 1] for representing certainty factors (CFs). The interval [0, 1] is more commonly used. In the latter case, a positive CF indicates confirming evidence for a fact, whereas negative CFs indicate opposing evidence or uncertainty. Obviously, CFs 1 and −1 represent absolute knowledge.

We will use a simple notation for representing CFs. The degree of certainty of each rule is written after the conclusion:

$$A, B \rightarrow C \text{ (with CF = 0.7)}$$

Such a rule may be naturally read, "If A and B hold, then there is suggestive evidence (0.7) that C holds." Some authors prefer a slightly different notation where the degree of certainty is written on the arrow. Thus the above rule may be written as

$$A, B \xrightarrow{CF = 0.7} C$$

In the evaluation of certainty factors, the following simple rules may be used. These guidelines are not unique, and several other variations have also been used.

- To compute the CF of the conjunction of a number of facts, find the minimum CF among those facts.

  Example: Suppose A, B, and C have CFs of 0.7, 0.5, and 0.8, respectively. The CF of their conjunction would be minimum(0.7, 0.5, 0.8) = 0.5

- To compute the CF of the conclusion produced by a rule, multiply the CF of the premise by the CF of the rule itself.

Example: Consider the rule

A, B, C → D (with CF = 0.9)

Suppose that A, B, and C have CFs of 0.7, 0.5, and 0.8, respectively. Then, as before, the CF of the premise is 0.5, and the CF of the conclusion D is 0.45 (i.e., the result of multiplication of 0.5 and 0.9).

- When the same conclusion is reached by more than one rule, then the CF of the conclusion is the maximum of the values obtained from all rules yielding that conclusion.

Example: Consider the rules

Rule R1:   A, B, C → D (with CF = 0.9)
Rule R2:   E, F → D (with CF = 0.8)

and suppose that A, B, C, E, and F have CFs of 0.7, 0.5, 0.8, 0.9, and 0.7, respectively. Then the CF of D is the maximum of the two values produced by rules R1 and R2. Rule R1 gives D the CF of 0.45, as we saw before. By using rule R2, the CF of D is

min(0.9, 0.7)*0.8 = 0.56

Thus D gets the CF of 0.56

Note that for the third step to apply, we need several rules with the same conclusion to *succeed,* each indicating a degree of certainty for that fact. The MYCIN inference engine uses a different method for combining certainty values. Suppose two rules succeed, both concluding fact A, one with a CF of 0.6 and one with a CF of 0.5. The two values may be combined in a cumulative manner as follows. By establishing a CF of 0.6 for A by the first rule, the amount of uncertainty is 0.4. The remainder, 0.4, is multiplied by the CF of 0.5 obtained from the other rule, to give the value 0.2 which is added to the previous value of 0.6. Therefore the combined CF is 0.8. In general, when a fact is concluded by two rules with certainty factors a and b, then the combined CF is

a + (1 − a)*b = a + b − ab

The reader may verify that when three rules conclude the same fact with values a, b, and c, then the combined value is given by

a + b + c − ab − ac − bc + abc

The idea behind this method is that as more positive information becomes available, the confidence in the conclusion increases.

8 BUILDING A RULE-BASED SYSTEM

8.1 Steps in the Process of Development

Artificial intelligence is considered to be an experimental science. Building AI systems, in general, involves many experimental steps and trials and errors. Expert systems are no exception. Although many criticize the evolutionary process of expert system development, the domains for which expert systems are considered good candidates, by their nature, resist an algorithmic process. Remember that expert systems are most helpful in complex domains where there is no systematic way of making decisions.

Generally, construction of an expert system goes through several stages and at each stage involves many iterations. The process requires patience from both knowledge engineers and experts. The following are some of the main steps.

1. Task definition. At this stage, the class of problems that the system is expected to solve, along with criteria for solutions, is identified.
2. Identifying resources. Such things as available expertise, computing facilities, and funds must be determined.
3. Formalization of knowledge and formulation of rules. The important issue here is finding key concepts and identifying the relationships between them.
4. Implementation and refinement of a prototype by
   a Creating the rule base
   b Implementing the method of inference
   c Experimenting with the system
   d Analyzing important issues
   e Generalizing the rules
5. Validation and testing. Bugs, redundancies, and inconsistencies must be detected and removed from the system. Comprehensive testing must be done on a number of diverse problems.

As we mentioned earlier, it takes several iterations over some of or all these steps before the system can be released for general use.

8.2 Other Components for Rule-Based Systems

A working expert system will need several other components that we have not yet discussed. So far, we have studied only the knowledge base, the working memory, and the inference engine. Figure 2-6 illustrates a more elaborate architecture. Although most of the elements are self-explanatory, we say a few words about each module.

The *user interface module* receives queries and requests for explanations from the user, and it activates the appropriate actions in the systems. It also ensures that results are presented in a readable and suitable manner.

Once the system has found a solution, the user may wish to ask for an explanation. The reason for such requests may be disbelief of the solution or

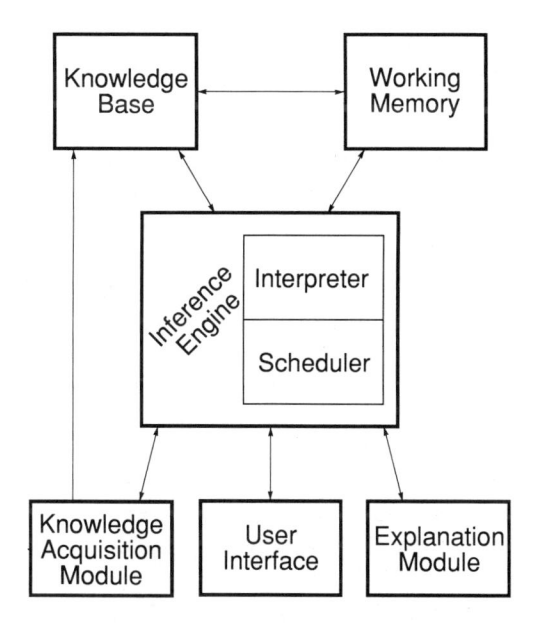

FIGURE 2-6
Architecture of a rule-based system.

simply testing of the inference process used by the system. The *explanation module* traces the rules invoked during the process of problem solving and explains why each rule was fired in that particular time. This serves several purposes, the most important of which is reassuring the user of the validity of the reasoning chain.

The *knowledge acquisition module* can be a simple program that receives data and rules from the knowledge engineer, carries out syntactic checks, and places the new addition into the knowledge base. More advanced versions may do validation checking before inserting the new rules.

The *scheduler* keeps an agenda (containing the activated rules) for the inference engine. By examining the priorities of rules and the strategies for rule selection, the scheduler chooses rules for the rule interpreter. The scheduler and the rule interpreter are considered to be the most important parts of the inference engine.

The modules shown in Fig. 2-6 vary from one system to another. In addition to these subsystems, several other modules have been proposed for rule-based systems.

9 EVALUATION OF RULE-BASED TECHNOLOGY

Although our study of rule-based systems has not been exhaustive, we have covered most of the important concepts. The rule-based technology has come a long way to arrive at its current state, and many additional techniques that are being investigated in research laboratories will affect its future. Rule-based systems are by far the most commercially successful AI products. This market demand will be an incentive for producers of these systems to improve on the shortcomings.

The rule-based approach has a number of clear advantages over other techniques. We will see that some of the positive features may lead to restrictions or inefficiencies. Here are some observations regarding the power and weaknesses of production systems.

The first and foremost advantage is the simplicity of rules. As we have seen from the examples, it is fairly easy to express knowledge, particularly what is known as the *know-how knowledge,* in the form of rules. In addition, this representation scheme has the advantage of uniformity. Rules are used for the representation of all kinds of knowledge, such as the operational knowledge and metaknowledge.

Rules are believed to be independent of one another. The independence allows new rules to be added to an existing rule base, thus making expansion possible. It is this feature that enables us to construct knowledge bases incrementally, step by step.

It is believed that the majority of experts have little difficulty stating their knowledge in the form of rules. Also many programmers find it easy to write code in this form. Because of their simplicity and ease of use, rule-based systems have been favored for software prototyping. Time is a crucial factor in prototyping, and speed becomes an important factor. Given a program specification (which should have all the information about *what* the program is supposed to do), the programmer may code that information in rules and run the rule base on an inference engine to get preliminary results. Obviously, such a product will be neither efficient nor general. However, prototypes are expected to be quick and easy rather than elaborate and complex.

Since in production systems the knowledge part is separated from the inference engine, the global control strategies are context-free. The selected control strategies may be applied regardless of the situation. This provides a sense of uniformity for the operations of the systems. However, bear in mind that generality imposes lack of flexibility. Therefore, if we want to make exceptions to the general rules, it may be either impossible or too inefficient.

Despite the strong features mentioned above, creation of a successful rule-based system may have several pitfalls. Most of these problems are being addressed by the research community, and it is hoped that future generations of production systems will avoid these complications. Note that most problems arise when we deal with large knowledge bases.

Recall that one of the advantages listed for rules was that they are modular in the sense that each one can be considered a separate entity. This level of modularity is adequate for small knowledge bases where the number of rules is not large. However, when we are dealing with large-scale systems, the granularity of rules is so fine that such a collection cannot be viewed as modular by any means. That is, a rule base containing, say, 200 rules may be considered a collection of 200 modules, but a rule base with, say, 100,000 rules is just a monolithic collection of unordered and unstructured tiny modules. A great deal of effort has been expended by both the software engineering community and the AI community to find techniques for the structuring of rule bases. One method employed by several AI tool-building companies is to use a combination of frames and rules for knowledge bases. In such hybrid systems, frames classify rules into groups, where each group represents a coherent concept or an object. Thus we take advantage of any explicit structures, such as taxonomic or cause-effect relationships, that the domain provides. Another method is to assign an index to those rules relevant to a particular task or goal. Then rules can be grouped into modules with respect to their indices.

Although so far we have viewed rules as independent entities, in reality, particularly when the volume of knowledge increases, undesirable interactions between rules cannot be avoided. This fact also relates to the problem of modularization. The existence of unnoticed interactions between rules makes the structuring of the rule base even more difficult.

While rules are very well suited for representation of empirical information, such as associations between cause and effect, they are not adequate for certain other types of knowledge. For example, in a medical expert system, histories of patients are very important. However, with current methodologies we need to rely on other systems, i.e., databases, for such information. In the past few years, the integration of expert system technology and database technology has received considerable attention. Often, the phrase *expert database system* is used to refer to this integration. One of the primary aims in this area is to give the expert system database capabilities. Note that there are some fundamental differences between the methods used in databases and those used in expert systems. One primary difference is that databases rely on online secondary storage, whereas current knowledge bases reside on main memory. There are three levels for the integration of expert systems and databases, as illustrated in Fig. 2-7. The simplest and most basic method is to create a communication channel between the two systems, whereby the expert system can send queries to the database and subsequently receive answers. The next level of integration is achieved when the two systems cooperate with each other. For example, the expert system, which knows the structure and the schema of the database, formulates its queries according to the modus operandi of the database. On the other hand, the database system returns its answers in a format that is readily usable by the expert system. The highest level of integration may be achieved when we can construct systems that have both database management capabilities (i.e., the ability to store, retrieve, and update large volumes of data) and expert system capabilities (the ability to reason with knowledge and produce more information based on existing information). Figure 2-7*a, b,* and *c* illustrates these three levels of integration. More on this topic can be found in Kerschberg (1986).

Debugging of rule-based systems is difficult for two primary reasons: obscure control and nontransparent behavior. The knowledge engineer con-

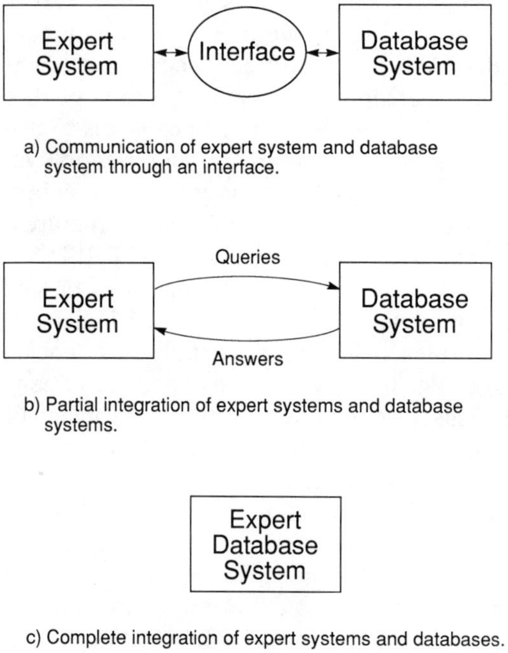

FIGURE 2-7
Integration of expert systems and database systems.

a) Communication of expert system and database system through an interface.

b) Partial integration of expert systems and database systems.

c) Complete integration of expert systems and databases.

stantly has to keep in mind the strategies used for conflict resolution, to ensure that the right rules will be executed at the right times. In particular, the addition of rules becomes a serious problem because the knowledge engineer has to find the correct location on the list of rules for each new addition.

Finally, the slowness of rule-based systems lessens their effectiveness for large applications. Specialized hardware for the processing of rules is being investigated, and any success in this area is bound to improve the speed of execution.

Despite these criticisms, the rule base technology will continue to be a primary methodology for the implementation of expert systems. The use of frames along with rules is one of the most promising avenues for further improvements. Once more sophisticated rule compilers and specialized hardware are developed, we can expect many of the current shortfalls to disappear.

ACKNOWLEDGMENTS

The author is thankful to the following people for their assistance. Comments from the editor, Prof. H. Adeli, improved the manuscript considerably. E. Cortes-Rello, T. Maghrabi, and W. Scott provided many helpful suggestions. Financial support from Bull H.N. Information Systems Inc. is gratefully acknowledged.

REFERENCES

Barr, A., and E. A. Feigenbaum, *The Handbook of Artificial Intelligence,* vols, 1, 2, Addison-Wesley, 1981.
Hayes-Roth, F., D. A. Waterman, and D. Lenat, *Building Expert Systems,* Addison-Wesley, 1983.
Jackson, P., *Introduction to Expert Systems,* Addison-Wesley, 1986.
Kerschberg, L., *Expert Database Systems, Proceedings of Workshop,* Benjamin Cummings, 1986.
Newell, A., and H. A. Simon, *Human Problem Solving,* Prentice-Hall, 1972.
Post, E. L., Finite combinatory processes, formulation I, *Journal of Symbolic Logic* 1:103–105 (1936).
Rich, E., *Artificial Intelligence,* McGraw-Hill, 1983.
Stachowitz, R. A., and J. B. Comb, Validation of expert systems, *Proceedings of Hawaii International Conference in Systems Sciences,* Kona, Hawaii, January 1987.
Winston, P. H., *Artificial Intelligence,* 2d ed., Addison-Wesley, 1984.

CHAPTER 3

KNOWLEDGE ACQUISITION SYSTEMS

BRIAN R. GAINES

1 KNOWLEDGE-BASED SYSTEMS

Knowledge acquisition is recognized as a major bottleneck in the development of knowledge-based systems. The first generation of knowledge acquisition techniques in the 1970s was based on human interviewing. The second generation in the 1980s has been based on computer-aided application of specific techniques such as repertory grids, behavior modeling, and text analysis. Current third-generation research is targeting tools for the 1990s which will provide highly integrated knowledge acquisition environments offering a wide range of complementary tools and techniques. This chapter surveys the state of the art in knowledge acquisition, providing a framework for the wide range of techniques in use, describing the major ones in detail, and illustrating them. It aims to give a solid foundation for tool developers and an indication of appropriate techniques for system developers.

This chapter focuses on the *acquisition* of knowledge for knowledge-based systems. It is primarily concerned with the *transfer* of knowledge from existing sources, such as experts and books, rather than the *learning* of knowledge from experience. *Machine learning* is an important aspect of knowledge-based systems, but it has a major literature of its own (Michalski and Carbonell, 1983; Michalski, Carbonell, and Mitchell, 1986) which is in large part distinct from issues of knowledge and expertise transfer from existing sources. In a broad sense, learning and transfer processes are both part of "knowledge acquisition" for knowledge-based systems and have many issues in common, particularly those of knowledge representation and inductive modeling. However, most practical knowledge-based systems currently acquire their knowledge base through the transfer of existent knowledge rather than learning from experience, and a major literature has developed about such transfer, its problems, and tools and techniques to overcome them (Gaines and Boose, 1988; Boose and Gaines, 1988). This literature is surveyed in this chapter.

One caveat should be added to the above remarks: The notion of *existent* knowledge is rather more complex than might appear at first. Knowledge transfer is rarely a straightforward problem of obtaining knowledge in some form and encoding it for the computer. For example, experts may be able to *perform* complex tasks and yet have major problems in *understanding* and *communicating* how they do this (Bainbridge, 1986; Gaines, 1987b). Understanding and communication are skills quite distinct from performance, and individuals may have widely differing capabilities in each of these three. The practitioner, the scientist, and the teacher are three distinct roles that are developed in differing degrees in different individuals. In knowledge acquisition it may be necessary to build a framework for the understanding and communication of expertise which did not previously exist. It is these types of issues that make knowledge acquisition for knowledge-based systems a complex and fascinating domain, of great practical significance, interacting with major research issues relating to human cognition, knowledge transfer processes in society, and computational problems of supporting and automating such processes.

Figure 3-1 shows the chain of information flow between expert performance of a task and client performance of the same task supported by an expert system.

The person in the role of an expert is shown as having three associated roles, or internal processes: performing the task; understanding the basis of the performance; and communicating the basis of the performance, termed the *practitioner, scientist,* and *teacher*, respectively, in the preceding paragraphs.

The person in the role of a knowledge engineer is shown as having three similar roles: communicating about the basis of the performance with the expert; understanding the basis of the performance at a level sufficient to specify it for a computer; and programming this specification in a form suitable for communication to an expert system on a computer.

The computer in the role of an expert system is shown as having three major internal processes: communicating about the basis of the performance with the knowledge engineer; "understanding" the basis of the performance in terms of its knowledge representation and inference processes at a level sufficient to generate advice for a client; and advising on specific task performance through communication with a client.

The person in the role of a client is shown as having three roles similar to those of the expert: communicating about the basis of specific performance with the expert system; understanding this at a level sufficient to carry out the performance; and performing the task based on the advice of the expert system.

The shaded arrow between the communication roles in the expert and client indicates the advisory role of the expert that is being emulated in part through the expert system. The actual flow of information involved in this ad-

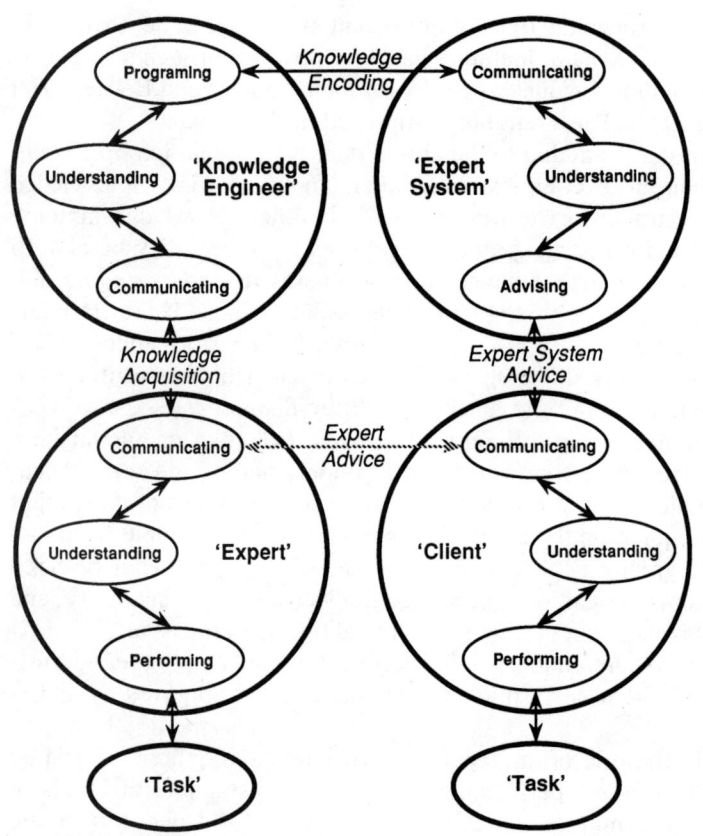

FIGURE 3-1
Information flow between expert performance and client performance.

vice consists of knowledge acquisition by the knowledge engineer from the expert; knowledge encoding by the knowledge engineer into the expert system; and then expert system advice from the expert system to the client.

The problem of expertise elicitation from a skilled person is well known in the literature of psychology (Nisbett and Wilson, 1977; Broadbent, Fitzgerald, and Broadbent, 1986). Hawkins (1983) has analyzed the nature of expertise and emphasizes its severe limitations and dependence on critical assumptions which are often implicit. Dixon (1981) has surveyed studies showing that much human activity is not accessible to awareness. Collins (1985) has studied knowledge transfer processes among scientists and suggests that some knowledge may not be accessible through the experts, not only because they cannot express it, but also because they may not be aware of its significance to their activity. Bainbridge (1979, 1986) has reviewed the difficulties of verbal debriefing and notes that there is no necessary correlation between verbal reports and mental behavior and that many psychologists feel strongly that verbal data are useless.

These are the main problems identified in accessing an expert's knowledge (Gaines, 1987b):

- "Expertise" may be fortuitous. Results obtained may be dependent on features of the situation which the expert is not controlling.
- Expertise may not be available to awareness. An expert may not be able to transmit the expertise by critiquing the performance of others because he is not able to evaluate it.
- Expertise may not be expressible in language. An expert may not be able to transmit the expertise explicitly because she is unable to express it.
- Expertise may not be understandable when it is expressed in language. An apprentice may not be able to understand the language in which the expertise is expressed.
- Expertise may not be applicable even when it is expressed in language. An apprentice may not be able to convert verbal comprehension of the basis of a skill to skilled performance.
- Expertise expressed may be irrelevant. Much of what is learned, particularly under random reinforcement schedules, is superstitious behavior that neither contributes to nor detracts from performance.
- Expertise expressed may be incomplete. There will usually be implicit situational dependencies that make explicit expertise inadequate for performance.
- Expertise expressed may be incorrect. Experts may make explicit statements which do not correspond to their actual behavior and lead to incorrect performance.

Thus, the many processes in the long chain of information flow shown in Fig. 3-1 are each a potential source of weakness in the transfer of expert performance from the expert to the client through a knowledge engineer and expert system. However, the fundamental issues underlying knowledge acquisition should not distract attention from the practical objectives of knowledge acquisition research: to develop effective knowledge-based systems. The concept of a knowledge-based system can be put in operational terms that do not involve any grand dreams of artificial intelligence or major advances in cognitive science.

- Computing technology is still far from the general intelligence and commonsense of people.
- However, in well-defined domains it is possible to build models of likely transactions and use them to aid and advise people.
- If these models are *explicit* (not embedded in code) and *high-level* (understandable in noncomputer terms), then it is reasonable to call the data structures involved *knowledge*.

- The advantages of such explicit, high-level knowledge-based systems are that they can be more easily *developed by end-user communities* and that the knowledge structures can be used in *explanations* such that the systems appear reasonably *intelligent*.

Put in these terms, the development of knowledge-based systems may be seen as part of the continuing evolution of data processing toward increased end-user understanding and control. This chapter encompasses both the fundamental and the practical aspects of this evolution.

2 KNOWLEDGE-BASED SYSTEMS IN COMPUTING

In the early days of systems, the nature of these systems appeared reasonably well defined. They were computer-based decision support systems that emulated the performance of a human expert based on knowledge obtained from that expert. However, as a variety of systems were developed and applied, some aspects of this definition became less clear.

- Does it matter if the system emulates an expert, if it performs well?
- Does it matter that the expertise is obtained from experts rather than other sources?
- Does it make sense to combine other decision support technologies, such as computation and simulation, and how do they interact effectively?
- Are expert systems intrinsically in a decision support role, or can they be used for direct control, for information retrieval, and so on?

What has occurred is an increasing integration of expert systems with other computational approaches, techniques, and methodologies, and a resultant blurring of distinctions.

It is useful to gain a perspective on this evolution of knowledge-based systems and their integration with conventional computing technology, by taking the viewpoint that computers were developed to support human knowledge processes, and all computing applications may be viewed in this light. What, then, differentiates expert and knowledge-based systems? Figure 3-2 classifies human knowledge processes and lists the relevant techniques and supporting systems, precomputing and postcomputing (Gaines, 1988c). The basic level is that of human *experience* as the generative substratum for knowledge processes. This has two major refinements, corresponding in engineering terms to *modeling*, allowing *anticipation* of potential future experience, and *control*, allowing the world being experienced to be changed through *interaction* with it.

In the survival of the species, it is important that experience and the lessons learned from it be shared. Others may be able to learn from experiences more effectively than those who have them. Individuals are finite and expendable, but their experience is part of the knowledge economy of the species. Hence the next level is *communication* of experience. This has two major refinements: accurate *description*, allowing others to model the experience, and

Human Knowledge Processes	Relevant Techniques	Supporting Systems Pre-Computing	Supporting Systems Post-Computing
Experience	Perception, Memory	Instruments, Data Recording	Instrumentation, Data Storage
Anticipation	Cognition, Modeling	Modeling Techniques, Statistics	Data Analysis
Interaction	Action, Control	Tools, Mechanisms	Simulation, Robotics
Communication	Mimicry, Rhetoric	Pamphlets, Novels, Drama, Movies	Games
Description	Representation, Language	Representational Painting & Writing, Phonograph, Television	Word Processing, Graphics, Information Retrieval
Discourse	Argument, Communicative Action	Books, Journals, Journalism, Telephone, Education System	Electronic Mail, Hypermedia, CAL, Natural Language
Reasoning	Induction, Abduction, Analogy	Jurisprudence Systems, Uncodified 'Commonsense'	Semantic Nets, Conceptual Graphs, 'Advice Systems'
Deduction	Mathematics, Set Theory, Arithmetic	Calculators, Diagraming Techniques	Number Crunchers, Relational Databases, 'Expert Systems'
Logic	Propositional & Predicate Calculi, Modal Logics	Proof Techniques, Sequent Calculi	Algebraic Manipulators, Theorem Provers
Category Theory	Universal Algebras, Topology, Homology	Metatheorems, Proof Techniques	Category-Theoretic Programming Languages

FIGURE 3-2
Human knowledge processes, relevant techniques, and supporting systems.

discourse, persuading others to adopt one's own model. Discourse allows for indirect action upon the world by persuading others to act in one's place. It involves the use of communication intentionally to affect the behavior of others and, in particular, to communicate not just experience but *understanding.*

Again, in the survival of the species, it is important that experience and the lessons learned from it be detached from the discourse processes of particular individuals, with whom interaction may not be widely available. Hence the next level is that of *reasoning,* concerned with the creation of knowledge from experience and its application to the generation of experience, i.e., determination of the meaning, significance, consequences, actions, and principles based on experience. Three major refinements are shown: the development of models and their consequences through *deduction* based on arithmetic and mathematics, formal reasoning based on *logic,* and metatheoretic foundations of mathematics and logic through *category theory*.

The techniques and support systems underlying these knowledge processes are shown in three columns in Fig. 3-2:

1 Experience
   a Relevant techniques: human interaction with worlds through perception, memory, cognition, and action
   b Supporting technology: instruments for enhancing perception; techniques for enhancing modeling; and tools for enhancing action
   c Computing technology: the support of natural and artificial experience, anticipation, and interaction through instrumentation, data analysis, simulation, and robotics
2 Communication
   a Relevant techniques: interpersonal interaction through mimicry and rhetoric; representation of experience through media and description of experience through language; mutual understanding of experience through argument
   b Supporting technology: literary forms for presenting experience; painting, writing, phonograph, and television as media for capturing experience; books, journals, and the education system supporting mutual understanding
   c Computing technology: games giving access to stylized experience; word processing and graphics as media representing and presenting experience; electronic mail, hypermedia, and computer-aided learning supporting mutual understanding
3 Reasoning
   a Relevant techniques: inductive and abductive inference and analogical pattern recognition; set theory, arithmetic, logical calculuses, topology, and homology
   b Supporting technology: surprisingly little—jurisprudence through the legal system to support justice in socially important practical reasoning, but primarily left to uncodified "common sense"; calculators, diagraming, and proof techniques at the formal level
   c Computing technology: knowledge representation and inference through semantic nets and conceptual graphs; numeric computing, relational databases, and expert systems; algebraic systems and theorem provers; category-theoretic programming languages

Looking down the right-hand column of Fig. 3-2, we see the full spectrum of computer-based techniques supporting human knowledge processes. Note how the development of computing has straddled the central region of discourse and practical reasoning. Computing commenced with number crunchers, moved into simulation and instrumentation, and then went into databases, graphics, and text processing. The support for the explanatory processes of discourse and the argument forms of practical reasoning is still very weak, even though these underlie our colloquial use of the term *expertise*. Current expert systems largely support formal reasoning and may be seen as a natural extension of database technology to allow retrieval of logically derived fields.

This weakness in the emulation of expert knowledge processes is clear in the knowledge acquisition literature where it appears as a gap between the knowledge elicited from experts and its accurate encoding for an expert system shell (Rappaport, Gaines, and Boose, 1988). The gap is being filed by research on knowledge representation schema better suited to discourse and practical reasoning, but so far these have had little impact on commercially available knowledge-based systems. Expert systems technology has developed primarily at the level of formal reasoning which is a very narrow subdomain of the human reasoning processes.

The diversity of human knowledge processes cannot be overemphasized—indeed, *anarchy* might be a better term than *diversity* (Feyerabend, 1975). "Knowledge" in the AI literature has become associated with formal structures from which inferences can be made. However, much human knowledge is not well structured for formal representation and computer processing (Gaines and Vickers, 1988). The philosophical dictionary definition of knowledge as "more than opinion, less than truth" (Runes, 1972) encapsulates what can be said. Classical data processing methods take it for granted that the data can be assumed to be "true" and base their inferences on this. In knowledge-based systems, we have data that is less than truth but more than opinion because it is "expert opinion." The implications of these notions of untruth (inconsistency, contradiction, change) and expertise (modeling, cognitive authority, reflective equilibrium) are critical to the design of knowledge acquisition systems.

The need for integration of computing technologies is very clear in Fig. 3-2. Practical knowledge support systems must be capable of providing combinations of graphics, text processing, hypermedia, number crunching, databases, communications, theorem proving, and semantic nets and expert system technologies. These are not highly separated functions but rather multiple aspects of human knowledge processing. The questions raised at the beginning of this section are also put into perspective. As part of the total infrastructure of information technology, expert systems provide a deductive reasoning capability that can be used in a variety of ways regardless of its source or relation to human expertise. However, the use of this reasoning capability in conjunction with overt, high-level data structures, or "knowledge," is a significant advance in the usability of information technology, and this does, in part, derive from the emulation of expert sources. However, once the concept and significance of overt, high-level knowledge become clear, it is less important how it is derived than how it is used.

The next section addresses the sources of expert knowledge and the role of the expert in society, giving a framework for the application of information technology and knowledge-based systems as part of the overall knowledge processes of society.

3 KNOWLEDGE ACQUISITION

The preceding section has suggested that knowledge-based systems should be regarded as an integral part of information technology and that the technology itself should be regarded as an integral part of human social processes. This section introduces a model of expert behavior and expertise which gives a practical framework for knowledge acquisition and itself suggests that acquisition should be regarded as an integral part of an expert system's performance.

The distinction between knowledge acquisition and performance systems is an invidious one. It can be taken to imply that expertise is a static collection of skills and knowledge, capable of being elicited and transferred to an expert system that will then be able to reproduce the expertise. This static model is an extremely dubious one in systems design, leading to systems that date rapidly, interfere with the growth of expertise, and have poor reliability and high maintenance costs. The following explains how these conditions arise and approaches for avoiding them.

Hawkins (1983) has abstracted from industrial experience in developing mineral exploration expert systems and proposed a model of human expertise relevant to expert systems. Figure 3-3 summarizes the essential features of this model:

- The expert first elicits data about the problem from the client.
- She or he develops a minimal model that accounts for the data provided.
- The expert generates advice based on the model and feeds this to the client.
- The client may accept the advice or query it, and possibly, the model.
- The queries lead to further data elicitation and repetition of the elicitation-modeling-advice-query cycle.

Thus, in Hawkins' model, the client plays an active role in further developing the model by providing more data until he or she is satisfied with the model and consequent advice. Expert advice giving and taking is part of a cycle of negotiation around a process of model formation.

The particular importance of this model of expert performance lies in the emphasis it places on the embedded knowledge acquisition process. The expert is managing a data collection, modeling, and decision cycle within a consultation. As Fig. 3-3 shows, this implies that the expert is able to enhance her of his collection of case histories, models, and advisory strategies through the consultation. Skilled performance is not the static utilization of expert knowledge and skills in Hawkins' model—it is, instead, a microcosm, and an essential part of, the wider processes of knowledge acquisition which are intrinsic to the role of an expert.

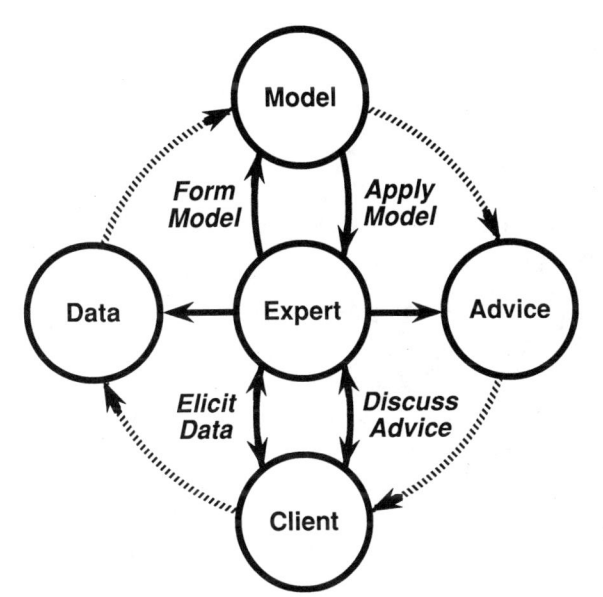

FIGURE 3-3
Negotiation cycle in expert-client interaction.

Access to the learning experiences of attempting to solve the problems of a client community is one major source of knowledge in the development of individual expertise. The client community provides access to practical experience, a variety of problems, particularly novel ones that go beyond existing expertise, and it also manages the growth of expertise through systems of reward, criticism, and access to resources. However, learning from experience is a slow and error-prone process, and socially significant areas of expertise become associated with professional communities that attempt to expedite learning and reduce errors through the sharing of experience. Figure 3-4 extends the model of Fig. 3-3 to include these other processes, such as apprenticeship, instruction, training, education, workshops, conferences, journals, and books. Professional communities usually play a major role in refining and directing the client community's reward, criticism, and resource allocation systems.

The role of the professional community can be seen as an indirect enhancement of the basic processes of knowledge acquisition in the negotiation cycle of expert performance. The individual can acquire the data of others through access to reported case histories and enhance his or her repertoire of models through the published theories developed by others to account for their data and the data of many others. The individual can enhance her or his repertoires of advice through the published advisory strategies developed by others and evaluated by them, or yet others, against past case histories, theoretical foundations, and experience in use.

This model of individual expertise as intrinsically a knowledge acquisition process at multiple levels intrinsically embedded in, part of, and sup-

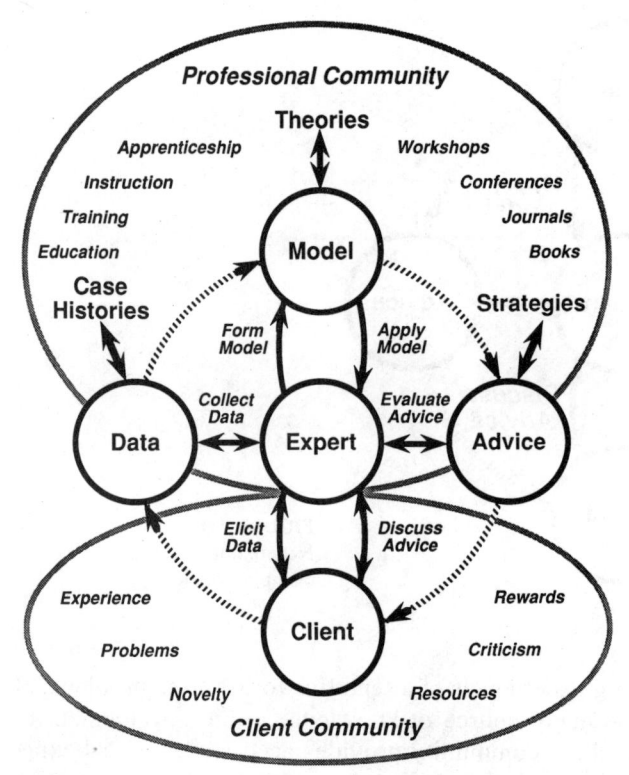

FIGURE 3-4
Extended knowledge acquisition process in expert-community interaction.

ported by essential social processes is consistent with Stich and Nisbett's (1984) view of the social role of experts as managers of the *reflective equilibrium* of the inductive process of knowledge generation in society. This model is also consistent with the emphasis in professional development books concerned with good practice on the importance of balancing experts' authoritarian roles as knowledge repositories and appliers with their entrepreneurial roles as knowledge acquirers (Schön, 1983).

Sociologists have noted the very strong positive feedback processes in the dynamics of knowledge acquisition in the scientific community (Hagstrom, 1965). Merton (1973) coined the term *Matthew effect* for those features of the reward system in research biased toward allocating greater credit for the same discovery to those with an established reputation. The qualitative effect of such positive feedback processes is to amplify random differences within a population, creating strong distinctions which have no other basis than the feedback process itself. That is, even if all scientists were created equal in their capabilities, some would become experts relative to others because random differences in early performance generated credit that affected resource allocation and gave them greater opportunities to acquire knowledge. The strength of this effect can be demonstrated by studies of the behavior of "experts" simulated by learning automata in a competitive learning environment with access to resources related to demonstrated expertise (Gaines, 1988b).

In many areas of expertise where expert systems are expected to play a major role, these positive feedback processes are exceptionally strong. In medicine, e.g., the key learning resource is access to medical problems, and the "owner" of such a problem has a strong personal interest in allowing only someone of very good reputation to handle it. The system, including considerations of legal liability, strongly funnels problems to those who are regarded as experts. It is, however, precisely from these problems that new knowledge is generated. Hence direct experience of case history data is made available primarily to those with established reputations for having capabilities to deal with it. Similar considerations apply to the award of scholarships, invitations to scientific congresses, and so on (Blume, 1974). They also apply not only to individuals but also to social units, such as a company subject to government procurement procedures that are heavily biased to contractors with prior experience and with whom the government agency has prior experience.

The introduction of knowledge-based systems in the processes shown in Fig. 3-4 may be seen as providing technological support to the expert role in the center of the figure. It corresponds to the change from expert advice to expert systems advice shown in Fig. 3-1. The system developer operates in the same environment, and Fig. 3-4 may be restructured to show the sources available to the developer of a knowledge-based system. Figure 3-5 illustrates the knowledge acquisition process with the system developer drawing on three major sources: knowledge encoded in *relevant media,* such as books, journals, and videotapes; knowledge available by observing and modeling the *relevant domain;* and knowledge available from discourse with, and observation of, the *relevant community.* It is this last source that is most associated with expert systems as such, although all three sources play significant roles in the development of any practical system.

Note that the *relevant community* has been split into requirers, experts, and client. Again, domain experts are the expected source of knowledge for an expert system, but detailed requirements specifications and involvement of end-users play as significant a role as they do in other computing system design. It is useful to think of those with these different roles as all being experts in some different aspect of the system functioning, since the knowledge acquisition techniques used with domain experts are often well suited to establishing requirements specifications and end-user perspectives.

The annotation of the *knowledge base* in Fig. 3-5 illustrates the range of technical features which an adequate knowledge-based system shell may be expected to support:

Relevant Media
Text, diagrams & other material

Relevant Domain
Instrumentation, observation & modeling

Relevant Community
Requirers, Experts, Clients

Knowledge Acquisition Process

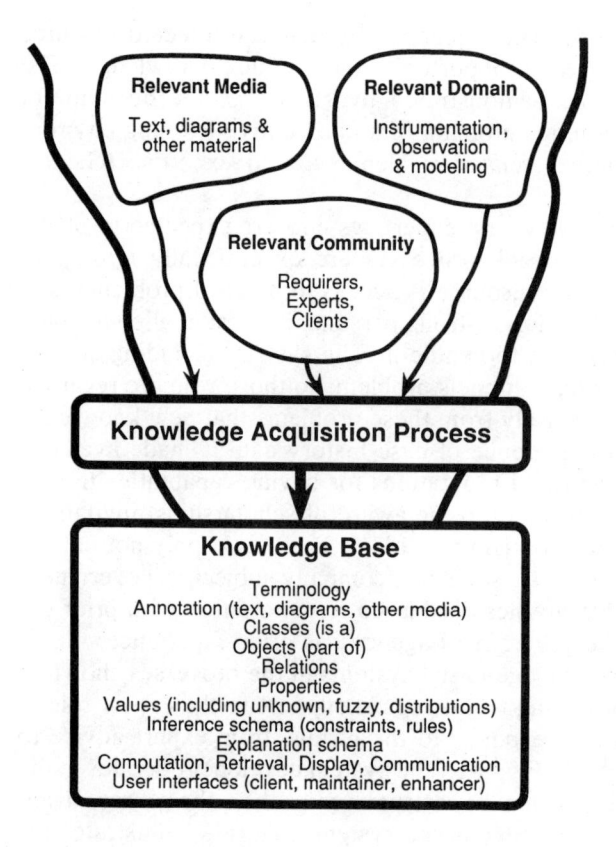

Knowledge Base
Terminology
Annotation (text, diagrams, other media)
Classes (is a)
Objects (part of)
Relations
Properties
Values (including unknown, fuzzy, distributions)
Inference schema (constraints, rules)
Explanation schema
Computation, Retrieval, Display, Communication
User interfaces (client, maintainer, enhancer)

FIGURE 3-5
Knowledge acquisition process for knowledge-based systems.

- *Terminology* is often neglected as computationally significant. However, it is particularly important in knowledge-based systems where it is critical that the terms used in questions, recommendations, and explanations be understood by clients in the same way as they are understood by experts. Obtaining the terminology direct from experts without distortion is an important design consideration for knowledge acquisition systems.
- *Annotation* is used here as a term for noncomputational knowledge such as text, diagrams, photographs, videos, and other media, which may be used to impart knowledge to people. Some of it may be suitable for computer storage but not for inferential processing. Providing appropriate access to such knowledge is important in many knowledge-based systems, and collecting and organizing such knowledge are part of the knowledge acquisition process.
- *Classes, objects, relations, properties, and values* form the abstract infrastructure of computational knowledge representation currently. Object-oriented knowledge bases provide a formal conceptual and computational schema which subsumes and supports a wide variety of knowledge representation schema (Shriver and Wegner, 1987). These primitives also correspond to natural cognitive operations of classifying, instantiating, relating, attribut-

ing, and measuring. An important aspect of knowledge acquisition is to determine these cognitive primitives in a domain (Rappaport, 1987a).

Knowledge representation goes beyond conventional data representation largely in the variety of types of value that must be supported. In a database, a numeric field may be expected to have a clearly defined value. In a knowledge base, because knowledge is "less than truth," it may be necessary to allow for the value's being unknown, constrained by a fuzzy hedge, a range, a distribution, or some combination of these. There may also be mutual constraints between such ill-defined values.

- *Inference schemata* are systems of constraints that express the relations between values in a knowledge base. These constraints may be thought of as rules expressing the relations, and many may necessarily be expressed by explicit rules in the knowledge-base. However, the effective organization of the knowledge base in terms of generic classes and relationships should allow the majority of constraints to be expressed implicitly in a natural way. This is the *framing problem* in knowledge representation for a domain, and an important aspect of knowledge acquisition is to determine a representation schema that minimizes the number of low-level rules required.
- *Explanation schemata* are forms of argument that generate satisfactory explanations of the inferences made from a knowledge base. Early expert systems generated explanations by displaying the chain of reasoning that was used to go from premises to conclusions. This approach is useful but does not necessarily "explain" the conclusions in a way that an expert would, or in a way satisfying a client's need for knowledge. The acquisition of knowledge necessary for adequate explanation goes beyond that adequate for performance.
- *Computation, retrieval, display, communication,* and other data processing activities play major roles in most knowledge-based systems. Modern expert system shells make provision for integration with data processing, simulation, database, graphics, and communications programs. The systems analysis for these needs to be integrated with the knowledge acquisition process.
- *User interfaces* are very important in knowledge-based systems for generally they are highly interactive and involve clients with little or no computer experience. The design of the interface in consultation with experts and clients is an important part of knowledge acquisition. Since most knowledge is fairly volatile, knowledge acquisition and updating the knowledge base are important aspects of system operation. The interface to system maintainers and enhancers also needs careful attention.

The emphasis on the complexity of the social processes underlying knowledge creation, dissemination, and application in society and the need to treat knowledge-based systems as an integral part of information technology make the development of such systems appear somewhat daunting. And so it is—it is easy to underestimate the effort required to develop effective expert systems. However, knowledge acquisition for knowledge-based systems has also tracked other trends in information technology, notably the use of the computer itself to support the system designer through computer-aided software engineering (CASE) (Gaines, 1988d). The following section describes these trends and their effects on the role of the knowledge engineer, and later sections describe and illustrate CASE tools for knowledge-based systems.

4 KNOWLEDGE ENGINEERING

Knowledge acquisition for practical system development has come to be termed *knowledge engineering,* following Feigenbaum's (1980) use of the term to describe the reduction of a large body of knowledge to a precise set of facts and rules. The term *knowledge engineer* describes the person responsible for such system development, and concise job descriptions for knowledge engineers have been given (Hayes-Roth, Waterman, and Lenat, 1983):

> Knowledge acquisition is a bottleneck in the construction of expert systems. The knowledge engineer's job is to act as a go-between to help an expert build a system. Since the knowledge engineer has far less knowledge of the domain than the expert, however, communication problems impede the process of transferring expertise into a program. The vocabulary initially used by the expert to talk about the domain with a novice is often inadequate for problem-solving; thus the knowledge engineer and expert must work together to extend and refine it. One of the most difficult aspects of the knowledge engineer's task is helping the expert to structure the domain knowledge, to identify and formalize the domain concepts.

Thus, the basic model for knowledge engineering has been that the knowledge engineer mediates between the expert and knowledge base, eliciting knowledge from the expert, encoding it for the knowledge base, and refining it in collaboration with the expert to achieve acceptable performance. Figure 3-6 shows this basic model with manual acquisition of knowledge from an expert followed by interactive application of the knowledge with multiple clients through an expert system shell:

- The knowledge engineer interviews the expert to elicit his or her knowledge.
- The knowledge engineer encodes the elicited knowledge for the knowledge base.
- The shell uses the knowledge base to make inferences about a particular case.
- Clients use the shell to obtain advice about particular cases.

Figure 3-6 may be compared with Fig. 3-4 to show the introduction of the knowledge engineer and expert system as intermediaries between the expert and client, as sociotechnical extensions to knowledge processes in society. Figure 3-1 illustrates in greater detail the many assumptions underlying the possibility of successfully introducing intermediaries in this way.

It seemed possible that knowledge engineering might develop as a profession on a par with systems analysis and programming, and that a major

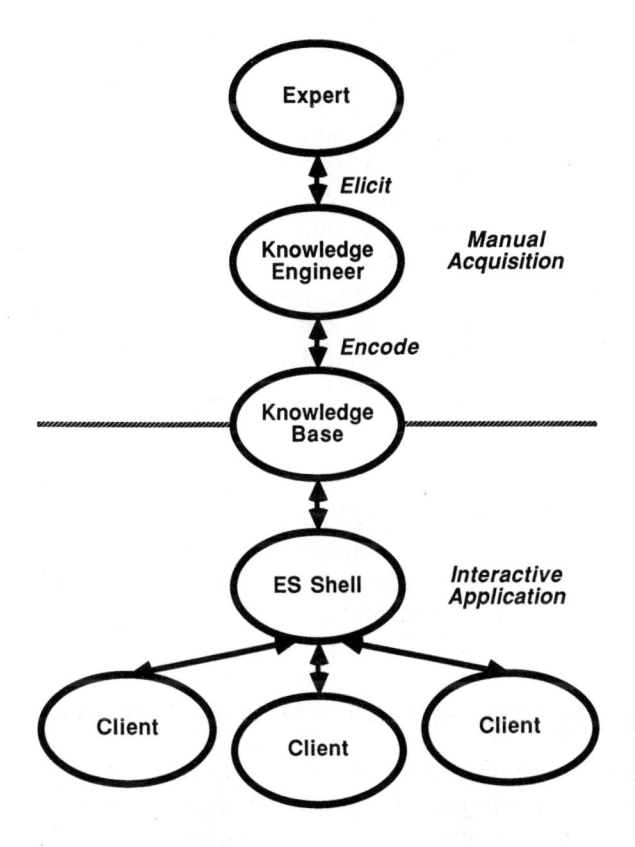

FIGURE 3-6
Basic model of knowledge acquisition by a knowledge engineer interviewing an expert.

shortage of skilled knowledge engineers would cause problems initially that would be overcome eventually as the profession developed. However, this scenario now appears less and less likely. There is certainly a shortage of knowledge engineers, and major problems exist in developing applications, but doubts have been cast on the notion that human labor is the appropriate solution to the knowledge engineering problem:

- The decline in costs of both hardware and software support for expert systems has brought the technology into a mass-market situation far more rapidly than originally envisioned.
- This has led to a growth in demand for expert systems that is proceeding far more rapidly than the growth in supply of trained and experienced knowledge engineers.
- The declining costs of expert system technology are also making the expense of human labor in tailoring the technology for particular applications appear to be the dominating constraint and an excessive cost.
- A move toward a labor-intensive activity such as knowledge engineering is contrary to all trends in industry.

- In particular, such a move is contrary to the trend toward automatic programing techniques in the computing industry.
- The role of the knowledge engineer as an intermediary between the expert and the technology is being questioned not only on cost grounds but also in relation to its effectiveness—knowledge may be lost through the intermediary, and the expert's lack of knowledge of the technology may be less of a detriment than the knowledge engineer's lack of domain knowledge.

These considerations aroused interest in the possibility of providing computer-based knowledge elicitation systems to automate knowledge engineering as a process of direct interaction between domain experts and the computer (Boose and Gaines, 1988). The tools and techniques used in such systems are described and illustrated in detail in later sections. Figure 3-7 shows how the role of the knowledge engineer has changed as such systems have come into use.

Interactive knowledge acquisition and encoding tools can greatly reduce the need for the knowledge engineer to act as an intermediary, but in most applications they leave a substantial role for the knowledge engineer. As shown in Fig. 3-7, knowledge engineers have responsibility for

- Advising the experts on the process of interactive knowledge elicitation
- Managing the interactive knowledge acquisition tools, setting them up appropriately
- Editing the unencoded knowledge base in collaboration with the experts
- Managing the knowledge encoding tools, setting them up appropriately
- Editing the encoded knowledge base in collaboration with the experts
- Validating the application of the knowledge base in collaboration with the experts
- Setting up the user interface in collaboration with the experts and clients
- Training the clients in the effective use of the knowledge base in collaboration with the expert by developing operational and training procedures

This use of interactive elicitation can be combined with manual elicitation and use of the interactive tools by knowledge engineers rather than, or in addition to, by experts. Knowledge engineers can directly elicit knowledge from the expert and use the interactive elicitation tools to enter knowledge into the knowledge base. Figure 3-7 shows multiple knowledge engineers since the tasks above may require the effort of more than one person, and some specialization may be appropriate. Multiple experts are also shown since it is rare for one person to have all the knowledge required, and even if this were so, comparative elicitation from multiple experts is itself a valuable knowledge elicitation technique (Boose, 1987; Shaw and Woodward, 1987; Shaw and Gaines, 1988).

Validation is discussed in greater detail in a later section. It is shown in Fig. 3-7 as a global test of the shell in operation with the knowledge base, i.e., of overall inferential performance. However, validation may also be seen as a

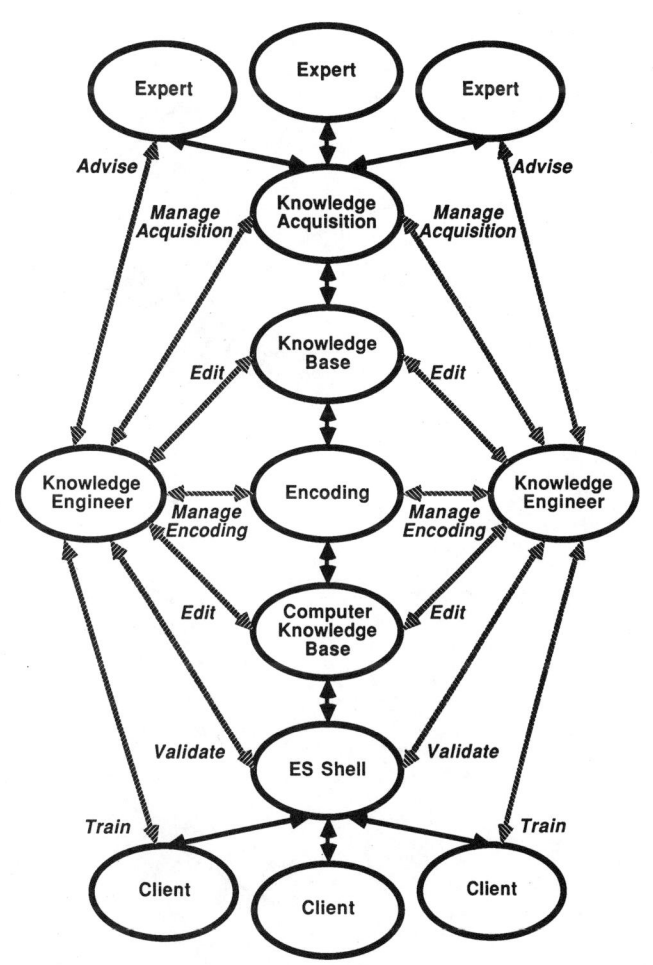

FIGURE 3-7
The knowledge engineers' roles in current system development using automated knowledge acquisition tools.

local feature of each step of the knowledge engineers' activities: the experts' proper use of the tools needs validation, the operation of the tools themselves needs validation, the resultant knowledge base needs validation, and so on. Attention to quality control through validating each step of the knowledge acquisition process is key to effective system development.

Figure 3-7 shows the complexity of the knowledge engineer's role and some of the support tools required. Figure 3-8 groups the support tools for editing, display, encoding, and validating the knowledge bases into a *knowledge-based system support environment* and lumps the various forms of knowledge base together. It shows the overall structure of a knowledge-based system as a

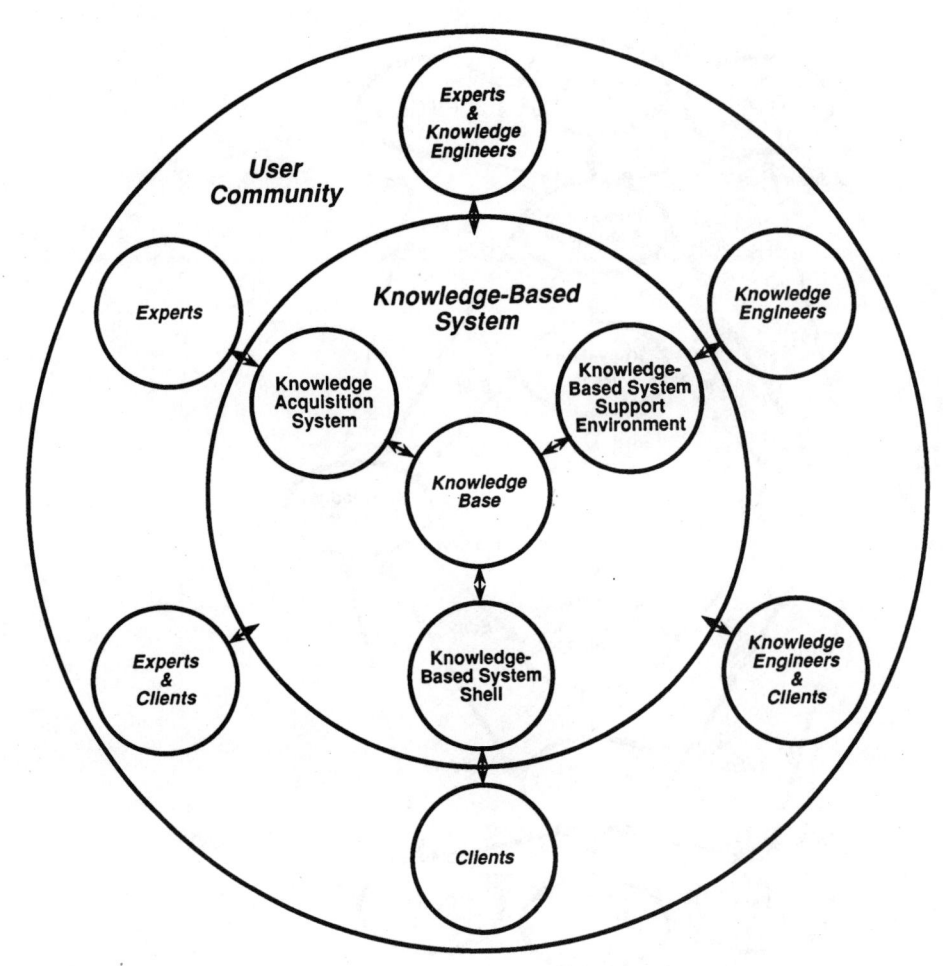

FIGURE 3-8
Major components of a knowledge-based system.

central knowledge base interacting through the knowledge acquisition tools, the expert system support environment, and the expert system shell with a user community of experts, knowledge engineers, and clients.

The notion of a *user community* is important in drawing attention to another trend in knowledge-based systems—that the distinction between experts, knowledge engineers, and clients is often one of variable roles rather than hard-and-fast job definitions. The expert may be a major client for the system or may play a major part in knowledge engineering or may do both. Some clients may customize or update the system in the role of knowledge engineers or extend its capabilities in the role of experts.

5 THE KNOWLEDGE-BASED SYSTEM SUPPORT ENVIRONMENT AND VIRTUAL MACHINE HIERARCHY

The significance of the knowledge-based system support environment shown in Fig. 3-8 in providing tools for the knowledge engineer is clear in the discussion above. The quality of this environment is very important in the development of expert systems, particularly in supporting the interaction between knowledge engineers and experts. The relation of this environment to knowledge acquisition is best seen by putting both in the context of the complete virtual machine hierarchy for knowledge-based systems. Figure 3-9 shows significant levels of this hierarchy:

1. *Machine architecture:* At the lowest level of the hierarchy is the machine on which the knowledge-based system runs. This is really irrelevant, but use of Lisp in the early days of expert systems has focused attention on Lisp machines. Now any good workstation is adequate.
2. *Operating system:* At the second level of the hierarchy is the operating system within which the implementation runs. This system needs to provide good interfaces to other programs, large databases, and communications.
3. *Implementation language:* At the third level of the hierarchy is the implementation language which actually interfaces to the computer. This tended to be Lisp in the early days of expert systems, but as speed and space efficiency have become significant and knowledge representation has become better understood, other languages that support dynamic data structures, such as C and Pascal, have become widely used.
4. *Knowledge-based system shell development language:* At the fourth level of the hierarchy is the language in which the knowledge-based system shell is written, generally a special-purpose environment for coping with knowledge representation and inference.
5. *Knowledge-based system shell:* At the fifth level of the hierarchy is the expert system shell as a run-time environment that elicits problem-specific information from the user, provides advice based on its knowledge base, and explains that advice in as much detail as required.
6. *Knowledge-based system support environment:* At the sixth level of the hierarchy is the equivalent of the application programming support environment (APSE) in conventional systems. The knowledge-based system support environment provides editing, browsing, and debugging tools for the knowledge base. This level encompasses the major differences between the high-quality knowledge-based system shells that support major system development through extensive support tools and the simpler shells that do not.
7. *Knowledge acquisition tools:* At the seventh level of the hierarchy are the tools for automating the systems analysis for expert systems, through automatic interview procedures, modeling of actual expert behavior, and analysis of knowledge in textual form. Automation of systems analysis is more significant for knowledge-based systems than for conventional data pro-

8 — Knowledge Support Systems
Integrate acquisition and performance in a multi-user environment supporting a variety of knowledge processes
AQUINAS, KSS0, NEXTRA

7 — Knowledge Acquisition Tools
Interview experts, model behavior, analyze text, form conceptual structures, generate inference rules
ETS, KITTEN, KNACK, KRITON, MOLE, OPAL, SALT

6 — Knowledge-Based System Support Environment
Display, edit, debug & validate the conceptual structures, inference rules and performance of the knowledge base
KEE, Knowledge Craft, ART, Nexpert

5 — Knowledge-Based System Shell
Elicit problem-specific data, apply and explain the application of the knowledge base
Run-time environments for above systems

4 — Shell Development Language
Provide suitable facilities for the effective representation of knowledge and efficient inferences from it
Prolog, OPS5, OPS83, special-purpose languages

3 — Implementation Language
Provide high-quality implementation environment for complex programs with dynamic data structures
Lisp, C, Pascal, Ada

2 — Operating System
Provide high-quality run-time environment for complex interactive programs with large knowledge bases
Unix, VMS, Aegis

1 — Machine Architecture
Provide cost-effective support for complex interactive programs with large dynamic data structures
Lisp machines, Prolog Machines, Workstations

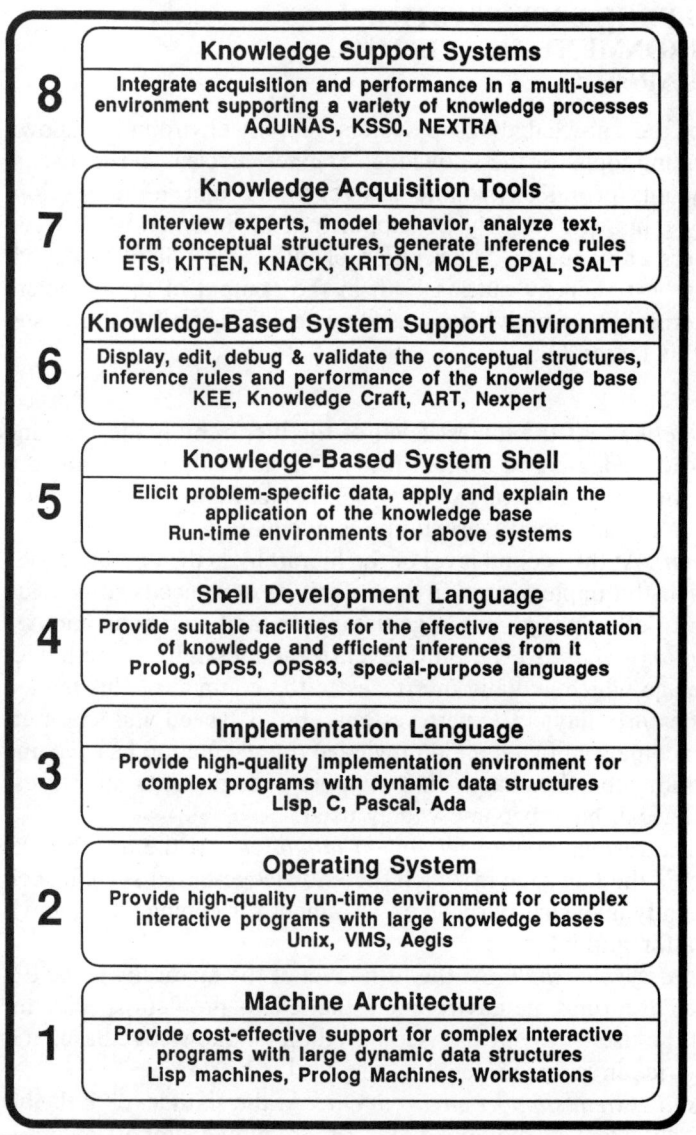

FIGURE 3-9
The virtual machine hierarchy for knowledge-based systems.

cessing because knowledge-based systems tend to be targeted on narrow areas of in-depth specialist knowledge, rather than broad general problems. In conventional terms, knowledge-based systems are satisfying the requirement for highly customized data processing based on idiosyncratic operations rather than general algorithms.

8. *Knowledge support systems:* At the top of the hierarchy are the objectives of recent research in knowledge acquisition which focuses on the integration of acquisition and performance systems to provide an environment supporting a wide variety of knowledge processes.

The hierarchy of Fig. 3-9 highlights some of the problems system managers often face in establishing development capabilities for knowledge-based systems: Should they use Lisp machines? Should they train their staff in Lisp or Prolog? How does an expert system shell compare with existing programming languages? What is knowledge engineering, and how does it differ from systems analysis? The problem is that knowledge-based systems are at the top of the virtual machine hierarchy which has developed in computing, and there are many layers of technology between a knowledge-based system and the computer on which it is running. In theory, system developers should not need to know about the lower levels of the hierarchy—machine architectures, operating systems, and implementation languages are remote from knowledge processing. But, in practice, these lower levels are the foundations on which systems are built, and any defects in them can undermine the functionality of the upper levels.

The virtual machine hierarchy for computing is very complex and has many branches. Figure 3-9 focuses on a path through the hierarchy that links the basic hardware and software to knowledge-based systems and knowledge acquisition issues. It omits other parts of the hierarchy that illustrate the integration issues discussed previously. Figure 3-10 shows some of these issues within a hierarchical framework (Gaines, 1987a).

The split between users and technology at the top of Fig. 3-10 emphasizes the importance of effective person-computer interaction in knowledge-based systems. As with much modern information technology, usability considerations dominate effectiveness—if the interface is poor, no amount of advanced technology in the rest of the system can compensate.

The tree below this to the top left depicts the distinctions between user roles and, in particular, the tasks of the knowledge engineer already discussed. The tree to the top right first splits the technology between knowledge-based and conventional components, depicting some of the integration issues already discussed, and then further splits the knowledge-based components into support environment, shell, and acquisition.

The remainder of this chapter focuses first on the tree to the bottom left, analyzing the knowledge-based system support environment and validation, and then on the tree to the bottom right, analyzing and exemplifying the methodologies, tools, and techniques used in knowledge acquisition.

6 THE KNOWLEDGE-BASED SYSTEMS SUPPORT ENVIRONMENT AND VALIDATION

The tree on the lower left of Fig. 3-10 shows some of the major issues in the knowledge-based system support environment at level 6 in Figs. 3-8 and 3-9 in terms of the distinctions involved. The basic tools required are for editing, dis-

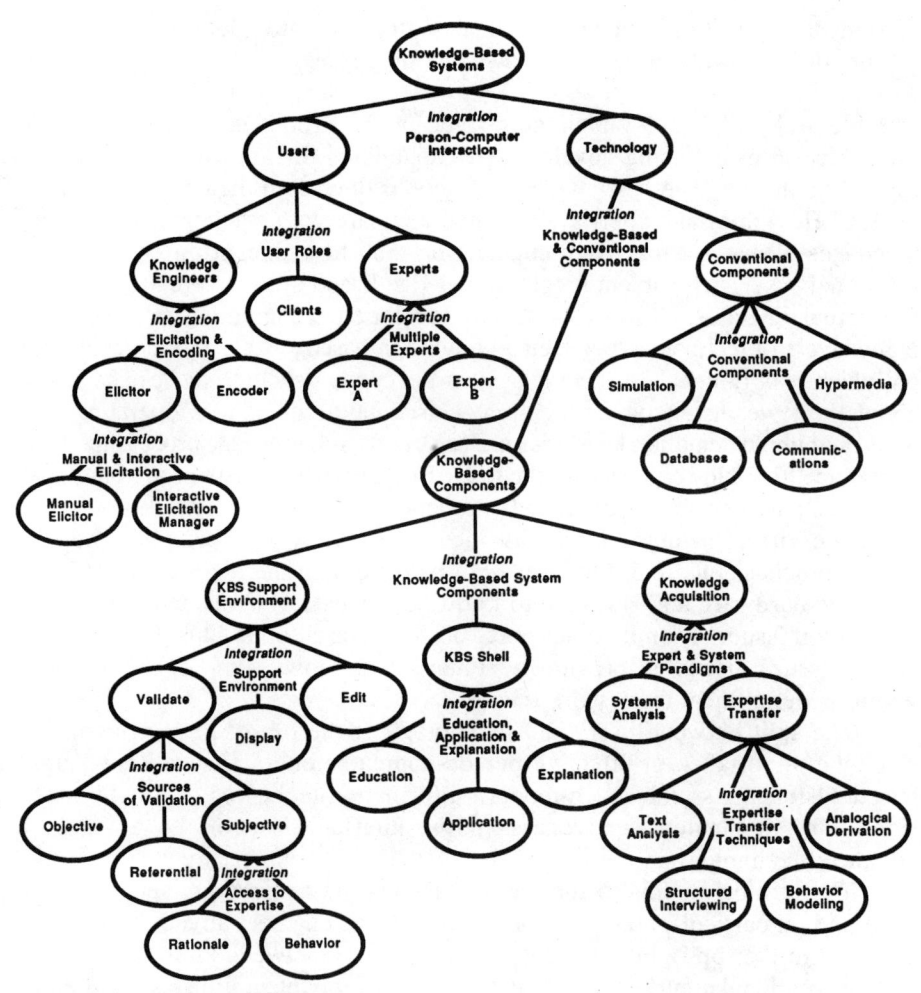

FIGURE 3-10
Integration issues in knowledge acquisition for knowledge-based systems.

playing, and validating the knowledge base. It is important that these tools present a uniform and consistent view of the knowledge base and that they aid consistency maintenance in the knowledge base during editing.

Editing is often regarded as a mundane requirement. However, the variety of related knowledge representations usually present in a knowledge-based system, particularly in the acquisition phase, makes the editing of knowledge bases very difficult to support. If representation A, e.g., text, is processed to give representation B, e.g., entities, attributes, and values, then the editing of the material in representation A should result in changes to that in representation B, and vice versa. However, the relation between representations is usually not functional but merely constrained; i.e., knowledge in representation A does not determine that in representation B, and vice versa, but only constrains it in some way, as shown in Fig. 3-11. Thus, the result of editing knowledge in one representation does not necessarily yield well-defined changes in other representations, only the detection of constraint violations. The editing system should minimally highlight the violations, but usually it can do more, e.g., it can adumbrate "likely" changes in other representations that will remove the violation.

Validation is a complex topic in its own right (Shaw and Woodward, 1987; Benbasat and Dhaliwal, 1988), and a variety of tools should be provided to support different forms of validation. The objective validation of the knowledge base in terms of its performance, its subjective validation in comparison with experts, and its referential validation in comparison with other knowledge sources should be supported. Objective validation may require significant case histories that can be collected only from an operational system in actual applications. Referential validation may require deep analysis of derivations of the knowledge base from other knowledge bases. There is a natural separation in subjective validation between experts' direct agreement that their rationale corresponds to that of the knowledge base and the observed correspondence between their behavior and that of the system based on the knowledge base. Both types of validation should be supported.

The validation of knowledge-based systems has all the dimensions of normal scientific validation together with those concerned specifically with the emulation of expertise (Gaines, 1988a). As shown in Fig. 3-12, the system can be validated through four primary pathways:

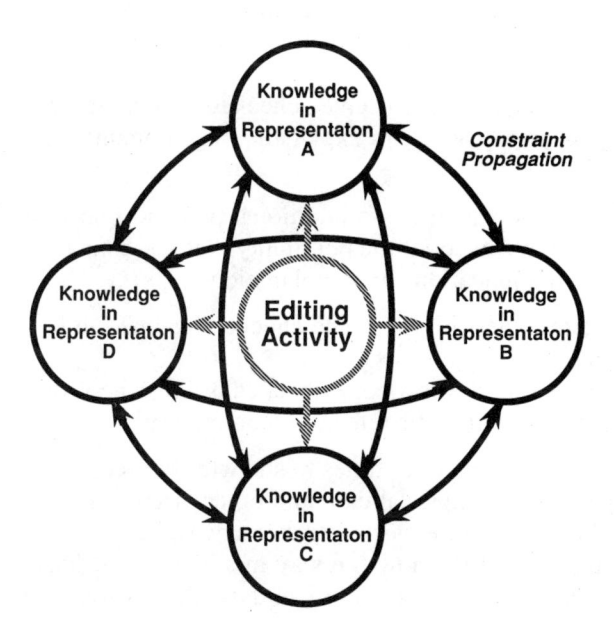

FIGURE 3-11
Constraint propagation in editing different representations of related knowledge.

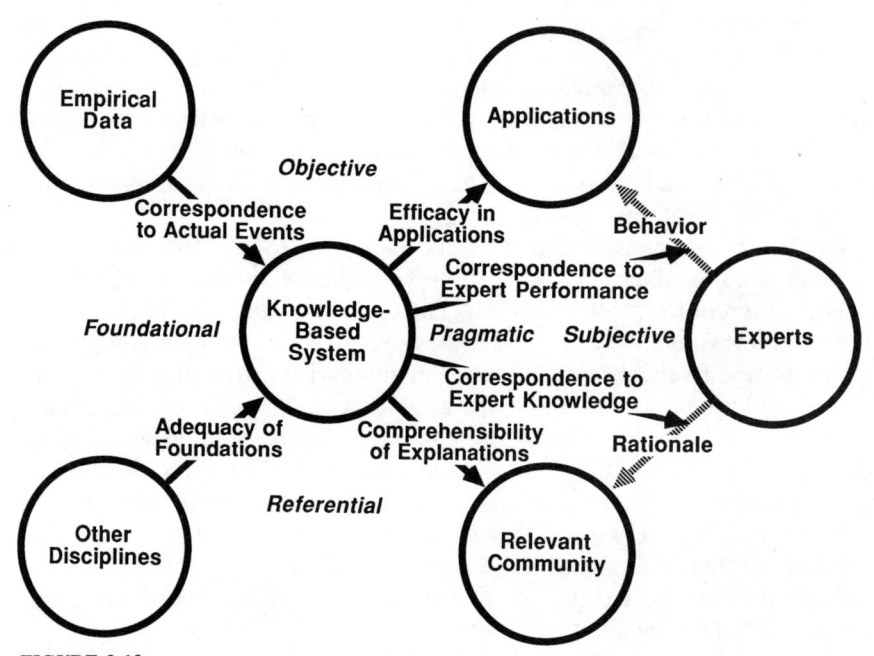

FIGURE 3-12
Dimensions of validation of knowledge-based systems.

- Through its correspondence to actual events checked against empirical data
- Through its efficacy in applications checked against actual applications
- Through the adequacy of its foundations checked against other disciplines corresponding to the derivation of presuppositions from other disciplines—this is sometimes termed *deep knowledge*
- Through the comprehensibility of explanations given checked against the relevant reference community—corresponding to explanation and communicative action

The expert systems paradigm introduces the two additional pathways shown on the right of Fig. 3-12 by taking into account the possibility that there may be "experts" who are competent in applications and explanations:

- Through its correspondence to expert performance checked against the behavior of experts in applications
- Through its correspondence to expert knowledge checked against the rationale of experts in communicating with their reference community

    This diagram may be iterated a number of times to show the source of the knowledge structures of experts from empirical data and other disciplines, introducing considerations of the validity of experts and reference communities. Similarly, the diagram may be iterated to show the way in which other disciplines, in turn, are validated against their own empirical data, deeper foundations, applications, and reference communities. There is no intrinsic conflict between the additional pathways of validation introduced and the positivist viewpoint of science emphasizing the primacy of empirical data. However, there is a major increase in the realism of this model in modeling the actual processes of knowledge acquisition rather than the normative criteria adopted in science as ultimate arbitrators.

7 MAJOR SOURCES FOR KNOWLEDGE ACQUISITION

The significant features of the knowledge acquisition component on the lower right of Fig. 3-10 are best analyzed by considering the natures of the classical and knowledge-based system design paradigms. Clearly knowledge-based systems use some new computational techniques, but perhaps less clear is that the system design approach involved is itself fundamentally changed. However, major paradigm shifts are involved (Gaines, and Shaw, 1985). The classical approach in decision and control system design involves the instrumentation, data collection, modeling, and optimization sequence shown on the left of Fig. 3-13:

- Knowledge of past case histories is used to select a class of system models.
- The information required to discriminate within this class determines how the system should be instrumented for data acquisition.
- Data is collected from the system through the instrumentation.
- A model is identified from the model class which best fits the data.
- This model is used to design a decision or control system for optimal performance.

This approach to system design underlies the methodologies of the physical sciences and technologies based on them. It has been extremely successful in engineering much of the technological infrastructure of our current society including our manufacturing industries. However, this approach is successful only to the extent that the systems under consideration are amenable to instrumentation and modeling. Its greatest successes have been where this amenability can be achieved normatively, i.e., in cases where the system to be controlled is itself a human artifact. For example, linear system theory has become a major tool in systems engineering but not because most natural systems are linear—they are not. The implication is in the opposite direction: Linear systems are mathematically tractable, and we design artificial systems to be linear so that we may model them readily.

The knowledge-based systems paradigm is particularly applicable when it is not possible to model the system but an alternative source of data is available, because human operators are able to performance the decision or control task. The right-hand column of Fig. 3-13 shows the use of knowledge sources in the expert system design paradigm:

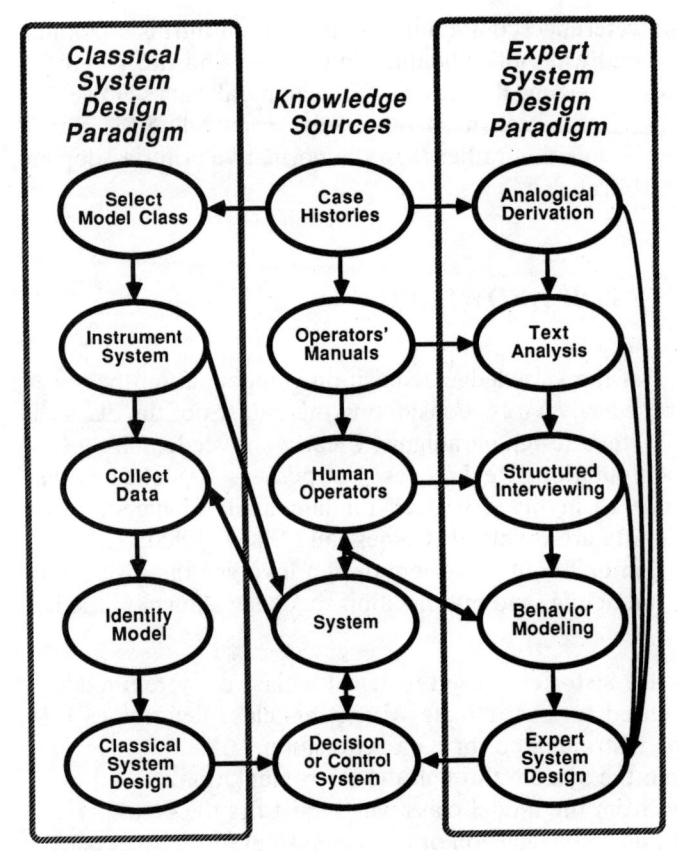

FIGURE 3-13
Classical and knowledge-based system design paradigms.

- Structured interviewing may be used to acquire knowledge directly from the operators.
- Behavior modeling may be used to identify the operators' strategies even if they are unaware of them or give incorrect ones in interviews.
- Text analysis may be used with instructional material such as the operators' manuals.
- Reasoning by analogy may be used based on the case histories without the data collection of the classical system design paradigm.

Note that the classical and knowledge-based system design paradigms, and the various acquisition techniques for knowledge-based systems, need not be regarded as competitive. They may all be used in the system design, and a collection of weak methods may be necessary because there is no overall strong method that can be applied.

The integration of the classical systems analysis with expertise transfer techniques is a major task. The primary problems currently involve the widely different forms of knowledge representation, and these are noted in a later section in terms of the integration of conventional and knowledge-based system components. However, at a knowledge acquisition level, the problem is one of imbalance in the types of knowledge structures studied in depth to date. There has been an imbalance favoring symbolic knowledge to redress the past emphasis on quantitative methods. This bias needs redressing with the knowledge structures underlying arithmetic being integrated with those underlying nonarithmetic symbol manipulation.

8 KNOWLEDGE ACQUISITION TOOLS

A number of tools have been developed for the rapid prototyping of knowledge-based systems automating some of or all the methodologies on the right of Fig. 3-13. The objective of systems such as PLANET (Shaw and Gaines, 1983, 1986a, 1987a), ETS (Boose, 1984, 1986), MOLE (Eshelman et al., 1986), SALT (Marcus, 1987), KITTEN (Shaw and Gaines, 1987b), KNACK (Klinker et al., 1986), KRITON (Diederich, Ruhmann, and May, 1986), OPAL (Musen et al., 1986), AQUINAS (Boose and Bradshaw, 1987) and KSSO (Gaines, 1987c) is to expedite the process of knowledge acquisition and transfer to knowledge-based systems.

Boose (1988) has surveyed a very wide range of knowledge acquisition tools and classified them along a variety of dimensions, analyzing the applicability of tools to tasks. He distinguishes three classes of tasks:

1 Analysis tasks
   a Classification: categorizing based on observables
   b Debugging: prescribing remedies for malfunctions
   c Diagnosis: inferring system malfunctions from observables
   d Interpretation: inferring situation descriptions from sensor data
2 Synthesis tasks
   a Configuration: configuring collections of objects under constraints in relatively small search spaces
   b Design: configuring collections of objects under constraints in relatively large search spaces
   c Planning: designing actions
   d Scheduling: planning with strong time and/or space constraints
3 Tasks combining analysis and synthesis
   a Command and control: ordering and governing overall system control
   b Instruction: diagnosing, debugging, and repairing student behavior
   c Monitoring: comparing observations to expected outcomes
   d Prediction: inferring likely consequences of given situations
   e Repair: executing plans to administer prescribed remedies

Boose notes that most knowledge acquisition tools currently are best suited to analysis tasks and are weak in acquiring expertise in synthesis.

These are the general requirements for a system to automate knowledge acquisition for knowledge-based systems:

- The knowledge acquisition system tools should be domain-independent.
- The tools should be directly applicable by experts without intermediaries.
- The tools should be able to access a diversity of knowledge sources including text, interviews with experts, and observations of expert behavior.
- The knowledge acquisition system should be able to encompass a diversity of perspectives including partial or contradictory input from different experts.
- The system should be able to encompass a diversity of forms of knowledge and relationships between knowledge.
- The system should be able to present knowledge from a diversity of sources with clarity as to its derivation, consequences, and structural relations.
- Users of the knowledge acquisition system should be able to apply the knowledge in a variety of familiar domains and freely experiment with its implications.
- The system should make provision for algorithmic expression of knowledge where appropriate.
- The system should make provision for validation studies.
- As much of the operation of the system as possible should be founded on well-developed and explicit theories of knowledge acquisition, elicitation, and representation.
- As the overall knowledge acquisition develops, it should converge to an integrated system.

It is particularly important in describing tools that they be presented in the context of user interaction, since their primary function is to operate directly with experts. It is impossible to describe all the knowledge acquisition tools currently in the literature, and the references cited at the beginning of this section and specialist collections should be consulted for details and examples of use (Gaines, and Boose, 1988; Boose and Gaines, 1988; Marcus, 1988). The next section takes one particular tool that covers the range of methodologies discussed in Section 7 and uses it to illustrate the major features of a wide variety of tools.

9 EXAMPLE OF AN INTEGRATED KNOWLEDGE ACQUISITION SYSTEM

The problems of eliciting knowledge from experts as skilled performers of a task are described in Section 1. Clinical psychologists see the problem as one of cognitive defenses that impede internal communication, and they have developed techniques of verbal interaction to identify underlying cognitive processes (Freud, 1914; Kelly, 1955; Rogers, 1967). These can be used to bypass cognitive defenses, including those resulting from automatization of skilled behavior, and Welbank (1983) has surveyed psychological techniques for doing this. In an industrial context, the application of techniques derived from clinical psychology to problems of hard system engineering is not an obvious or probable step to take, and the early approach to knowledge engineering was to "manage" it by inventing the profession of knowledge engineer as discussed above. However, for the reasons given above, this is not a realistic approach, and the focus shifted to computer-based tools, many of which have been derived from techniques of clinical and social psychology.

One major class of tools for knowledge acquisition has been developed around the "personal construct psychology" of George Kelly (1955) and, in particular, his repertory grid technique for eliciting conceptual systems as individual models of the world. In computational terms, the grid may be seen as small entity-attribute database in which a domain is characterized by an expert in terms of critical objects within it and the properties of these objects relevant to problem-solving within the domain. The elicitation of an expert's conceptual system necessarily involves approximation since a complete system may involve indefinitely complex relations. However, as a start, one wishes to elicit the major distinctions that an individual uses in a domain, the terminology used for them, and the relation of such distinctions and terminology to those of others. Entity-attribute grid elicitation is an effective method for eliciting major distinctions and terminology in a domain (Shaw, 1981).

The grid technique is an extensional approach in which individuals are asked to specify a set of entities in a domain and then make distinctions among them, naming the distinctions and classifying all the specified entities in terms of them. The extension of a distinction determined in this way is only an approximation to the underlying concept since critical entities may be missing in the classification. However, both manual and computer-based elicitation techniques attempt to prompt the individual for additional entities to discriminate between extensionally related distinctions (i.e., making the same, or similar, classifications). Knowledge acquisition is essentially a negotiation process leading to approximations to conceptual structures that are adequate for some practical purposes.

The application of interactive repertory grid elicitation to expert system development was first proposed by Gaines and Shaw (1980), taken up and demonstrated effective in industrial applications by Boose (1984), and is now in widespread use through a number of derivative system developments (Gammack and Young, 1985; Diederich, Ruhmann, and May, 1986; Wahl, 1986; Garg-Janardan and Salvendy, 1987). Figure 3-14 gives a family tree for some of the major developments in repertory grid knowledge acquisition tools and the extensions of Kelly's original concepts.

KSS0: Knowledge Support System Zero

Figure 3-15 shows the architecture of one system that integrates a wide range of knowledge acquisition methodologies. KSS0 is written in Pascal and runs on the Apple Macintosh family of computers to provide a highly interactive and graphical knowledge acquisition environment. At the heart of KSS0 is an object-oriented knowledge base in which knowledge is formally represented as a multiple-inheritance structure of classes, objects, properties, values, and re-

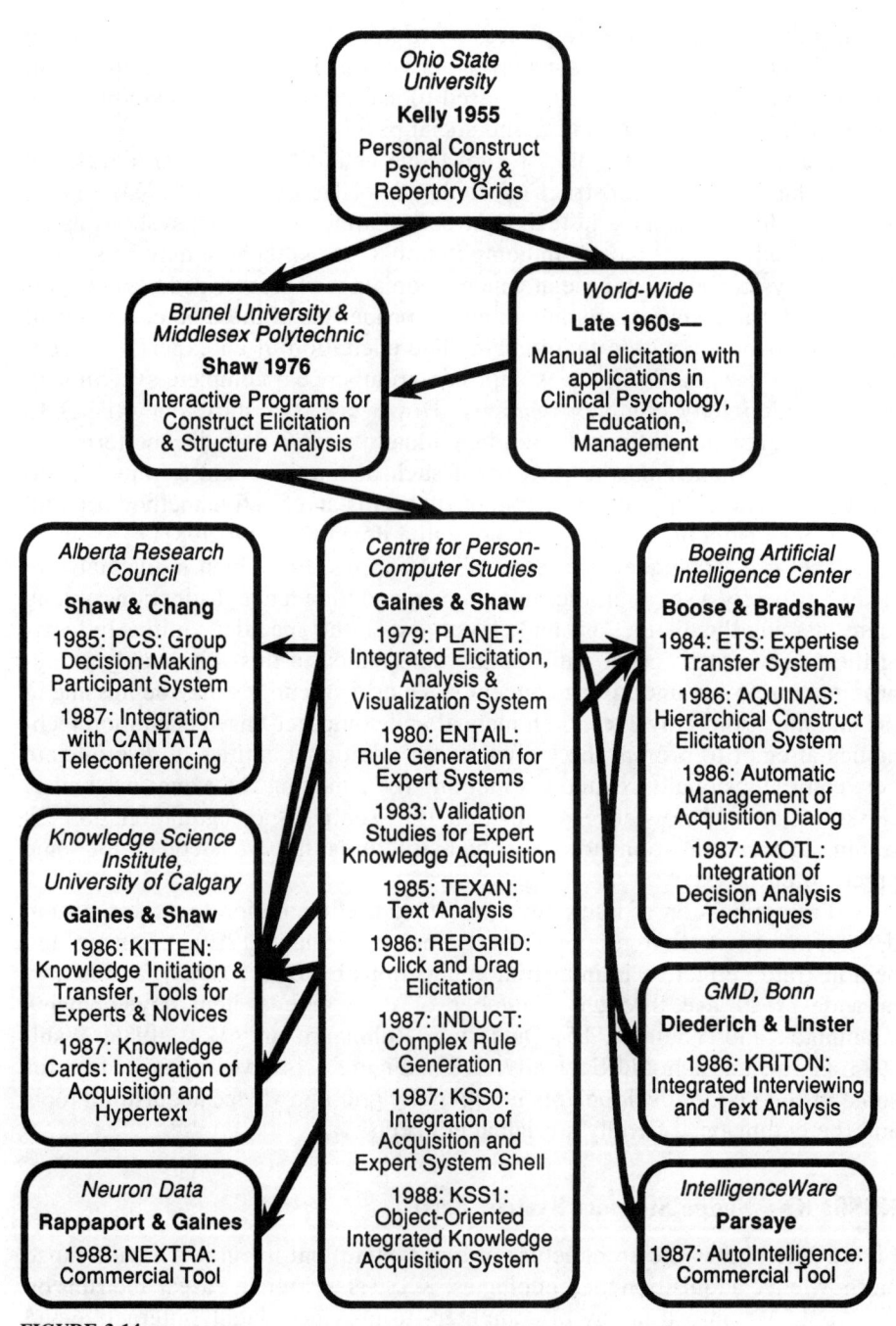

82 KNOWLEDGE ENGINEERING: FUNDAMENTALS

```
                    Ohio State University
                    Kelly 1955
                    Personal Construct
                    Psychology &
                    Repertory Grids
                   /                    \
  Brunel University &              World-Wide
  Middlesex Polytechnic            Late 1960s—
  Shaw 1976                        Manual elicitation with
  Interactive Programs for         applications in
  Construct Elicitation            Clinical Psychology,
  & Structure Analysis             Education,
                                   Management
```

Alberta Research Council	Centre for Person-Computer Studies	Boeing Artificial Intelligence Center
Shaw & Chang	**Gaines & Shaw**	**Boose & Bradshaw**
1985: PCS: Group Decision-Making Participant System	1979: PLANET: Integrated Elicitation, Analysis & Visualization System	1984: ETS: Expertise Transfer System
1987: Integration with CANTATA Teleconferencing	1980: ENTAIL: Rule Generation for Expert Systems	1986: AQUINAS: Hierarchical Construct Elicitation System
	1983: Validation Studies for Expert Knowledge Acquisition	1986: Automatic Management of Acquisition Dialog
	1985: TEXAN: Text Analysis	1987: AXOTL: Integration of Decision Analysis Techniques

Knowledge Science Institute, University of Calgary		
Gaines & Shaw	1986: REPGRID: Click and Drag Elicitation	
1986: KITTEN: Knowledge Initiation & Transfer, Tools for Experts & Novices	1987: INDUCT: Complex Rule Generation	**GMD, Bonn**
1987: Knowledge Cards: Integration of Acquisition and Hypertext	1987: KSS0: Integration of Acquisition and Expert System Shell	**Diederich & Linster**
		1986: KRITON: Integrated Interviewing and Text Analysis

Neuron Data		IntelligenceWare
Rappaport & Gaines	1988: KSS1: Object-Oriented Integrated Knowledge Acquisition System	**Parsaye**
1988: NEXTRA: Commercial Tool		1987: AutoIntelligence: Commercial Tool

FIGURE 3-14
Family tree of applications and developments of repertory grids in knowledge acquisition.

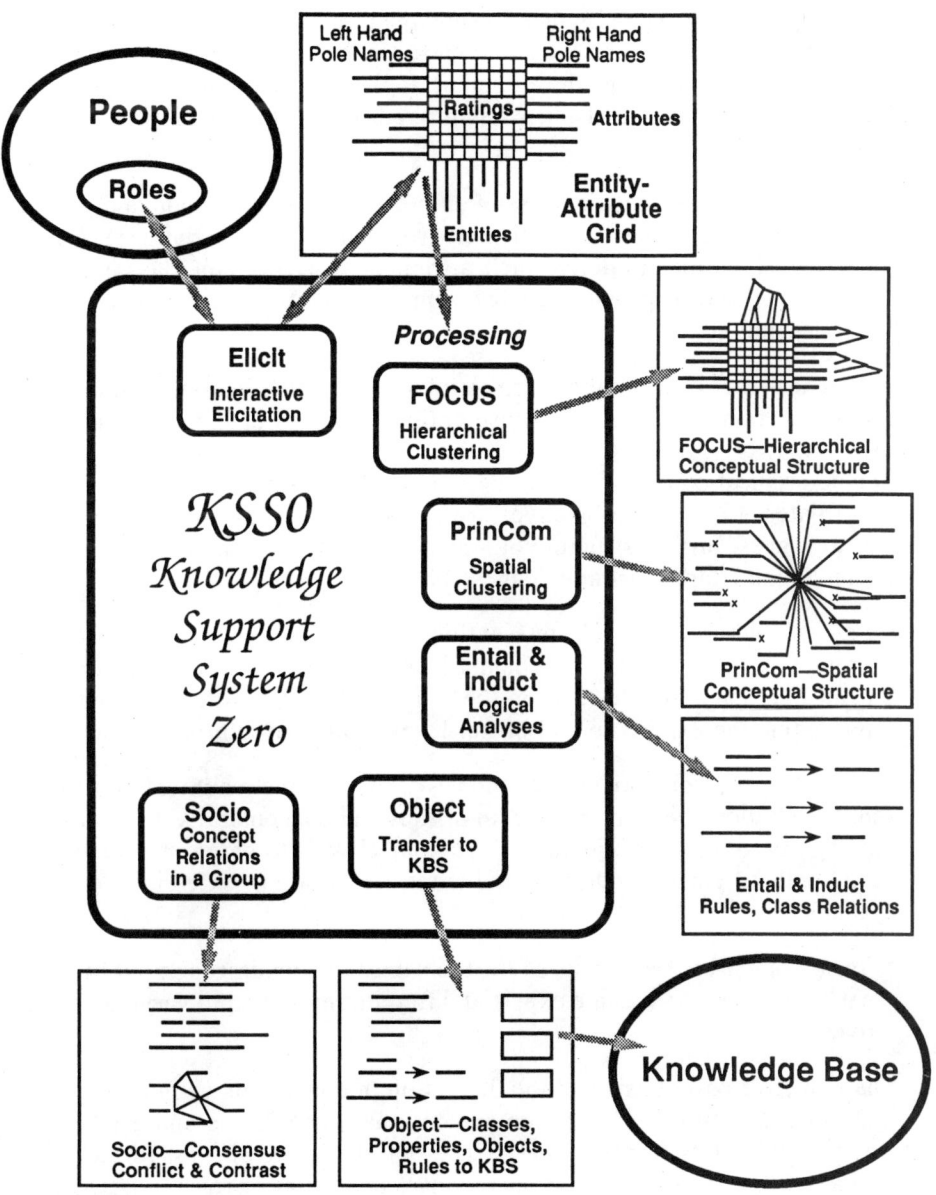

FIGURE 3-15
KSS0, an integrated knowledge acquisition and support system.

lations. Such a structure generalizes the entity-attribute grids used in several early knowledge acquisition systems and has proved both general and powerful in a variety of applications.

The elicitation tools in KSS0 are based on Shaw's (1980) computer-based interviewing techniques extended through the use of graphical rather than numerical data entry. The visualization and the direct manipulation of knowledge structures through the graphical and click-and-drag facilities of modern workstations are an important development at the user interface giving experts, knowledge engineers, and clients improved access to the knowledge structures.

- *Interview* accepts specifications of entities within a subdomain and provides an interactive graphical elicitation environment within which the experts can distinguish entities to derive their attributes. The resultant class is continuously analyzed to provide feedback, prompting the expert to enter further entities and attributes.

The visual analysis tools consist of an interactive interface, to represent the abstractions derived from those entities in terms of hierarchical clusters using Shaw's (1980) FOCUS algorithm, and relational diagrams such as nonhierarchical conceptual maps derived through principal components analysis (Slater, 1976). The objectives are to validate the raw domain knowledge and to suggest further structure at a higher level through interactive topological induction (Rappaport and Gaines, 1988):

- *FOCUS* hierarchically clusters entities and attributes within a subdomain, prompting the experts to add higher-level entities structuring the domain.
- *PrinCom* spatially clusters entities and attributes within a subdomain, prompting the experts to add higher-level entities structuring the domain.

The group comparison tools consist of an interactive interface to represent the relations between the terminologies and conceptual systems of different experts, or experts and clients. The objective is to determine the consensus, conflict, correspondence, and contrast between different conceptual systems (Shaw and Gaines, 1988):

- *Socio* compares the structures for the same subdomain generated by different experts or by the same expert at different times or from varying perspectives.

The inductive part consists in the derivation of constraints within the conceptual structures through logical entailment analyses (Gaines and Shaw, 1985; Quinlan, 1987; Cendrowska, 1987). The objective is to suggest further structure at a higher level that translates to class inclusions or rules in NEXPERT:

- *Entail* and *Induct* induce logical entailments enabling the attributes of an entity, or the evaluations of a decision-making situation in a domain, to be derived from other attributes.

The generative part consists in the transformation of the knowledge analysis made by the previous tools into a formalism understandable by the NEXPERT inference mechanisms (Rappaport, 1987b):

- *Object* formats the specifications of subdomains as classes, of entities as objects, of attributes as properties, and of entailments as methods, and it transfers them to the performance tool.

Knowledge Acquisition in Action: Defining the Domain

Figure 3-16 is the start-up screen for KSS0. The system is designed to be very clean and simple to use with a substantial entertainment value. The objective is to make interaction with system attractive and fun while retaining a professional working environment. Help is available at all times through context-sensitive, interactive "advice buttons." The expert should never be at a loss, need help that she or he cannot get, or have to access a manual. The default action at all times is to click on the button at the top right of the lower control panel—usually "OK," "Yes," "More," or the most likely action. During elicitation, when the user has a choice of actions, this default action will select one through biased random choice from a reasonable set generated by rules which can be edited by the knowledge engineer. Thus, experts can begin to make effective use of the system very quickly with no need for computer experience and very little hand holding.

Figure 3-17 is an entity screen from KSS0 in a knowledge elicitation sequence concerned with factory layout planning (Adler and Matheson, 1986). KSS0 allows its vocabulary to be tailored to the problem to maximize expert involvement, and the expert and knowledge engineer have agreed on a basic terminology. The expert has entered various "factors" which affect layout—

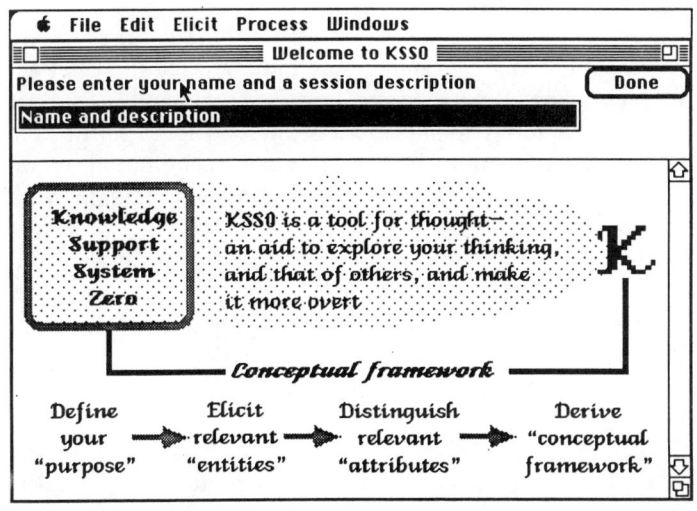

FIGURE 3-16
KSS0 start-up screen.

86 KNOWLEDGE ENGINEERING: FUNDAMENTALS

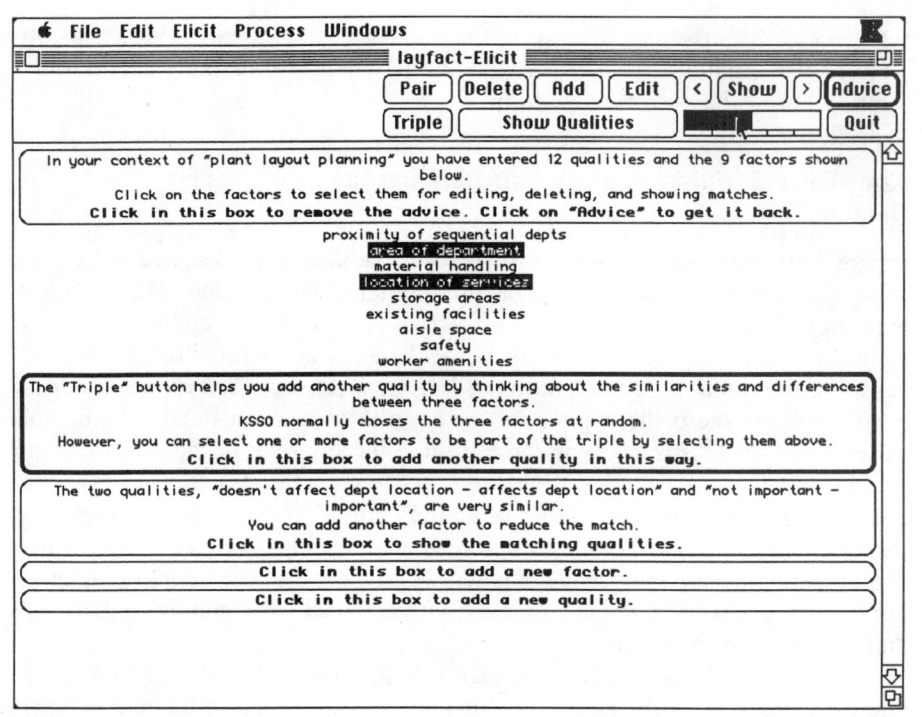

```
 File  Edit  Elicit  Process  Windows
═══════════════════════ layfact-Elicit ═══════════════════════
                    [ Pair  ][Delete][  Add  ][ Edit ][<][ Show ][>][Advice]
                    [Triple ][ Show Qualities ][      ]         [ Quit  ]
┌──────────────────────────────────────────────────────────────┐
│ In your context of "plant layout planning" you have entered 12 qualities and the 9 factors shown │
│                                below.                                                            │
│     Click on the factors to select them for editing, deleting, and showing matches.              │
│     Click in this box to remove the advice. Click on "Advice" to get it back.                    │
│                        proximity of sequential depts                                             │
│                              area of department                                                  │
│                              material handling                                                   │
│                              location of services                                                │
│                                storage areas                                                     │
│                              existing facilities                                                 │
│                                 aisle space                                                      │
│                                    safety                                                        │
│                              worker amenities                                                    │
├──────────────────────────────────────────────────────────────┤
│ The "Triple" button helps you add another quality by thinking about the similarities and differences │
│                              between three factors.                                              │
│             KSSO normally choses the three factors at random.                                    │
│     However, you can select one or more factors to be part of the triple by selecting them above.│
│            Click in this box to add another quality in this way.                                 │
├──────────────────────────────────────────────────────────────┤
│ The two qualities, "doesn't affect dept location – affects dept location" and "not important –   │
│                              important", are very similar.                                      │
│                   You can add another factor to reduce the match.                                │
│            Click in this box to show the matching qualities.                                     │
├──────────────────────────────────────────────────────────────┤
│                Click in this box to add a new factor.                                            │
├──────────────────────────────────────────────────────────────┤
│                Click in this box to add a new quality.                                           │
└──────────────────────────────────────────────────────────────┘
```

FIGURE 3-17
Entity screen showing matches.

these are the entities—and we will see the system help the expert to elicit relevant ''qualities''—these are the attributes.

The interactive advice buttons are also apparent in this figure. They are generated dynamically from the current context and may be used to select actions without use of the control panel above, or the pull-down menus. ''Power users'' may turn them off completely to allocate more screen space to the data structures—this works well because by the time the user has entered a substantial number of entities and attributes, she or he has become familiar with the system and no longer needs the tutorial advice.

The system keeps track of matches between entities and between attributes which indicate structures in the knowledge, and the expert has decided to investigate one of these. The indicator at the upper right of Fig. 3-17 shows the value of the highest match between entities. The expert has clicked on it to highlight the matching entities and can use the '' < '' and '' > '' buttons to step backward and forward through the matches.

The expert next clicks on the ''Show'' button to see whether he or she can input an attribute that will reduce the match, i.e., a counterexample to the inferred structure. KSS0 goes to the screen shown in Fig. 3-18 which makes explicit the question that needs to be answered and shows how the match has come about. The horizontal bars show the ratings of the entities on the attributes, with those of the upper entity (location of services) shown by the upward markers and those of the lower entity (area of department) shown by the downward markers.

The expert can edit these ratings by dragging the markers along the scales or modify the entity names by clicking on them to open up a text edit window. At no time in KSS0 is the user constrained to follow a specific course of action since the system is designed to facilitate free exploration and development of knowledge structures. In this case the expert decides to complete the action by answering yes to the question asked. KSS0 then goes to the screen shown in Fig. 3-19 where the expert is prompted to define a new attribute and then rate all the entities on it by clicking and dragging them to the vertical rating bar in the center.

Figure 3-20 shows the start of the complementary sequence in which the expert starts from the attribute screen and clicks on the indicator to see the highest matched attributes. Then the expert clicks on the "Show" button again and moves to the screen shown in Fig. 3-21, where the basis of the match is shown and again a very specific question is asked about the type of entity required to break the match. The expert can again edit the ratings by clicking

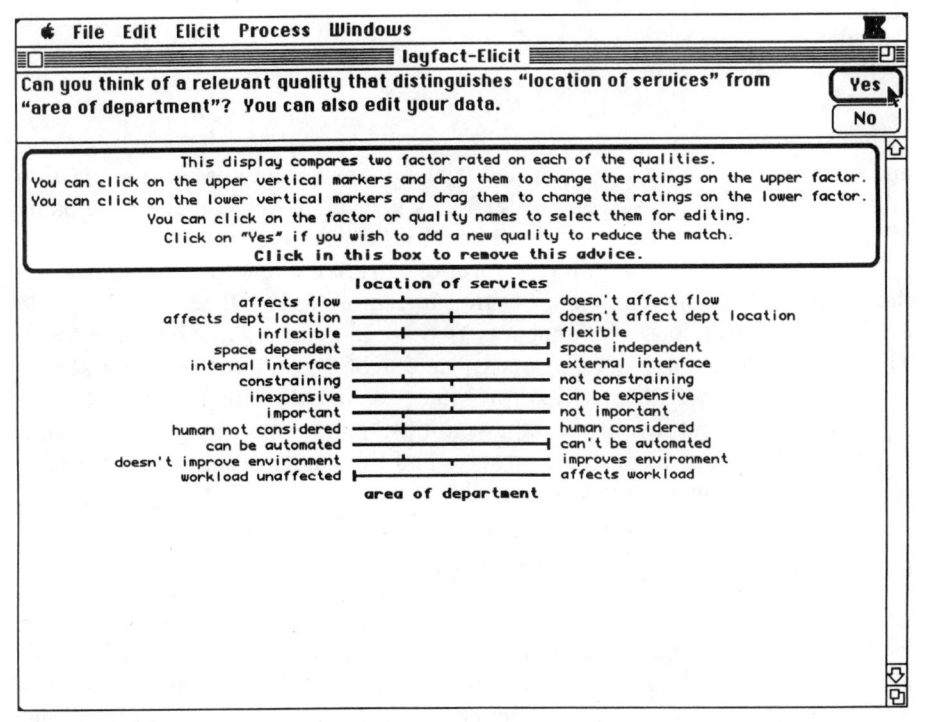

FIGURE 3-18
Entity match screen.

88 KNOWLEDGE ENGINEERING: FUNDAMENTALS

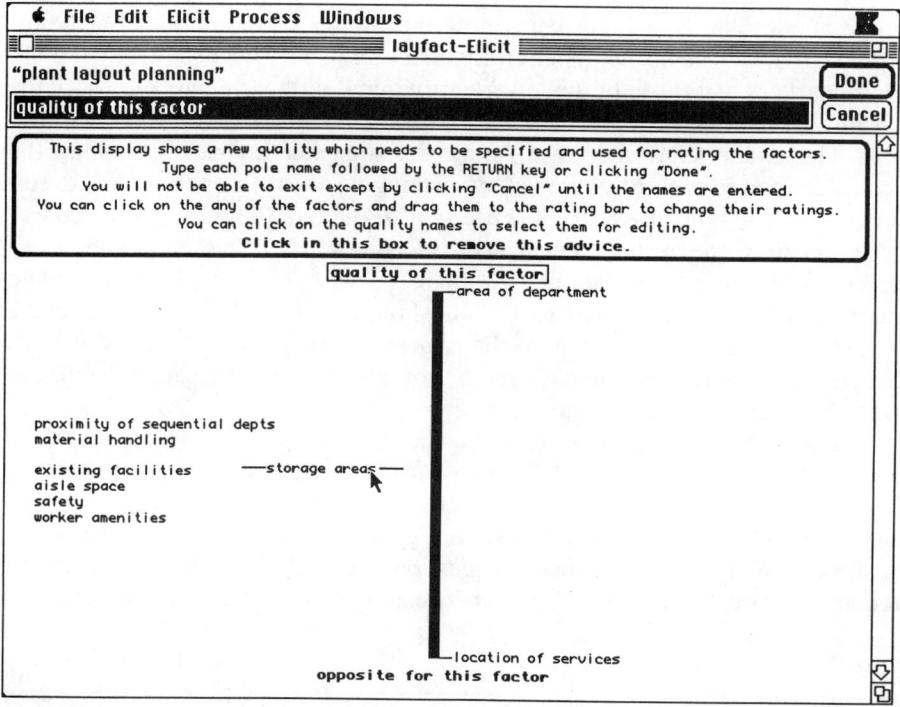

FIGURE 3-19
Attribute rating screen.

and dragging the entities if he or she wishes. Answering yes to this question takes her or him to the screen in Fig. 3-22 where he or she can enter the name of the new entity and then rate it on each of the existing attributes.

Figure 3-23 shows a different form of attribute elicitation initiated when the expert clicks on the "Triple" button in Fig. 3-17. KSS0 generates a triple of three entities and asks in what way two are alike and differ from the third. The expert clicks on "storage areas," and KSS0 goes to the rating screen shown in Fig. 3-24 with the triple prerated, so that the expert may enter the attribute and the ratings of the other entities. The triples are generated on a random basis biased to promote variety, and this mode of elicitation tends to trigger recollections of significant attributes otherwise forgotten.

Knowledge Acquisition in Action: Visualization

The elicitation screens shown above all use the graphical capabilities of the workstation to encourage and support the experts' visualization of the data being entered. The rating of an entity on an attribute is seen in relation to the ratings of other entities to enable immediate consistency checks to be made. The analysis tools continue this emphasis on visualization and validation by

KNOWLEDGE ACQUISITION SYSTEMS **89**

presenting the conceptual structures underlying the data entered in graphical form.

At any time during the elicitation, experts can ask for the structure of their entity-attribute conceptual system to be analyzed and displayed. Figure 3-25 shows the basic data structure of the grid that has been entered with entity names, attribute names, rating values, and shading to indicate low, medium, and high values. Figure 3-26 shows a hierarchical clustering of the entities and attributes in Fig. 3-25 based on the matches between them by using Shaw's (1980) FOCUS algorithm. Note how this has tended to pull similarly shaded areas together. The clusters can be used to suggest the introduction of hierarchical relations in the conceptual framework, e.g., that location of services, area of department, and aisle space are all spatial constraints.

Figure 3-27 shows an alternative mapping into two dimensions based on a principal components analysis (Slater, 1976). Again this can be used to identify groupings leading to more abstract attributes. Both forms of clustering are also important in allowing the expert to examine the semantic consistency of the data that has been entered—if the way that entities and attributes cluster does not make sense, then it is a clear indication that something is wrong in the data entered or the expert's conceptualization of the domain.

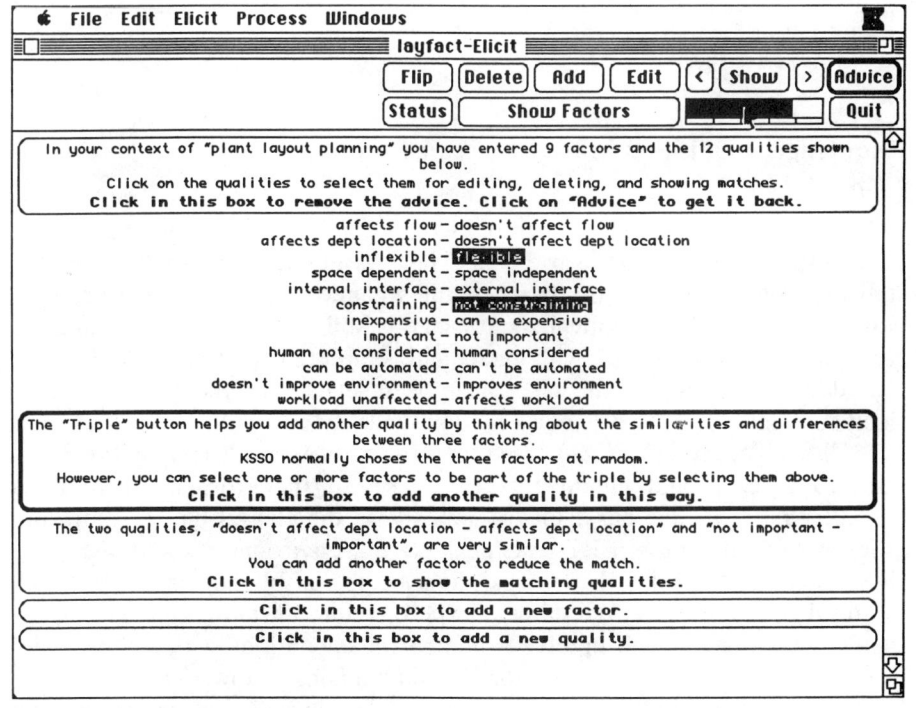

FIGURE 3-20
Attribute screen showing matches.

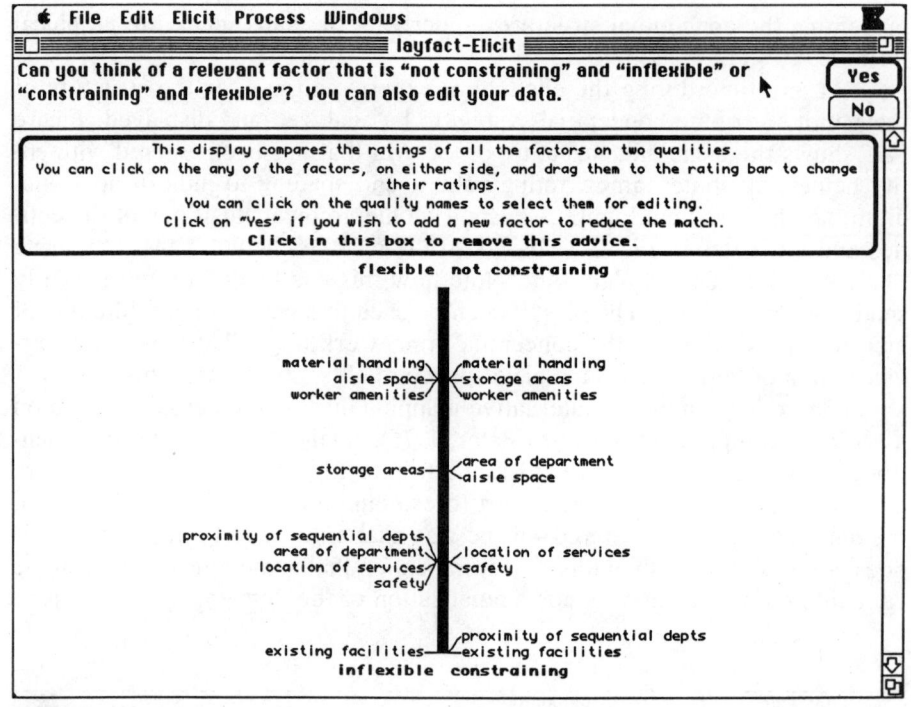

FIGURE 3-21
Attribute match screen.

Knowledge Acquisition in Action: Transfer to the Shell

Rules relating attributes are generated from the rule induction algorithms Entail and Induct applied to the grid data. The dataset elicited is small so that the statistical significance of the induced rules cannot be expected to be high. However, sorting by statistical significance is still effective in preventing the generation of spurious rules with inadequate support. The small dataset is usually adequate because all the cases have been selected by the expert as being significant and correct—it is not a random collection of noisy data.

The Object tool in KSS0 converts attribute sets to class definitions, entities to objects, and entailments to rules and then formats them as a NEXPERT knowledge base. Figure 3-28 shows the class, objects, and properties within the NEXPERT graphical editing environment, and Fig. 3-29 shows some of the rules. If the task might be solved by a decision tree, then the result is a complete, operational expert system. A more complex expert system development will involve a number of elicitations with KSS0 to develop the conceptual structures for the subdomains followed by the use of NEXPERT editing tools to link the subdomains through appropriate control structures.

The technical details of the processes going on in Figs. 3-16 through 3-28 are less important than their psychological impact on the expert. Users of KSS0 generally find that it gives fascinating insights into their own thought processes which both makes it attractive and puts users into an introspective frame of mind conducive to knowledge elicitation. Most of the dialogue that users see on the screen is generated from their own inputs and hence appears natural in terms of their professional terminology. Most importantly, this translates immediately to the structures generated for use in the expert system shell—the structures use the expert's jargon—and their derivation from the data entered has already been shown to the expert and changes negotiated.

Thus, the initial prototype knowledge structures, such as those shown in Figs. 3-27 and 3-28, are comprehensible to the expert in their derivation and terminology. Also by this stage the expert has become accustomed to the use of the computer—windows, menus, and mice are familiar tools—and will often move over to editing the rules in the shell without really being aware that the environment has changed.

In most problem environments, there are multiple knowledge islands, each requiring a separate elicitation process specific to the knowledge structures within the island. These islands tend to form a hierarchy; e.g., "safety"

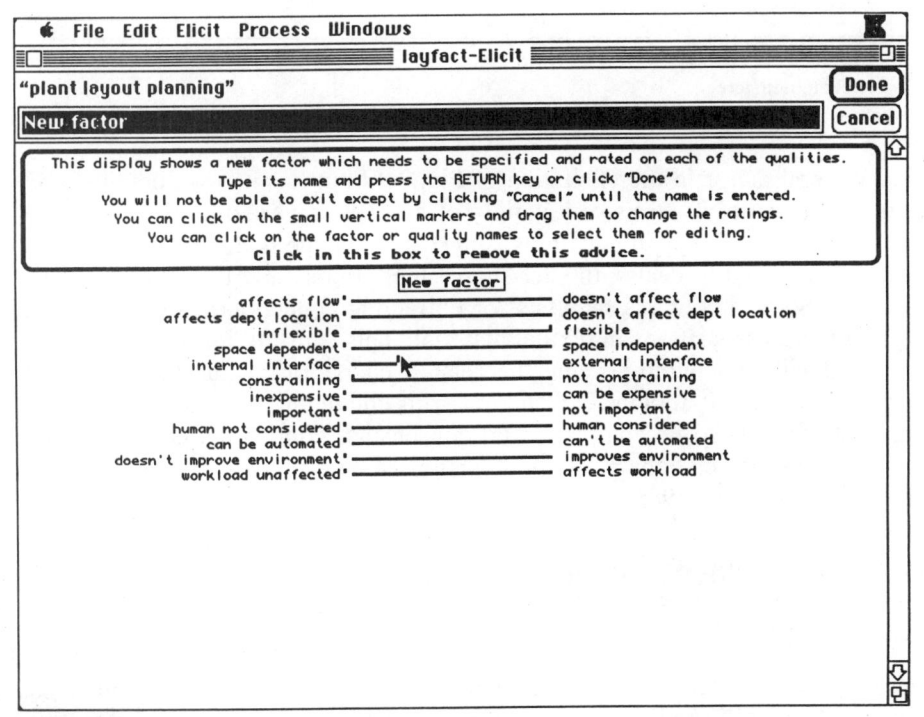

FIGURE 3-22
Entity rating screen.

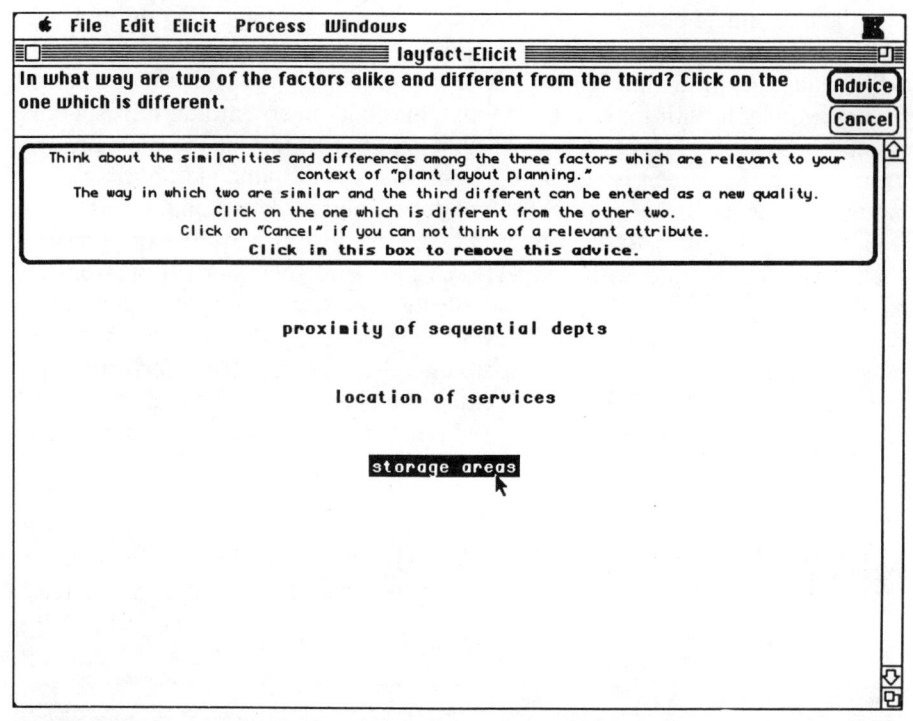

FIGURE 3-23
Triple elicitation screen.

is a general factor in the elicitation of Fig. 3-17 but has its own detailed conceptual structure in terms of a variety of hazards, a variety of forms of accident, a variety of legal implications, and so on. The methodology shown above can be extended to deal with such hierarchies (Boose and Bradshaw, 1987).

KSS0 also contains facilities for extracting conceptual clusters by analyzing word associations in textual material. This is useful for defining knowledge islands and pump-priming the related elicitations with entities and attributes. The Socio tool in KSS0 allows grids elicited in the same domain from multiple experts to be compared and consensual, conflicting, corresponding, and contrasting relations between attributes to be displayed in graphical form (Shaw and Gaines, 1988).

10 CONCLUSIONS: IMPEDIMENTS TO KNOWLEDGE-BASED SYSTEM DEVELOPMENT

Expert systems technology, like many other computing developments, has been "oversold," and the expectations of what may be achieved through the use of knowledge-based system technology have been raised too high. The bottleneck of knowledge engineering discussed in Section 4 still remains. The quotation at the head of that section dates from 1983; meanwhile major effort has swung into research on knowledge acquisition techniques, and there has been a substantial increase in the experience of developing expert systems. Notwithstanding this research and experience, there are still major problems in the development of knowledge-based systems, and it is appropriate to conclude this chapter by summarizing these problems. Figure 3-30 gives a synopsis of the major impediments to knowledge-based system development. The top box in Fig. 3-30 summarizes what may be termed the problem of *professionalism* in developing knowledge-based systems:

- *Theory:* The lack of scientific and engineering foundations for knowledge-based systems is a major impediment to the development of professional standards for expert system development. We understand remarkably little about the nature of expertise even though it is one of the most valued commodities in our society. We also have little fundamental understanding of knowledge, the ways in which it is represented in human thought processes, used in human problem solving, and transmitted between people.

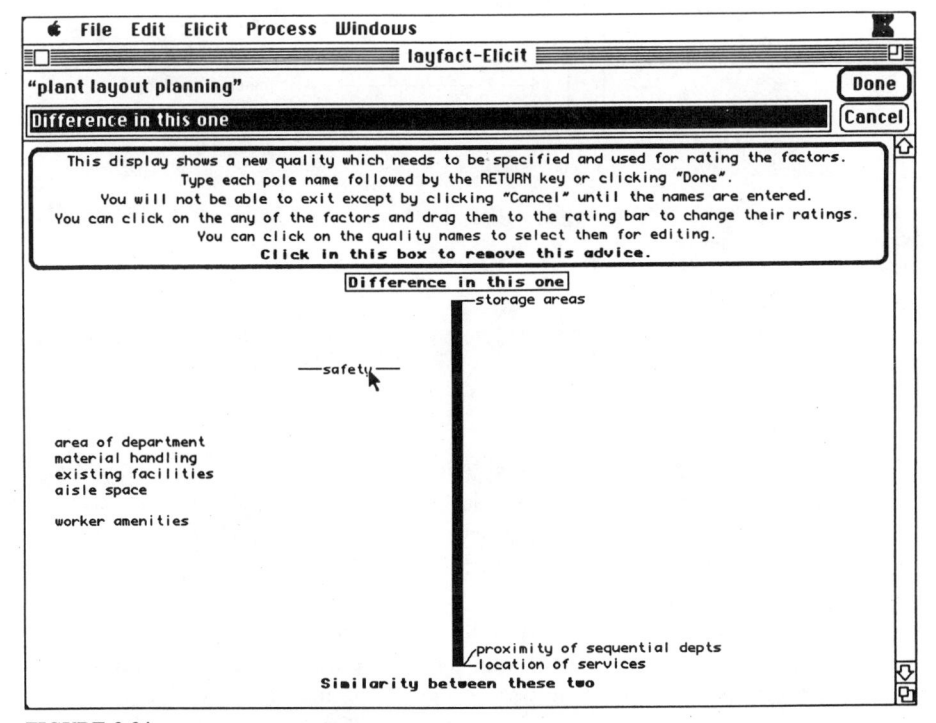

FIGURE 3-24
Attribute entry and rating after a triple.

94 KNOWLEDGE ENGINEERING: FUNDAMENTALS

Display: layfact
Entities: 9, Attributes: 12, Range: 1 to 5, Context: plant layout planning

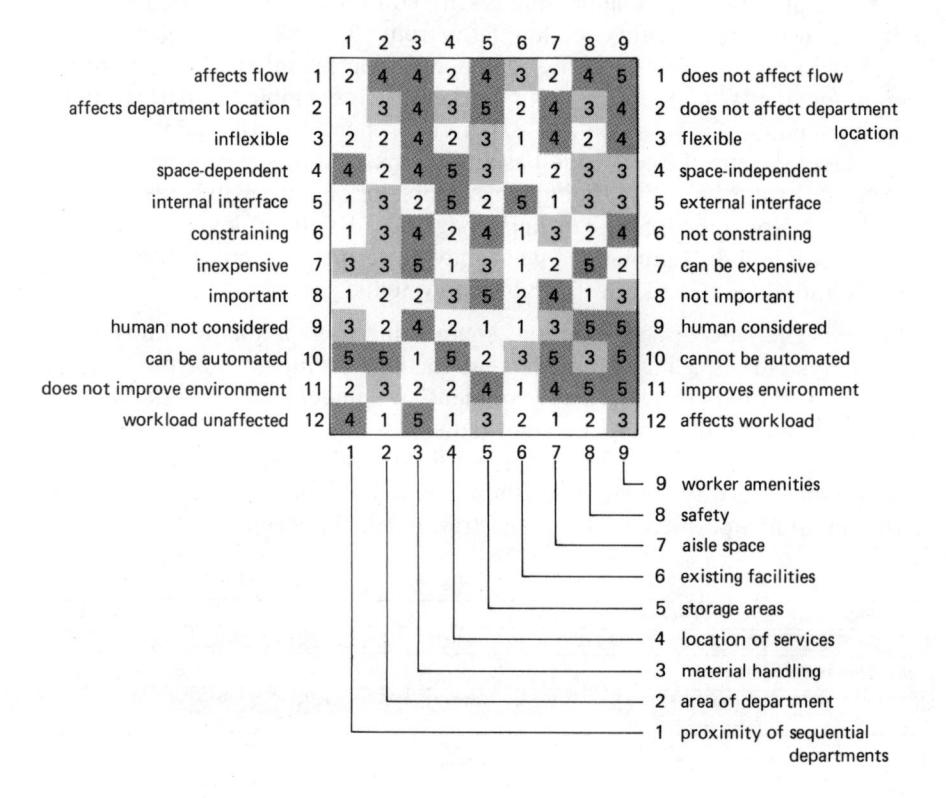

		1	2	3	4	5	6	7	8	9		
affects flow	1	2	4	4	2	4	3	2	4	5	1	does not affect flow
affects department location	2	1	3	4	3	5	2	4	3	4	2	does not affect department location
inflexible	3	2	2	4	2	3	1	4	2	4	3	flexible
space-dependent	4	4	2	4	5	3	1	2	3	3	4	space-independent
internal interface	5	1	3	2	5	2	5	1	3	3	5	external interface
constraining	6	1	3	4	2	4	1	3	2	4	6	not constraining
inexpensive	7	3	3	5	1	3	1	2	5	2	7	can be expensive
important	8	1	2	2	3	5	2	4	1	3	8	not important
human not considered	9	3	2	4	2	1	1	3	5	5	9	human considered
can be automated	10	5	5	1	5	2	3	5	3	5	10	cannot be automated
does not improve environment	11	2	3	2	2	4	1	4	5	5	11	improves environment
workload unaffected	12	4	1	5	1	3	2	1	2	3	12	affects workload

1. proximity of sequential departments
2. area of department
3. material handling
4. location of services
5. storage areas
6. existing facilities
7. aisle space
8. safety
9. worker amenities

FIGURE 3-25
An entity-attribute grid.

- *Techniques:* The lack of well-established techniques for knowledge engineering is a major problem for new development teams. There is some folklore, generally mutually contradictory, and a few rules of thumb based on experience, but so far no in-depth studies of what is actually done and how successful it has been.
- *Tools:* The lack of commercially available tools for knowledge acquisition and transfer is also a major problem. There are now many research reports of such tools, but few of those developed are available as products.
- *Training:* The training situation is improving all the time, largely because it has become the major secondary industry to the supply of expert system shells. A range of courses are being offered by commercial and university

FOCUS: layfact
Entities: 9, Attributes: 12, Range: 1 to 5, Context: plant layout planning

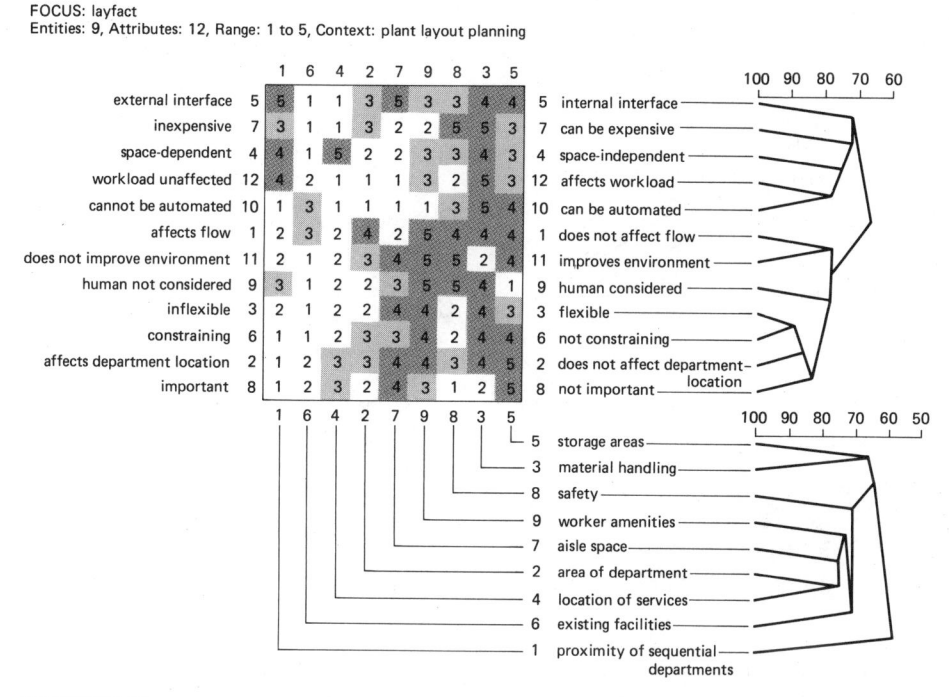

FIGURE 3-26
Hierarchical cluster analysis.

organizations. It is still a problem to get good-quality courses targeted to specific requirements and on the schedules needed.

The middle box in Fig. 3-30 summarizes problems of *knowledge acquisition and transfer* in developing knowledge-based systems. Knowledge acquisition and transfer processes in human society are complex and poorly understood. Neurology, psychology, linguistics, education, sociology, anthropology, philosophy, and systems theory all make significant contributions. However, integrating them, clarifying jargon, and combining different objectives and perspectives are major problems. A variety of short-term problems arise from these weaknesses in foundations:

- *System specification:* The formal specification of software that has made software engineering a professional discipline which can ensure quality within well-defined budgets and schedules is notably lacking for expert system developments. We seem to be at the stage of conventional software development of twenty years ago. This should change as expert system developments move out of the laboratories into the hands of experienced software development groups trained in software management. There is little in the literature to support and expedite this change currently, and we need to develop specification disciplines for expert systems.

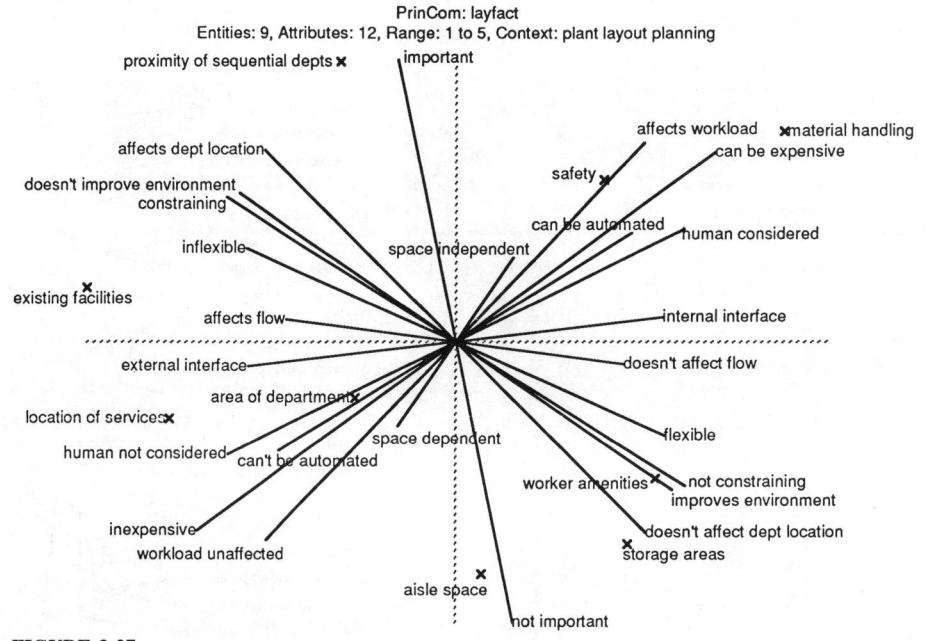

FIGURE 3-27
Spatial cluster analysis.

- *Access to expertise:* The problems of getting experts to make the time commitments necessary to the development of expert systems should never be underestimated. Successful projects inevitably seem to have involved the experts in such a way that they become personally committed to the success of the system.
- *Acquisition of skills:* Communicating the basis of skilled performance between people, and between people and computers, is very difficult.
- *Representation of knowledge:* Encoding knowledge within the representation schemes of current shells can present major difficulties. Production rules as used in early expert systems offer very limited facilities for encoding simple knowledge structures at a single level. Frames and their generalization through objects allow conceptual hierarchies to be better expressed. Neither rules nor frames are adequate to encode the procedural or temporal knowledge which is often a key component of expertise.

The bottom box in Fig. 3-30 summarizes problems of *system performance* in developing knowledge-based systems. Rapid prototyping is a significant practical objective, and techniques for accelerated initial development without loss of long-term quality are now common in conventional data processing. However, expert system development emphasizes weaknesses in some areas of performance which we tend to take for granted in standard information technology. There are problems with:

- *Validation:* Because we can rarely define and quantify expertise, the validation of expert systems can be very difficult. Do we subjectively validate against the opinions of the expert, or do we attempt objective validation of performance against performance standards, accepting that the experts may not themselves satisfy such criteria? How do we determine whether the knowledge base captures the expert's relevant knowledge accurately, consistently, and completely? Very few validation studies exist in the literature.
- *Usability:* The usability of expert systems raises many problems. If we accept the jargon of the expert and the rules of inference based on usability, how do we ensure that the operators will use this jargon in the same way as the expert, so that the inferences remain valid? It is such considerations that make nonsense of the popular misconception that access to encoded expertise can make experts of us all. Understanding the questions, advice, and explanations from an expert system generally requires an operator with the same professional background as the expert.
- *Maintainability:* The superficiality of the knowledge encoded in current expert systems raises major problems of maintenance. Keeping the knowledge base up to date may require an ongoing effort at the same level as the original development, a situation very different from most conventional data

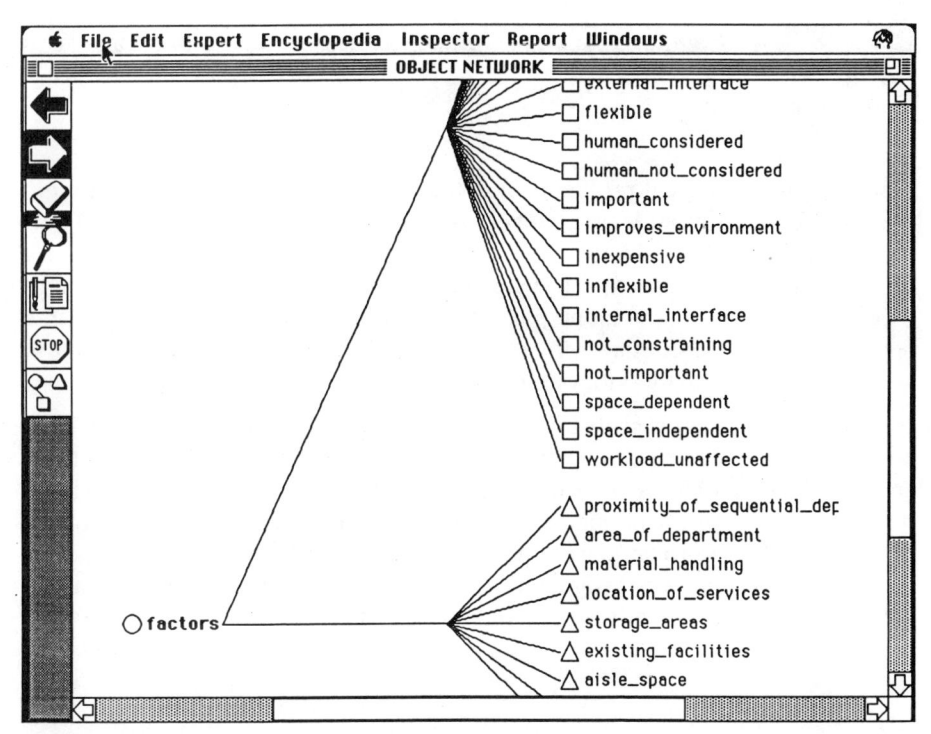

FIGURE 3-28
Class, objects, and properties exported to NEXPERT.

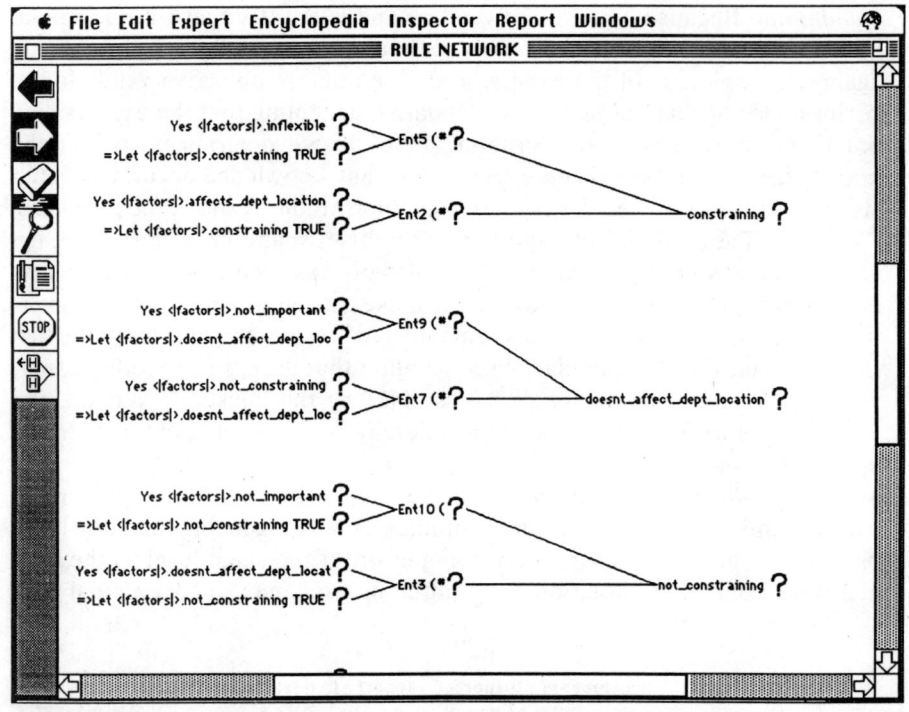

FIGURE 3-29
Rules exported to NEXPERT.

processing. The usual problems of extensive maintenance introducing problems of its own are at least as severe for expert systems as for standard data processing.

- *Upgradability:* Many expert systems and the shells supporting them have been designed as stand-alone packages with little scope for integration into large-scale data processing environments. These systems will be difficult to upgrade and integrate with other software.

This emphasis on problems impeding knowledge-based system development is not intended to be negative. In any major technological advance there are significant problems, and it is the recognition and classification of these that indicates the growing maturity of the field. There is intensive, worldwide research in all the problem areas noted above and growing experience in how to overcome them. By recognizing the problems, project managers may be able to target their objectives to avoid some and to allocate their resources to overcome others.

ACKNOWLEDGMENTS

I am grateful to many colleagues for discussions over the years that have influenced this chapter. In particular, I would like to thank John Boose, Jeff

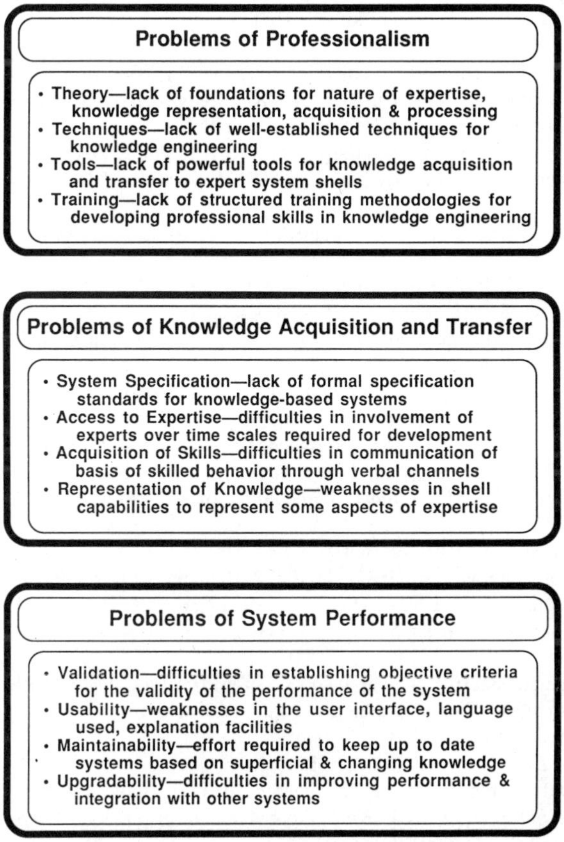

Problems of Professionalism

- Theory—lack of foundations for nature of expertise, knowledge representation, acquisition & processing
- Techniques—lack of well-established techniques for knowledge engineering
- Tools—lack of powerful tools for knowledge acquisition and transfer to expert system shells
- Training—lack of structured training methodologies for developing professional skills in knowledge engineering

Problems of Knowledge Acquisition and Transfer

- System Specification—lack of formal specification standards for knowledge-based systems
- Access to Expertise—difficulties in involvement of experts over time scales required for development
- Acquisition of Skills—difficulties in communication of basis of skilled behavior through verbal channels
- Representation of Knowledge—weaknesses in shell capabilities to represent some aspects of expertise

Problems of System Performance

- Validation—difficulties in establishing objective criteria for the validity of the performance of the system
- Usability—weaknesses in the user interface, language used, explanation facilities
- Maintainability—effort required to keep up to date systems based on superficial & changing knowledge
- Upgradability—difficulties in improving performance & integration with other systems

FIGURE 3-30
Impediments to knowledge-based system development.

Bradshaw, Alain Rappaport, and Mildred Shaw for access to their own research and many stimulating discussions. Much of the research reported here is part of a long-term collaborative study with Mildred Shaw of the cognitive foundations of human knowledge processes.

REFERENCES

Adler, J., and J. Matheson (1986). Expert knowledge elicitation using personal construct theory: Obtaining plant layout expertise with "PLANET." B.A. Sc. thesis, Department of Industrial Engineering, University of Toronto./ Bainbridge, L. (1979). Verbal reports as evidence of the process operator's knowledge. *International Journal of Man-Machine Studies* 11(4):411–436 (July).

Bainbridge, L. (1986). Asking questions and accessing knowledge. *Future Computing Systems* 1(2):143–149.

Benbasat, I., and J. S. Dhaliwal (1988). A framework for the validation of knowledge acquisition. *Proceedings of the Third AAAI Knowledge Acquisition for Knowledge-Based Systems Workshop*, pp. 1-1–1-18. Banff (November).

Blume, S. S. (1974). *Towards a Political Sociology of Science*. New York: Free Press.

Boose, J. H. (1984). Personal construct theory and the transfer of human expertise. *Proceedings AAAI-84*, pp. 27–33. Menlo Park, Calif.: American Association for Artificial Intelligence.

Boose, J. H. (1986). *Expertise Transfer for Expert System Design.* New York: Elsevier.
Boose, J. H. (1987). Rapid acquisition and combination of knowledge from multiple experts in the same domain. *Future Computing Systems* 1(2):191–216.
Boose, J. H. (1988). A survey of knowledge acquisition techniques and tools. In: J. H. Boose and B. R. Gaines (Eds.), *Proceedings of the Third AAAI Knowledge Acquisition for Knowledge-Based Systems Workshop,* pp. 3-1–3-23. Banff (November).
Boose, J. H., and J. M. Bradshaw. Expertise transfer and complex problems: Using AQUINAS as a knowledge acquisition workbench for knowledge-based systems. *International Journal of Man-Machine Studies* 26:3–28.
Boose, J. H., and B. R. Gaines (Eds.) (1988). *Knowledge Acquisition Tools for Expert Systems.* London: Academic Press.
Broadbent, D. E., P. Fitzgerald, and M. H. P. Broadbent (1986). Implicit and explicit knowledge in the control of complex systems. *British Journal of Psychology* 77:33–50.
Cendrowska, J. (1987). An algorithm for inducing modular rules. *International Journal of Man-Machine Studies* 27(4):349–370 (October).
Collins, H. M. (1985). *Changing Order: Replication and Induction in Scientific Practice.* London: Sage.
Diederich, J., I. Ruhmann, and M. May (1986). KRITON: A knowledge acquisition tool for expert systems. In: J. H. Boose, and B. R. Gaines (Eds.), *Proceedings of the First AAAI Knowledge Acquisition for Knowledge-Based Systems Workshop,* pp. 12-0–12-12. Banff (November).
Dixon, N. (1981). *Preconscious Processing.* Chichester: Wiley.
Eshelman, L., D. Ehret, J. McDermott, and M. Tan (1986). MOLE: A tenacious knowledge acquisition tool. In: J. H. Boose and B. R. Gaines (Eds.), *Proceedings of the First AAAI Knowledge Acquisition for Knowledge-Based Systems Workshop,* pp. 13-0–13-12. Banff (November).
Feigenbaum, E. A. (1980). Knowledge engineering: The applied side of artificial intelligence. Report STAN-CS-80-812, Department of Computer Science, Stanford University.
Feyerabend, P. (1975). *Against Method.* London: NLB.
Freud, S. (1914). *Psychopathology of Everyday Life.* London: Benn.
Gaines, B. R. (1987a). Advanced expert system support environments. In: J. H. Boose and B. R. Gaines (Eds.), *Proceedings of the Second AAAI Knowledge Acquisition for Knowledge-Based Systems Workshop,* pp. 8-0–8-14. Banff (October).
Gaines, B. R. (1987b). An overview of knowledge acquisition and transfer. *International Journal of Man-Machine Studies* 26(4):453–472 (April).
Gaines, B. R. (1987c). Rapid prototyping for expert systems. In: M. D. Oliff (Ed.), *Intelligent Manufacturing: Proceedings from First International Conference on Expert Systems and the Leading Edge in Production Planning and Control,* pp. 45–73. Menlo Park, Calif.: Benjamin Cummins.
Gaines, B. R. (1988a). Knowledge acquisition and technology. In: J. H. Boose and B. R. Gaines (Eds.), *Proceedings of the Third AAAI Knowledge Acquisition for Knowledge-Based Systems Workshop,* pp. 8-1–8-21. Banff (November).
Gaines, B. R. (1988b). Positive feedback processes underlying the formation of expertise. *IEEE Transactions on Systems, Man and Cybernetics,* to appear.
Gaines, B. R. (1988c). Second generation knowledge-acquisition systems. *Proceedings of the Second European Workshop on Knowledge Acquisition for Knowledge-Based Systems (EKAW 88).* GMD-Studien Nr. 143. pp. 17-1–17-14. Bonn, Germany: Gellsellschaft für Mathematik und Datenverarbeitung (June).
Gaines, B. R. (1988d). Software engineering for knowledge-based system. *Proceedings of CASE 88, Second International Workshop on Computer-Aided Software Engineering,* pp. 14-1–14-7.
Gaines, B. R., and J. H. Boose (Eds.) (1988). *Knowledge Acquisition for Knowledge-Based Systems.* London: Academic Press.
Gaines, B. R., and M. L. G. Shaw (1980). New directions in the analysis and interactive elicitation

of personal construct systems. *International Journal of Man-Machine Studies* 13(1):81–116 (July).
Gaines, B. R., and M. L. G. Shaw (1985). From fuzzy sets to expert systems. *Information Science* 36(1, 2):5–16 (July).
Gaines, B. R., and M. L. G. Shaw (1986). Induction of inference rules for expert systems. *Fuzzy Sets and Systems* 8(3):315–328 (April).
Gaines, B. R., and J. N. Vickers (1988). Design considerations for hypermedia systems. *Microcomputers for Information Management* 5(1):1–27 (March).
Gammack, J. G., and R. M. Young (1985). Psychological techniques for eliciting expert knowledge. In: M. Bramer (Ed.), *Research and Development in Expert Systems*, pp. 105–116. New York: Cambridge University Press.
Garg-Janardan, C., and G. Salvendy (1987). A conceptual framework for knowledge elicitation. *International Journal of Man-Machine Studies* 26(4):521–531 (April).
Hagstrom, W. O. (1985). *The Scientific Community*. New York: Basic Books.
Hawkins, D. (1983). An analysis of expert thinking. *International Journal of Man-Machine Studies* 18(1):1–47 (January).
Hayes-Roth, F., D. A. Waterman, and D. B. Lenat (Eds.) (1983). *Building Expert Systems*. Reading, Mass.: Addison-Wesley.
Kelly, G. A. (1955). *The Psychology of Personal Constructs*. New York: Norton.
Klinker, G., J. Bentolila, S. Genetet, M. Grimes, and J. McDermott (1986). KNACK—report-driven knowledge acquisition. In: Boose, J. H. and B. R. Gaines (Eds.), *Proceedings of the First AAAI Knowledge Acquisition for Knowledge-Based Systems Workshop*, pp. 23-0-23-13. Banff (November).
Marcus, S. (1987). Taking backtracking with a grain of SALT. *International Journal of Man-Machine Studies* 26(4): 383–398 (April).
Marcus, S. (Ed.) (1988). *Automating Knowledge Acquisition for Expert Systems*. New York: Kluwer.
Merton, R. K. (1973). *The Sociology of Science: Theoretical and Empirical Investigations*. Chicago: University of Chicago Press.
Michalski, R. S., and J. G. Carbonell (Eds.) (1983). *Machine Learning: An Artificial Intelligence Approach*. Palo Alto, Calif.: Tioga.
Michalski, R. S., J. G. Carbonell, and T. M. Mitchell (Eds.) (1986). *Machine Learning: An Artificial Intelligence Approach*, vol. 2. Los Altos, Calif.: Morgan Kaufmann.
Musen, M. A., L. M. Fagan, D. M. Combs, and E. H. Shortliffe (1986). Using a domain model to drive an interactive knowledge editing tool. In: J. H. Boose, and B. R. Gaines (Eds.), *Proceedings of the First AAAI Knowledge Acquisition for Knowledge-Based Systems Workshop*, pp. 33-0-33-11. Banff (November).
Nisbett, R. E., and T. D. Wilson (1977). Telling more than we can know: Verbal reports on mental processes. *Psychological Review* 84:231–259.
Quinlan, J. R. (1987). Simplifying decision trees. *International Journal of Man-Machine Studies* 27(3): 221–234 (September).
Rappaport, A. (1987a). Cognitive primitives. In: J. H. Boose and B. R. Gaines (Eds.), *Proceedings of the Second AAAI Knowledge Acquisition for Knowledge-Based Systems Workshop*, pp. 15-0-15-13. Banff (October).
Rappaport, A. (1987b). Multiple-problem subspaces in the knowledge design process. *International Journal of Man-Machine Studies* 26(4):435–452 (April).
Rappaport, A., and B. R. Gaines (1988). Integration of acquisition and performance systems. *Proceedings of the Third AAAI Knowledge Acquisition for Knowledge-Based Systems Workshop*, pp. 25-1-25-20. Banff (November).
Rappaport, A. T., B. R. Gaines, and J. H. Boose (Eds.) (1988). *Proceedings of the First AAAI Knowledge Acquisition for Knowledge-Based Systems Workshop*, St. Paul, Minn. (August).
Rogers, C. R. (1967). *On Becoming a Person: A Therapist's View of Psychotherapy*. London: Constable.

Runes, D. D. (1972). *Dictionary of Philosophy*. London: Peter Owen.
Schön, D. A. (1983). *The Reflective Practitioner*. New York: Basic Books.
Shaw, M. L. G. (1980). *On Becoming a Personal Scientist*, London: Academic Press.
Shaw, M. L. G. (Ed.) (1981). *Recent Advances in Personal Construct Technology*. London: Academic Press.
Shaw, M. L. G., and B. R. Gaines (1983). A computer aid to knowledge engineering. *Proceedings of British Computer Society Conference on Expert Systems,* pp. 263–271. Cambridge (December).
Shaw, M. L. G., and B. R. Gaines (1986a). Interactive elicitation of knowledge from experts. *Future Computing Systems* 1(2):151–190.
Shaw, M. L. G., and B. R. Gaines (1986b). Techniques for knowledge acquisition and transfer. In: J. H. Boose, and B. R. Gaines (Eds.), *Proceedings of the First AAAI Knowledge Acquisition for Knowledge-Based Systems Workshop,* pp. 39-0–39-13. Banff (November).
Shaw, M. L. G., and B. R. Gaines (1987a). An interactive knowledge elicitation techniques using personal construct technology. In: A. Kidd (Ed.), *Knowledge Elicitation for Expert Systems: A Practical Handbook,* pp. 109–136. New York: Plenum Press.
Shaw, M. L. G., and B. R. Gaines (1987b). KITTEN: Knowledge initiation and transfer tools for experts and novices. *International Journal of Man-Machine Studies* 27(3):251–280 (September).
Shaw, M. L. G., and B. R. Gaines (1988). A methodology for recognizing consensus, correspondence, conflict and contrast in a knowledge acquisition system. In: J. H. Boose and B. R. Gaines (Eds.), *Proceedings of the Third AAAI Knowledge Acquisition for Knowledge-Based Systems Workshop,* pp. 30-1–30-19. Banff (November).
Shaw, M. L. G., and J. B. Woodward (1987). Validation of a knowledge support system. In: J. H. Boose and B. R. Gaines (Eds.), *Proceedings of the Second AAAI Knowledge Acquisition for Knowledge-Based Systems Workshop,* pp. 18-0–18-15. Banff (October).
Shriver, B., and P. Wegner (1987). *Research Directions in Object-Oriented Programming*. Cambridge, Mass.: M.I.T. Press.
Slater, P. (Ed.) (1976). *Dimensions of Intrapersonal Space*, vol. 1. London: Wiley.
Stich, S. P., and R. E. Nisbett (1984). Expertise, justification and the psychology of inductive reasoning. In: T. L. Haskell (Ed.), *The Authority of Experts,* pp. 226–241. Bloomington: Indiana University Press.
Wahl, D. (1986). An application of declarative modeling to aircraft fault isolation and diagnosis. In: P. A. Luker and H. H. Adelsberger (Eds.), *Intelligent Simulation Environments,* pp. 25–28 (January). La Jolla, Calif.: Society for Computer Simulation.
Welbank, M. (1983). *A Review of Knowledge Acquisition Techniques for Expert Systems*. BTRL, Ipswich, U.K.: Martlesham Consultancy Services.

CHAPTER 4

MODEL-BASED KNOWLEDGE ACQUISITION

MEHDI T. HARANDI
RENSE LANGE

INTRODUCTION

The success of current expert system technology is due in large part to the decision to enforce a strict separation between a domain-dependent knowledge base and a domain independent inference engine to reason with this knowledge. Over the past years, the development of inference engines has shown continuous progress, and today's expert system builders can select an appropriate inference engine from several "shell" systems (Duda et al., 1979; van Melle et al., 1984; Forgy, 1981). It has become increasingly clear, however, that the power of an expert system depends more on the quality of its knowledge base than on the nature of its inference engine. Yet, although the inferencing aspect of expert systems has largely been solved, the development and maintenance of knowledge bases continue to be major problems in expert system development. Feigenbaum's assessment in 1978 (Feigenbaum, 1978) that knowledge acquisition is the "bottleneck" in expert system development is still accurate today.

Given this state of affairs, it is not surprising that many research attempts have been made to streamline the knowledge acquisition process. These efforts differ in the degree to which they adhere to the traditional **expert → knowledge engineer → knowledge base** paradigm. Some approaches stay squarely within this framework and focus mainly on reducing the clerical burden on the knowledge engineer, through the development of intelligent editing systems designed to facilitate the updating of knowledge bases. On the other end of the spectrum one finds research in similarity-based learning aimed at eliminating the involvement of knowledge engineers as well as experts. For example, work by Michalski and Chilausky (1980) has shown that expert system rules can, in principle, be learned from a library of previously solved problems.

In this chapter we discuss approaches which lie between these extremes with respect to their reliance on the traditional acquisition paradigm. Clearly, sufficient progress has already been made to proceed beyond the development of knowledge base editing tools. It seems natural, therefore, to reflect on approaches aimed at decreasing the roles of experts and knowledge engineers. The most promising avenue of research at this point seems to lie in the development of automated knowledge acquisition systems which decrease the involvement of knowledge engineers but continue to rely on domain experts. In essence, then, we focus on systems that follow an **expert → acquisition system → knowledge base** approach in which the knowledge engineer is largely replaced by an automated acquisition system.

A decrease in the involvement of knowledge engineers—let alone their complete elimination—means that a vital link in the knowledge acquisition chain is now missing. For this reason, in the first section we analyze the consequences of this situation in terms of the role and contribution of the knowledge engineer in the traditional knowledge acquisition paradigm. Naturally, the omission of the knowledge engineers' contributions to the knowledge acquisition process will have to be compensated for elsewhere. Therefore, in the next sections we present a "model-based" paradigm. In this paradigm, knowledge acquisition is guided by an acquisition system's current level of domain knowledge, as represented by a "domain model." Different types of domain models are discussed, together with some applications of this approach.

ROLE OF THE TRADITIONAL KNOWLEDGE ENGINEER

The traditional knowledge engineering approach follows the **expert → knowledge engineer → knowledge base** paradigm. According to this paradigm, a knowledge base is created by knowledge engineers who extract domain knowledge from (one or more) experts and translate this information to a format suitable for the inference engine. Experts then conduct trial consultations to verify that the knowledge base is correct and complete. In addition, experts check whether the style of reasoning of the inference engine agrees with their problem-solving approach. If any errors or omissions are found, these are reported to the knowledge engineer. Since an initial knowledge base tends to contain many errors or omissions, several cycles may be required before it meets the experts' final approval. This trial-and-error approach to knowledge acquisition is well known, and it has been described extensively in standard texts on expert systems (see, e.g., Hayes-Roth et al., 1983; Buchanan and Shortliffe, 1984; Waterman, 1985). In this section we limit the discussion to the role of the knowledge engineer in this process.

Knowledge engineering is a formidable, expensive, and time-consuming task in the traditional paradigm. We list five major reasons, in order of increasing complexity:

(1). *Vocabulary:* To communicate with a domain expert, a knowledge engineer has to become conversant with the basic vocabulary of a domain (the names of basic components, structures, standard problem-solving strategies, etc.).
(2). *Completeness:* A knowledge engineer has to be able to identify pieces of information that are missing from the knowledge base. Being able to do so requires an understanding of the domain as well as a good overview of the contents of the current knowledge base.
(3). *Integration:* As new information becomes available, it has to be determined how this fits into the current knowledge base. Because new information can interact with already available information in undesirable ways, it may be difficult to find the appropriate way to encode new information.
(4). *Analysis:* Experts often find it difficult to explain exactly how and why they arrived at certain conclusions. Consequently, knowledge engineers may have to conduct lengthy interviews with experts. In addition to analytical skills, this requires considerable social skills on the part of the knowledge engineer.
(5). *Transparency:* A major factor in the evaluation of a knowledge base concerns the extent to which it leads to problem-solving behavior that agrees with experts' styles of reasoning. A knowledge engineer should incorporate this reasoning style in the knowledge base or, in special cases, in the functioning of the inference engine.

Depending on the domain in question, other requirements could be added to this list. Our main point is, however, that the traditional acquisition approach causes knowledge engineers to learn extensively about the domain, either through their own efforts (by reading books, manuals, etc.) or through direct interaction with experts. Knowledge engineering requires a good "feel" for the application area, and experts are often faced with the task of teaching knowledge engineers very much as they would do with an apprentice. In other words, the traditional approach forces knowledge engineers to become mini-experts themselves. This assessment is consistent with the finding that applications in highly technical domains require knowledge engineers with backgrounds similar to those of the domain experts (Lange et al., 1986). Further, the best known and most successful expert systems such as MYCIN (Buchanan and Shortliffe, 1984) and XCON (McDermott, 1982; Bachant and McDermott, 1984) were built by highly skilled knowledge engineers who had a formal education in the area of application (i.e., medicine and computer science, respectively).

MODELS OF DOMAIN KNOWLEDGE

Any system intended to automate the acquisition process to a significant extent needs to perform the same types of tasks as a knowledge engineer does. That is, such a system needs to know the basic vocabulary of the domain; it should identify missing pieces of information, integrate new information into partially developed knowledge bases, and elicit explanations and clarifications when experts' information is incomplete or inconsistent. Finally, the knowledge base should be organized so as to maximize the transparency for the end user. To perform these tasks, the acquisition system needs a model of domain of application, and it should bring this model to bear on the acquisition process.

Even a cursory review of the AI literature indicates a great diversity in the definition, use, content, and scope of domain models (e.g., Weiss et al., 1978; Boose, 1983; Eshelman et al., 1987; de Kleer and Brown, 1984; Forbus, 1984; Swartout, 1983). Rather than describing the various approaches in detail, we develop a categorization based on the required completeness of the domain models being used. The extremes of this dimension are formed by the two basic types of machine learning, i.e., similarity-based learning and explanation-based learning. Similarity-based learning requires no explicit domain model and thus forms the low end of the completeness dimension. In contrast, explanation-based learning requires a complete and consistent domain model and thus forms the high end of the dimension. We discuss these two forms of machine learning in some detail before dealing with approaches that use domain models of intermediate levels of completeness.

NO DOMAIN MODEL REQUIRED: SIMILARITY-BASED LEARNING

The general goal of similarity-based learning is to produce rules for classifying objects as belonging to one or more classes, based on the objects' properties (Michalski et al., 1983). In essence, classification rules state which minimal combinations of properties are sufficient to make a classification unambiguous. Most learning systems deal with two-valued classifications (discrimination learning), whereas other systems can also deal with more complex cases where several classifications are possible. Classification rules are not unlike the rules in the knowledge base of a typical expert system. For example, a medical expert system might contain rules that describe the set of observations (test results, etc.) sufficient to arrive at the correct diagnosis of a particular illness. A similarity-based learning system can produce such classification rules without the direct involvement of experts. It does so by analyzing known example classifications, thereby focusing on properties that cover positive instances while excluding negative instances.

Strictly speaking, the resulting classification rules cover only the examples on which they are based. Further generalization may be warranted, however, to the extent that the example classifications are representative of the entire domain. In addition, great care has to be taken in the selection of properties to describe the example classifications. If one or more essential attributes are lacking, classification rules become needlessly complex and, in extreme cases, no rules can be formulated.

Several similarity-based learning algorithms have been described in the literature (see, e.g., Michalski et al., 1983, 1986). One of the earliest and best known algorithms is ID3 (see Quinlan, 1983), a descendant of Hunt's CLS (Hunt, 1962). As this algorithm is comparatively simple, it is described here in some detail.

An Illustrative Example: ID3

ID3 assumes that each object is described in terms of a fixed number of attributes, each with some set of attribute values. For example, a test result (the attribute) might have the outcomes of *positive* and *negative* (the attribute values). In addition, each object must be identified as belonging to some class of objects. For simplicity, we assume that only two classifications are possible: positive (i.e., the object is a member of the class) or negative (the object is not a member of the class). ID3 discovers minimal classification rules that correctly assign each object to one of these two categories. As is shown in the following example, classification rules take the form of a decision tree.

Assume that we want to form rules to classify patients as having a particular disease X, and for each patient the results on four medical tests, A, B, C, and D, are known. It is further assumed that the results of each test as well as the classification are deterministic and binary; i.e., a test is either positive (+) or negative (−), and either patients have disease X (+) or they do not have disease X (−). Table 4-1 shows the measurements and the classifications for five hypothetical patients.

In general, a decision tree is formed by selecting an attribute and creating a branch for each value of this attribute. This process continues until the classification of the objects within a branch is homogeneous (i.e., all objects have the same classification) or until there are no objects left to be classified (i.e.,

TABLE 4-1
Patient classification based on measurements

Patient	Test A	Test B	Test C	Test D	Disease X
1	+	+	−	+	+
2	−	+	−	−	−
3	+	−	+	−	+
4	−	−	−	+	−
5	−	−	+	−	+

the branch is empty). For example, taking test C as the root attribute, we get the partial decision tree shown in Fig. 4-1.

In this tree, the left branch is homogeneous because patients 3 and 5 both have disease X; consequently, no further splitting is possible, and we have arrived at a leaf of the classification tree. The right branch is not homogeneous because patient 1 is classified as having disease X (+), whereas patients 2 and 4 do not have the disease (−). Closer inspection reveals that further splitting on the values of Test A would result in two homogeneous subbranches: a value of + for Test A leads to a positive classification, and a value of − leads to a negative classification. Thus, by ignoring redundant information (Tests B and D) we have simplified the original classification based on four tests (A through D) to one based on two tests only (C and A). If all redundant attributes are removed, the decision tree of Fig. 4-2 is obtained. In this figure, the leaves of the tree have been labeled by their appropriate classifications (i.e., Disease X = + or −).

ID3 attempts to minimize the expected number of tests needed to arrive at a classification. Thus, the choice of a branching attribute is crucial. ID3 solves this problem by selecting at each choice point the attribute that leads to the greatest increase in homogeneity. Although this approach does not guarantee that the size of the entire tree is minimal, the approach is satisfactory for most purposes and leads to a computationally efficient algorithm. (For additional details, the interested reader is referred to Quinlan, 1983.)

It is relatively straightforward to translate the information in a decision tree such as the one above into the more familiar production rule format by the following procedure. For each leaf, make the label of the leaf the consequent of the rule, and then follow the tree up to the root while forming a conjunct of the labels found on each branch. In the example this yields the following three rules:

IF Test C = + THEN Disease X = +
IF Test C = − & Test A = − THEN Disease X = −
IF Test C = − & Test A = + THEN Disease X = +

AN APPLICATION: SOYBEAN DISEASE DIAGNOSIS. Illustrative of the application of similarity-based learning in knowledge acquisition is the work by Michalski and Chilausky (1980). By using the AQ11 induction algorithm, clas-

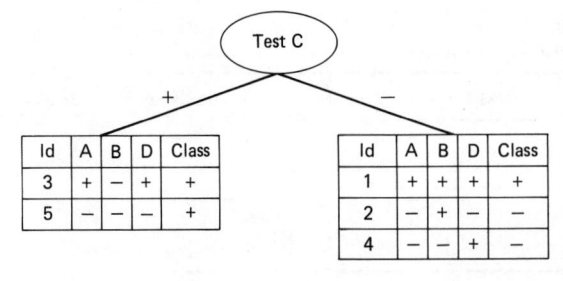

FIGURE 4-1
Partial decision tree for Test C.

FIGURE 4-2 Complete decision tree.

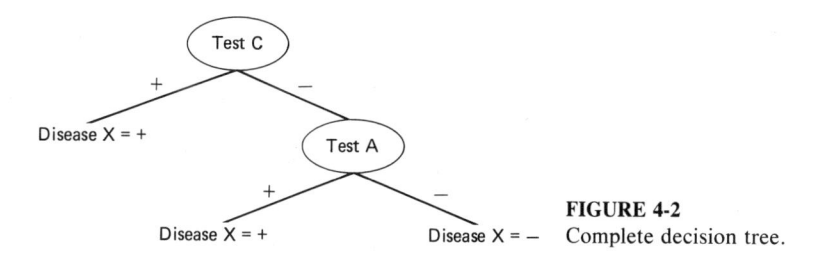

sification rules were induced for the diagnosis of 15 different soybean diseases based on a sample of 290 diagnoses of diseased plants. Next, on a second batch of 340 new cases, an experiment was performed to compare the performance of the induced rules with that of rules induced by a highly respected plant pathologist. The machine-generated rules ranked the correct diagnosis first 97.6 percent of the time, whereas the expert's rules ranked the correct diagnosis first only 71.8 percent of the time. In addition, the machine-generated rules never failed to list the correct diagnosis among the possible ones, whereas the expert's rules failed to mention it in 3.1 percent of the cases. So the experiment showed that, at least under certain conditions, induced rules can outperform rules generated by domain experts.

COMPLETE DOMAIN MODEL REQUIRED: EXPLANATION-BASED LEARNING

Although explanation-based learning can be traced to early robot problem-solving systems such as STRIPS (Fikes et al., 1972), the current interest in this approach is mainly due to the work by Mitchell (1983) and DeJong (1981). Explanation-based learning is applicable to several types of problem solving (see DeJong and Mooney, 1986; Mitchell et al., 1986), and in the following we outline its foundations and some applications relevant to knowledge acquisition.

In the terminology of DeJong and Mooney (1986), explanation-based learning can be thought of as a way to generalize the solutions found in a particular problem-solving situation. In this context, a problem is specified by some initial state, and the solution consists of a sequence of steps which transform the problem state to a state that matches the goal. A solution can be given directly (e.g., it can be obtained through observation of experts' problem-solving behavior), or it may have to be produced first by some problem-solving system (for instance, a theorem prover). In either case, an explanation-based learning system takes the complete solution as its input and transforms it into a more general explanation that solves not only the original problem but possibly others as well. This generalization can be stored and retrieved for later use in the solution of new and similar problems.

Any solution to a problem can be thought of as a proof that a set of given initial conditions leads to achieving a particular goal. To understand such proofs, an explanation-based learning system requires complete knowledge of all relevant contingencies, relations, and properties of the domain. Without loss of generality, it is assumed that such domain knowledge can be represented in the form of inference rules and that the proof consists of applications of these rules. Thus, the problem of generalizing a solution is transformed to the problem of generalizing proofs. Several methods to generalize proofs have been proposed (e.g., Waldinger, 1977; Mitchell et al., 1986). However, in the following we use the method described in DeJong and Mooney (1986).

Generalizing a Solution

To contrast similarity-based learning and explanation-based learning, we continue the hypothetical example used earlier concerning the diagnosis of disease X here. Explanation-based learning requires a domain theory, and so we are forced to state some detailed properties and relations governing the domain. Such information can conveniently be expressed in the notation of first-order logic. By convention, lowercase letters are used to write constants, whereas variable names start with a capital letter. Predicate names can be in either lowercase or uppercase letters.

For purposes of illustration we assume the following:

- Disease X is actually the result of an imbalance between two hormones a and c. This fact can be expressed by an inference rule:

    Imbalance(a, c) $\Rightarrow$ Disease X

- An imbalance, in turn, is defined as

    level(H1, L1) & level(H2, L2) & different(L1, L2) $\Rightarrow$ imbalance(H1, H2)

- Tests A and C are indicators of the levels of hormones a and c, respectively. This assumption is represented by two inference rules:

    Test A = X $\Rightarrow$ level(a, X)   and   Test C = Y $\Rightarrow$ level(c, Y)

- It is further given that a person tested low on test A and high on test C, or

    Test A = −   Test C = +

Together, the above information allows the deduction of the fact that the person suffers from the hormonal imbalance underlying disease X. An outline of the explanation is shown in Fig. 4-3. In this outline, the consequents and the antecedents of the same rule are connected by single lines, whereas the antecedents and consequents of different rules are connected by double lines.

By applying the substitution {a/H1, c/H2, +/La, +/L1, −/Lc, −/L2}, the above structure explains why the results on Tests A and C indicate an imbalance between hormones a and c. That is, this substitution causes the anteced-

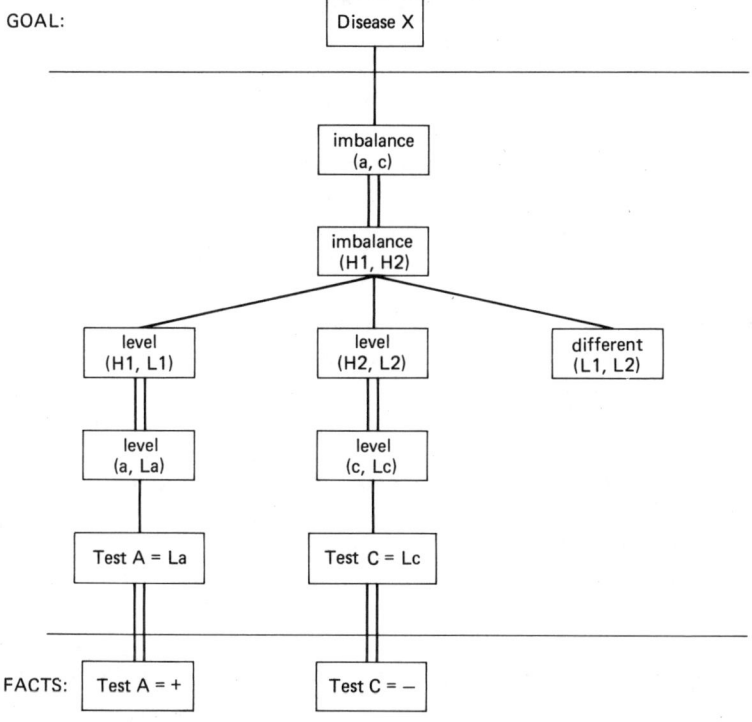

```
GOAL:                    ┌──────────┐
                         │ Disease X│
                         └────┬─────┘
─────────────────────────────┼──────────────────────────────
                         ┌───┴──────┐
                         │ imbalance│
                         │  (a, c)  │
                         └────┬─────┘
                         ┌───┴──────┐
                         │ imbalance│
                         │ (H1, H2) │
                         └────┬─────┘
          ┌───────────────────┼────────────────────┐
     ┌────┴─────┐        ┌────┴─────┐         ┌────┴─────┐
     │  level   │        │  level   │         │ different│
     │ (H1, L1) │        │ (H2, L2) │         │ (L1, L2) │
     └────┬─────┘        └────┬─────┘         └──────────┘
     ┌────┴─────┐        ┌────┴─────┐
     │  level   │        │  level   │
     │ (a, La)  │        │ (c, Lc)  │
     └────┬─────┘        └────┬─────┘
     ┌────┴─────┐        ┌────┴─────┐
     │Test A = La│       │Test C = Lc│
     └────┬─────┘        └────┬─────┘
─────────────┼───────────────┼──────────────────────────────
     ┌──────┴───┐        ┌───┴──────┐
FACTS:│Test A = +│        │Test C = −│
     └──────────┘        └──────────┘
```

FIGURE 4-3
Explanation outline for "Patient has Disease X."

ents and consequents of all the rules and facts to match, and the resulting structure forms a proof for the presence of Disease X. The generalization of this explanation focuses on the middle part of the figure; i.e., the goal and the facts are excluded. The substitution list in this case is {a/H1, c/H2, La/L1, Lc/L2}. Application of this substitution to the middle part of the explanation structure, ignoring all the internal nodes of the explanation structure, yields the following generalized explanation:

Test A = La & Test C = Lc & different(La, Lc) ⇒ imbalance(a, c)

This generalized explanation for Disease X agrees with the example data for ID3 (Table 4-1) and with the induced decision tree. For example, Patients 1 and 5 have different test results on Tests A and C, they are both classified as having disease X, and this classification is consistent with the generalized explanation. Thus, explanation-based learning yielded a rule—derived from a single case only—that correctly applies to additional cases also. This net gain may appear relatively insignificant, and it certainly is in the present example. However, in more realistic domains with long proofs involving many inference rules, facts, etc., the derivation and subsequent reuse of a generalized explanation can lead to considerable savings in computation time. For example, research by Steier and Newell (1988) showed that under certain conditions the use of a form of explanation-based learning provided a 28 to 69 percent speedup in the execution time of their automated algorithm design system.

The example also points to one of the major necessities for explanation-based learning. Notice that the generalized explanation does not classify Patient 3 as having Disease X (since both Test A and Test C are positive), whereas this patient is classified positively in Table 4-1. This does *not* mean that the generalized explanation is faulty. Rather, the domain theory used in our example simply does not cover the data for Patient 3, and no proof (let alone a generalization) can be found for this patient's classification. This shows that the power of explanation-based learning depends directly on the scope and power of the domain theory.

APPLICATIONS OF EXPLANATION-BASED LEARNING. Mitchell et al. (1985) applied explanation-based learning in the development of a learning apprentice system called LEAP which directly assimilates new knowledge by observing experts' problem-solving behavior in the domain of VLSI circuit design. This system offers advice in designing circuits, but it allows the user to override this advice if so desired. In cases where users modify or refine the circuits proposed by LEAP, the system records the refinement as a training example. Later such training examples are refined by using a form of explanation-based learning.

PARTIAL DOMAIN MODELS

In the preceding section we dealt with the two extremes in the use of models in knowledge acquisition. Taken in isolation, neither approach is entirely satisfactory in the context of knowledge acquisition for expert systems. Similarity-based learning is not universally applicable. First, the approach works best when one is dealing with applications requiring a classification into a limited number of categories. Second, similarity-based learning requires the availability of suitable data. Third, similarity-based learning groups examples into categories based on their attributes and attribute values only. In most cases, however (and almost certainly in expert system applications), additional background information is available about a domain. Such background knowledge may be theoretical (e.g., part of the domain may be formalized) or pragmatic (e.g., experts may follow a certain approach because they know that it works in practice). Most similarity-based learning systems cannot take advantage of such information in the induction process. Consequently, the rules produced by a rule induction system may appear unnatural and counterintuitive to domain experts and expert system users (Michalski and Chilausky, 1980).

In many ways, explanation-based learning poses the reverse picture. Explanation-based learning is not particularly useful for classification problems, and it cannot take advantage of the same type of available data as similarity-based learning can. Most importantly, however, explanation-based learning requires a complete and consistent domain theory. As pointed out earlier, without such a complete domain theory the method simply cannot be applied. This decreases the suitability of explanation-based learning for knowledge acquisition purposes since knowledge acquisition takes place in a context where domain information is incomplete and (possibly) inconsistent.

In the following we focus on three approaches: Teiresias (Davis 1984), the BLIP system as designed in the Lerner project (Emde and Morik, 1986; Wrobel, 1987), and the distributed knowledge acquisition system (DKAS) (Lange and Harandi, 1986; Buchner, 1988). These three were selected because they highlight some of the major differences in this area. In particular, it will be shown that Teiresias bases its models mainly on syntactic properties of the knowledge base. In contrast, the BLIP system is directed at modeling semantic, domain-independent knowledge. The DKAS uses a causal model of the domain, combining both syntactic and semantic information.

Syntactic Models: Teiresias

The Teiresias system as conceived and developed by Davis (see Davis and Lenat, 1982; Davis, 1984) is one of the earliest and most influential attempts to automate important aspects of the knowledge acquisition process. Teiresias should perhaps best be thought of as an intervening step between a domain expert and an expert system. Its approach to knowledge acquisition is "failure-driven," in that it follows a continual debugging and acquisition cycle. The debugging part of this cycle is initiated whenever an expert discovers an error in the expert system's reasoning. At this point, Teiresias will assist the expert in tracking down the bug by providing detailed information concerning the sequence of inference steps that lead to the error. The system basically does so by retracting and displaying all the rules involved in reaching the point at which the error was detected. The system then queries the expert to determine if the invocation of these particular rules was justified, given the context and prevailing set of circumstances. Analogous to the use of standard program tracing facilities, "seeing" the rules in action greatly facilitates the debugging task. However, the importance of Teiresias greatly transcends its function as a rule-checking mechanism. Instead, the primary importance of the system lies in its use of metaknowledge in the form of so-called rule models. As is described below, rule models provided the basis for Teiresias' knowledge acquisition functions.

Rule models serve three major functions (Davis, 1984, p.189). First, they capture the structure, organization, or content of the experts' reasoning. Second, they can be used to guide the interpretation of information provided by experts. Third, rule models express expectations concerning the likely content of new rules. One of the major assumptions made in the development of Teiresias is that these functions can, in large part, be derived from syntactic properties of the rules in a knowledge base. That is, it is assumed that rules about a particular topic have common characteristics that reflect experts' ways of reasoning about this topic. More specifically, the central idea behind rule models is that a rule base can be divided into meaningful subsets of rules, each of which is characterized by some combination of "typical" syntactic items. In practice, typicality was defined in terms of the frequency of occurrence of syntactic items over a subset of rules such that more frequently occurring items are more likely to be included in a rule model than items with a lower frequency of occurrence.

Figure 4-4 shows an example of a rule model (from Davis, 1984). As indicated, rule models consist of five major parts. First, the EXAMPLES list states the rules from which the model was constructed. Next, the PREMISE list describes typical premises in the antecedents of the rule mentioned in EXAMPLES. For instance, the rules' premises refer to attributes such as **gram** and **morph,** which have as typical values **same** and **notsame**. Likewise, the ACTION list describes typical action attributes such as **category** with the typical value **conclude**. The numbers at the end of sublists indicate the total strength or reliability of the associations between an attribute and its possible values. The higher the number, the more reliable the association. For instance the example set of rules is more likely to contain references to the attribute **gram** (premise 1, strength = 3.83) than to the attribute **site** (premise 8, strength = 1.23). In addition, the attribute-value(s) combinations are clustered into subgroups according to their patterns of co-occurrence. For example, the model shows that

CATEGORY-IS	
EXAMPLES	((RULE116 .33) (RULE050 .70) (RULE037 .80) (RULE095 .90) (RULE152 1.0) (RULE140 1.0))
P-ADVICE	((GRAM SAME NOTSAME 3.83) (MORPH SAME NOTSAME 3.83) ((GRAM SAME) (MORPH SAME) 3.83) ((MORPH SAME) (GRAM SAME) 3.83) ((AIR SAME) (NOSOCOMIAL NOTSAME SAME) (MORPH SAME) (GRAM SAME) 1.50) ((NOSOCOMIAL NOTSAME SAME) (AIR SAME) (MORPH SAME) (GRAM SAME) 1.50) ((INFECTION SAME) (SITE MEMBF SAME) 1.23) ((SITE MEMBF SAME) (INFECTION SAME) (PORTAL SAME) 1.23))
R-ADVICE	((CATEGORY CONCLUDE 4.73) (IDENT CONCLUDE 4.05) ((CATEGORY CONCLUDE) (IDENT CONCLUDE) 4.73))
MORE-GENL	(CATEGORY-MOD)
MORE-SPEC	NIL

FIGURE 4-4
A rule model.

whenever the attribute **gram** occurs, the attribute **morph** is likely to occur also. Finally, the MORE-GENL and MORE-SPEC fields refer to more general subsets of rules and more specific sets of rules.

Note that Teiresias' rule models are not entered a priori by experts. Rather, the system uses simple statistical techniques (e.g., frequency computations) to analyze its own knowledge base, finding new rule models or updating existing ones. Thus, as new information becomes available, the system's performance improves because the system continues to learn better and more appropriate generalizations.

Teiresias uses its rule models for a number of purposes. In the acquisition process, rule models serve as a guide for the interpretation and understanding of new information. Using relatively straightforward pattern-matching mechanisms and a dictionary of connotations, the system extracts key words from the text input by an expert and attempts to construct an appropriate internal representation of this text. Usually, several alternative interpretations are possible. Rule models provide information concerning the level of specificity of the input (based on the MORE-GENL and MORE-SPEC fields, see Fig. 4-4) as well as the probable content of rules, etc. Thus, rule models can be used to order the interpretations according to their plausibility. The system can then report the most plausible interpretation of the input and query the expert for confirmation.

The system plays an even more active role during the rule-editing process. When an expert edits existing rules, Teiresias uses its rule models to determine how well the result conforms to its own expectations concerning the topic. For example, certain expected items may appear to be missing, whereas other may appear superfluous or simply unfamiliar to the system. Thus, the system attempts to "second-guess" experts, and it may suggest plausible improvements in the experts' inputs as well as query the experts for additions and clarifications. In a sense, Teiresias acts as a (good) student who continuously tries to match her understanding against the input provided by her teachers (i.e., the domain experts).

Semantic Models: The Lerner Project

The Lerner project at the University of Berlin was started in 1985 with the aim of developing a convenient system to assist knowledge engineers in building an expert system. This section discusses the basic principles behind the development of the BLIP system (Emde and Morik, 1986).

Influenced by the work by researchers such as Clancey (1983) and Swartout (1983), the Lerner approach assumes that qualitatively different types of knowledge can be distinguished. Specifically, in the context of building a knowledge acquisition system, a sharp distinction is made between consultation-dependent knowledge and consultation-independent knowledge. Consultation-dependent knowledge is task-oriented, domain-specific, and typically precompiled in order to be used efficiently by some inference engine. In other words, this type of knowledge corresponds approximately to the notion of a standard knowledge base. In contrast, consultation-independent knowledge is application- and implementation-independent, it is useful over several (closely) related applications, and it is independent of a particular consultation system.

Consultation-independent knowledge is structured in so-called domain models which represent terminological and empirical knowledge as it is presented by encyclopedia, accompanied by facts or observations. In their basic form, domain models consist of declarative statements in a sorted, higher-order, semantic representation language. Thus, this language allows for not only the definition of first facts in first-order logic but also the expression of (higher-order) "metafacts." In essence, this language provides a convenient and economical way to represent much of the fundamental conceptual and terminological information needed throughout the knowledge acquisition process. The notation used for representation of constants and variable names is the same as before.

For example, the following three statements could be considered simple first-order facts in a drug domain (the equivalent descriptions in English are shown in brackets):

1. contains(aspirin, ass)   [aspirin contains ASS]
2. strengthen-1(ass, asthma)   [ASS strengthens asthma attacks]
3. strengthen-2(aspirin, asthma)   [aspirin strengthens asthma attacks]

In addition, a metafact like

4. transitive(contains, strengthen-1, strengthen-2)

indicates that the contains, strengthen-1, and strengthen-2 relations are transitive. Metafact 4 is a shorthand version of the following first-order rule:

$$\text{contains}(X, Y) \,\&\, \text{strengthen-1}(Y, Z) \Rightarrow \text{strengthen-2}(X, Z)$$

The representation language also allows for the use of meta-metapredicates to express relations between various metapredicates. No special distinction is made between the various metalevels, and thus the same metapredicates can be used at various levels. However, the language is sorted, and this feature restricts the sorts (types) of arguments for each predicate. This, in turn, affects the nature and content of the rules that can be derived.

An application of the above is found in the BLIP system (Emde and Morik, 1986). BLIP is a learning system concerned with improving initial ("sloppy") domain models which are likely to be incorrect and incomplete. A major function of the system concerns consistency checks of the input. For example, given the meta-metafact

5. opposite-2(opposite-1, inclusive)

and the metafacts

6. opposite-1(monosubstance, multisubstance)
7. inclusive(monosubstance, multisubstance)

the system would detect an inconsistency. This is because fact 5 implies the metarule

8. opposite-1(...) ⇒ not inclusive(...)

and this rule contradicts metafact 7.

In addition, BLIP translates domain models to rules for a target expert system called TWAICE (Mescheder, 1985). BLIP can transform first-order facts as well as higher-order facts (metafacts). As a simple example, the first-order fact

9. contains(aspirin, ass)

would be translated to the following TWAICE rules:

10. IF drug · name = aspirin THEN drug · contains = ass
11. IF substance · name = ass THEN substance · is-contained-in = aspirin

Notice that two rules resulted from fact 9. In general, although many rules are implied by a set of facts, probably not all these rules are useful to the expert system. For example, if the knowledge base never used the is-contained-in relation, rule 11 would be useless. Also rules assigning values to attributes should not be generated if the attribute values must always be provided by the users of the expert system (e.g., the name of a patient). BLIP analyzes the consultation knowledge contained in the actual knowledge base to guide the transformation process and to avoid the generation of useless rules.

Causal Models: The ACDM Approach

The *abstract causal domain model* (ACDM) is essentially a method for encapsulating expert domain knowledge into small, hierarchical, independent, composable, and decomposable units of knowledge. These units of knowledge, when composed, are able to interact with each other in such a way as to model the high-level processes of expert reasoning. The basic units of the domain model are termed *components*. The interactions between components are called *causal flows*.

A domain model is constructed by connecting components so as to represent the topology of the application domain. An output from one component may be connected to the input of another, creating a causal flow. A completely connected set of components represents an ACDM, which can be utilized for reasoning as well as knowledge acquisition purposes.

This modeling technique is similar in some ways to the domain theories used in qualitative physics, in particular, the work of de Kleer and Brown (1984). The ACDM, however, is structurally a much simpler model than is de Kleer's. Specifically, the notion of the component state is avoided and made up for by allowing some of the causal flows to indicate a component's state and thus provide the required flexibility. A principal distinction with de Kleer's domain theory is that the modeling described here explicitly represents the direction of causality, thus providing an absolute temporal ordering of domain states when change within the domain is being investigated.

COMPONENT DESCRIPTION. The basic unit of knowledge in this technique is the component. A component is an encapsulated representation of the functionality of some abstract domain entity in terms of its causal interactions with other domain entities, what we call its *causal inputs* and *causal outputs*. For example, a component may represent the functionality of a pump in terms of the fluid intake, fluid output, and power source of the pump. The power source and fluid intake are causal inputs since they are provided to the pump from other components. The fluid output of the pump is a causal output because it is provided to some other component. Conceptually, a component is a blackbox procedure. It has a functional process for determining the values of its causal outputs, given the values of its causal inputs. Figure 4-5 exemplifies this property for the pump component.

A component contains knowledge of the names, types, sources (or sinks), and values of its causal inputs and outputs. This is the extent of its external knowledge. And, in general, a component does not know, or need to know, from where its causal inputs are coming or to where its causal outputs are going. This information is needed, however, when the component is to be integrated into a higher-level component.

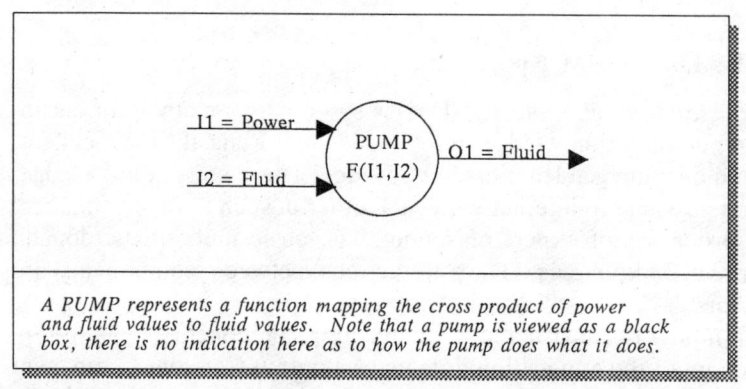

A PUMP represents a function mapping the cross product of power and fluid values to fluid values. Note that a pump is viewed as a black box, there is no indication here as to how the pump does what it does.

FIGURE 4-5
Representation of a pump component.

Causal inputs and outputs are the mechanism for hooking components together. The connection of one component's causal output to another component's causal input is the only path by which the components may interact. This pathway between components is termed a *causal flow*. The binding of values to causal flows and their subsequent propagation, via component causal equations, to other causal flows are the central reasoning mechanism within the domain model.

COMPONENT HIERARCHIES AND TYPES. An important aspect of expert reasoning is the ability to reason about a domain at varying degrees of detail. By pursuing a top-down reasoning approach, an expert can avoid the myriad details which might otherwise impede thinking. For example, in the domain of automobile repair, first an expert will isolate a problem to be in a very general component of the automobile, perhaps the transmission. Then he may reason about the transmission as if it were, itself, the entire domain. Details about tires, radiators, and various other parts may be safely neglected.

To represent these levels, the ACDM allows each component to contain knowledge of the component's internal structure in terms of other, lower-level domain components and their causal interconnections. There is no structural difference between a top-level domain model and a particular component's decomposition detail; both are causal domain models. This turns out to be useful for two reasons. First, it allows the same reasoning algorithms to work at all levels of the domain; second, when this model is applied to knowledge acquisition, it will allow experts to deal in the same manner with widely different domain granularities. For example, an expert on carburetors could work exclusively on carburetors while other experts treat the carburetor as a bottom-level component.

Components with similar internal structure and functionality can be abstracted as submodels. Figure 4-6 shows the topology of a subdomain model for a car's fuel system (top) and how this structure has been abstracted into a single component (bottom). The external inputs and outputs and the functional relationship between the inputs and outputs of both the detailed and abstracted models are the same. One or more instances of this submodel can then be incorporated into a higher-level model.

By maintaining a type hierarchy on the set of domain components, it is possible to facilitate inheritance of component properties and properties of their inputs and outputs. It would allow more effective search of domain components by the expert when the expert needs to incorporate a certain type of component into a domain model being constructed. Figure 4-7 is a hierarchy for pump components; it represents the knowledge that a water pump, a heatpump, and a heart are all examples of pumps.

The domain model also maintains a type hierarchy over the set of causal input and causal output types. Two types are compatible if either type is an instance of the other type (i.e., there is a path between them in the type lattice). A simple-type hierarchy is shown in Fig. 4-8. In this hierarchy water is a type of fluid, but liquid is not a type of gas.

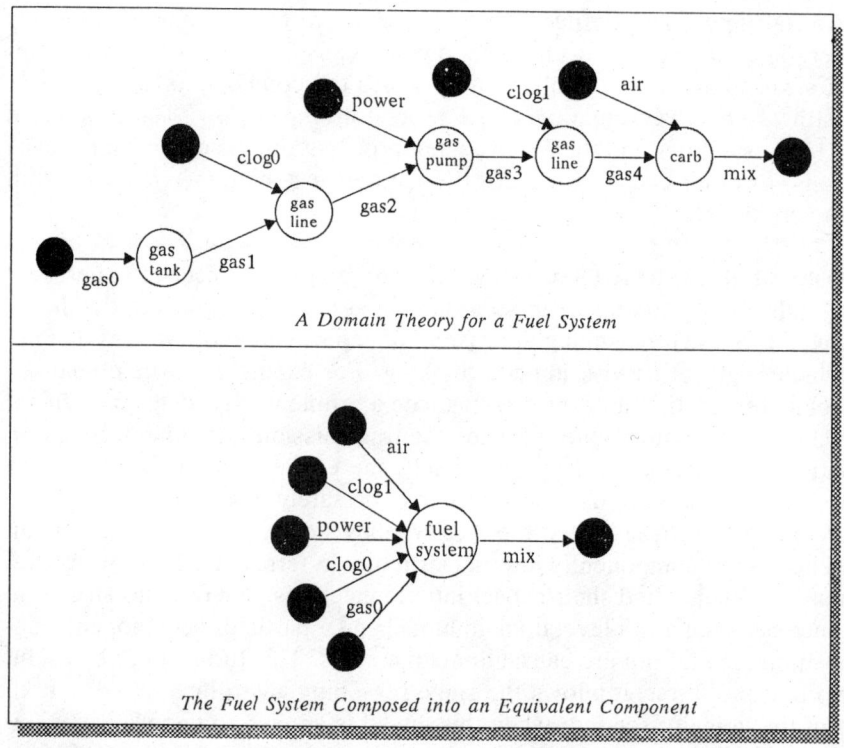

FIGURE 4-6
Composition of a fuel system.

By providing typed causal inputs and outputs, the user is prevented from making mistakes such as hooking a water pump up to a gas line. This would be prevented because the types of the water pump's output and the gas line's input, water and gas, respectively, are not compatible. Additionally, when components are connected, compatible types may provide the user (and the system) with more specific descriptions of an input or output. For example, in Fig. 4-8, when a pump (which transports fluids) is connected to a gas line (which transports gasoline), the pump is coerced, if possible, to be a transporter

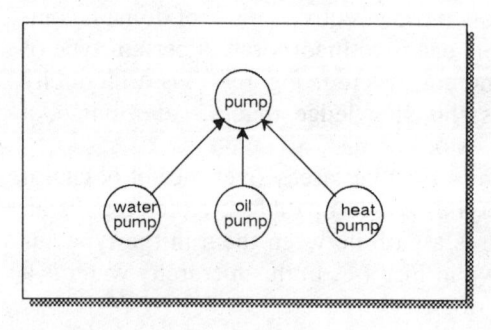

FIGURE 4-7
A component hierarchy.

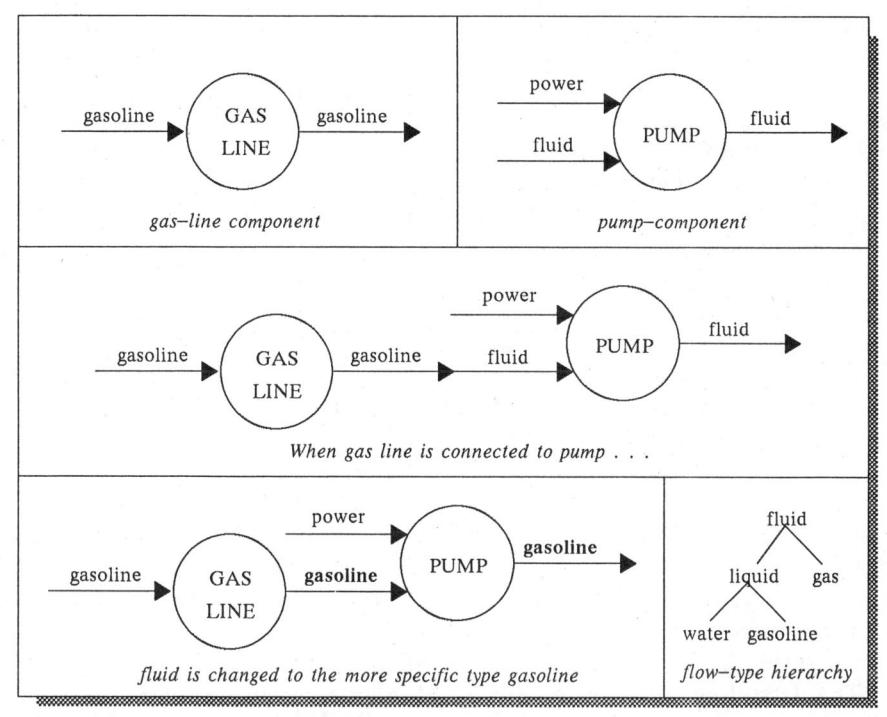

FIGURE 4-8
Type consistence in connecting components.

of gasoline. This is possible in this example because as the flow-type hierarchy of Fig. 4-8 shows, gasoline is a type of liquid which in turn is a type of fluid.

CAUSAL EQUATIONS (COMPONENT FUNCTIONALITIES). A component must be able to determine values of its causal outputs, given values of its causal inputs. For this purpose each component maintains a *causal equation*. In general, a causal equation is a partial specification of the component's functional behavior in terms of its causal input and causal output values. It is partial because for some input conditions only a subset of the causal output values is specified. Thus, in essence, a component with *m* causal inputs and *n* causal outputs is mapping from *m*-tuples of causal input values to *n*-tuples of causal output values, assuming we allow the unknown value to be incorporated into the causal input and causal output value sets.

The functionalities of the components of the decomposed fuel system of Fig. 4-6 are as follows:

gas-tank:	gas1 = gas0
gas-line:	gas2 = gas1 − clog0
gas-pump:	gas3 = gas2 + power
gas-line2:	gas4 = gas3 − clog1
carb:	mix = gas4 + air

By composing these functionalities, we arrive at the causal equation for the composed fuel system component: mix = air − clog1 + power − clog0 + gas0.

The values of causal inputs and causal outputs are expressed qualitatively, by using values from the set { − , 0, +, ?}. The meanings are given in Table 4-2. Since we desire to model change in the domain, the most typical use of these values is in conjunction with causal flows representing first derivatives of domain flows. If we have a causal flow which represents the first derivative of some domain variable, then the above values may be viewed in the language of change as described by Table 4-3. For example. if we let velocity be a variable, whose first derivative is represented by a flow, then we can partition the variable values into three categories: decelerating, constant, and accelerating.

In general, an expert will not know or be able to specify the complete causal equation for a component. Despite this, we wish our domain model to be useful and incrementally modifiable with respect to its causal equations. We therefore take the approach of initially representing causal equations by some easily specified, easily obtained set of default equations. For each causal output of a component we obtain an algebraic equation which can be used to partly specify the functionality for that causal output. The equation relates the causal output to each causal input in one of three ways: directly related (the input is preceded by a +), inversely related (the input is preceded by a −), or not related (the input does not appear in the equation).

The default equation, when evaluated in an environment of causal-flow bindings, results in a determination of the net effect (+, −, 0, or ?) of each causal input on the causal output in question. Table 4-4 is used to calculate causal output values. Note that if all the nonzero effects of an equation are the same, i.e., all minuses or all pluses, then the causal output can be determined as 0, +, or −. Otherwise, the output value is indeterminate.

There are two deficiencies associated with default equations. First, an assumption made in utilizing these default equations is that the causal equations are linear. This, of course, is incorrect for many components but is a reasonable model to begin with. Second, due to their qualitative nature, these equations will often yield the unknown value and hence be indeterminate as to the causal output value. While these problems preclude the use of such equations for a robust expert system, they can be utilized to facilitate knowledge acquisition in the early development stages of an expert system. As experts interact

TABLE 4-2
Interpretation of causal values for a quantity represented by X

| Causal flow representing X ||
Value	Meaning in terms of X
0	X is 0
−	X is negative
+	X is positive
?	X is unknown

TABLE 4-3
Interpretation of causal values for the first derivative of a quantity X

Value	Causal flow representing δX
	Meaning in terms of X
0	X is constant
−	X is decreasing
+	X is increasing
?	δX is unknown

with the domain model and encounter failures (either indeterminate equations or incorrect equations), they will be able to correct these functional shortcomings and have their corrections remembered as annotations to the original default equations. The system's ability to obtain more and more complete causal equations is actually an important aspect of knowledge acquisition since it will incrementally produce a more accurate model of the domain.

REASONING WITHIN THE ABSTRACT CAUSAL DOMAIN MODEL. There are two basic methods of reasoning available within the ACDT: reasoning from causes to effects and reasoning from effects to causes. In the first method an initial domain state **d** (some binding of values to causal flows) is given, and the causal equations of the components are applied to each component to calculate the resulting domain state. This process iterates until it reaches a final or a repeated state, when further application of the equations results in a loop. The results of this process is a queue of unique domain states.

Reasoning from effects to causes differs in that any given effect might have multiple causes. Each predecessor of a domain state is a potential cause for that state. The result of this process, therefore, is a tree of domain states, with the effect at the root. Each node in this tree is a potential cause of the effect at the root. More generally, any state which is a descendant of

TABLE 4-4
Addition table for causal flow values

	Value 2			
Value 1	−	0	+	?
−	−	−	?	?
0	−	0	+	?
+	?	+	+	?
?	?	?	?	?

some other state in the tree is a potential cause of that other state. The causal explanation for any particular state can be obtained by following the links of the tree, in reverse, from the state's node up to the root of the tree.

Consider, for example, the model represented in Fig. 4-9. It represents a simple domain model which consists of a gas line and a gas pump. The gas line may be clogged. The gas pump requires power to operate. The equation for the gas line declares that the value for gas2 is directly affected by the gas1 value and inversely affected by the clog value. The equation for the gas pump declares that the value of gas3 is directly affected by the values of both gas 2 and power. The small black circles indicate connections to higher-level components that are not of direct interest here.

We represent a state of the above domain by listing the causal flows and their values. If a causal flow has a value of 0, we omit it. For example, the state of the above domain in which there is a clog in the gas line and an increase in power is represented as ([clog +], [power +]).

Now consider the cause-to-effect reasoning with a given state of ([clog +]). The algorithm first adds ([clog +]) to an empty queue. The result of applying the causal equations to this state produces the state ([clog +], [gas2 −]), and the result of applying the causal equations to this new state produces the state ([clog +], [gas2 −], [gas3 −]). An interpretation of the results is then that

1. The presence of clog in the gas line had the effect of a decrease in gas2 from the gas line.
2. The decrease in gas2 to the gas pump had the effect of a decrease in gas3 from the gas pump.

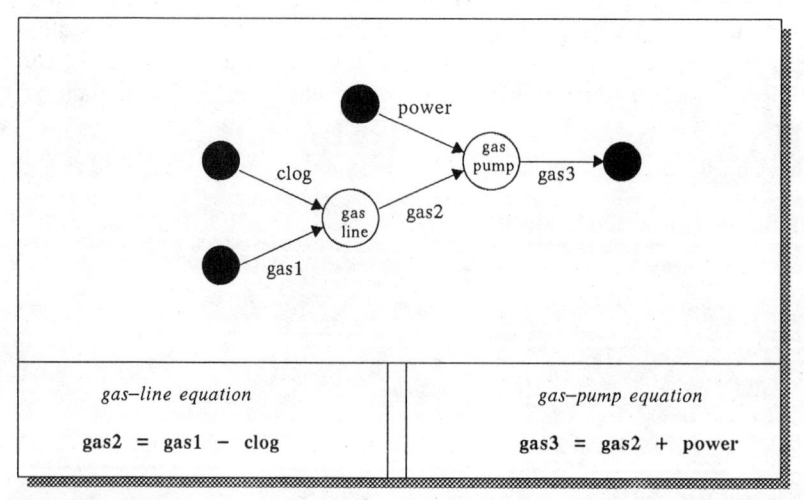

gas–line equation

gas2 = gas1 − clog

gas–pump equation

gas3 = gas2 + power

FIGURE 4-9
A domain theory for a gas pump connected to a gas line.

Now let us examine what happens when the effect-to-cause reasoning is applied with the above model and a state of ([gas3 −]). This is equivalent to asking what can cause a decrease in the gas output of the gas pump. Now ([gas3 −]) is made the root of the cause tree. The inverse causal equations indicate that the state can be caused by one of three possible states: power is −, gas2 is −, or both power and gas2 are −. Each of these states becomes a child of the root, and each is recursively investigated to determine its causes. The final tree is shown in Fig. 4-10 (only the changes in states have been displayed at each node).

Each node of this tree represents a different sequence of events that could result in the reduction of gas3, the state at the root of the tree for which a cause is to be determined. For example, starting from the leftmost leaf of the tree, one could obtain the following interpretation:

A clog causes a decrease in gas2 which causes a decrease in gas3.

UTILIZATION OF THE DOMAIN MODEL. Two of the required abilities of ACDM are to justify and to complete expert input. By justifying expert input we mean verifying the possibility of an expert's claim that a cause is related to an effect, i.e., making sure the expert's claim is consistent with the knowledge base. By *completing* expert input we mean providing, if necessary, a less general description of the expert's claims; i.e., we expect that although the expert's input may be consistent, it is likely that it overgeneralizes. We assume that expert's input is in the general form *A causes B*. Here A and B are state descriptions of the domain that, in general, are true for some subset D′ of the set of all domain states D.

In justifying the expert's input, we wish to ensure that the expert's statement of causality is consistent with the current domain model. It is justified if there exists at least one domain state specified by condition A which causes at least one domain state specified by condition B. If no such states

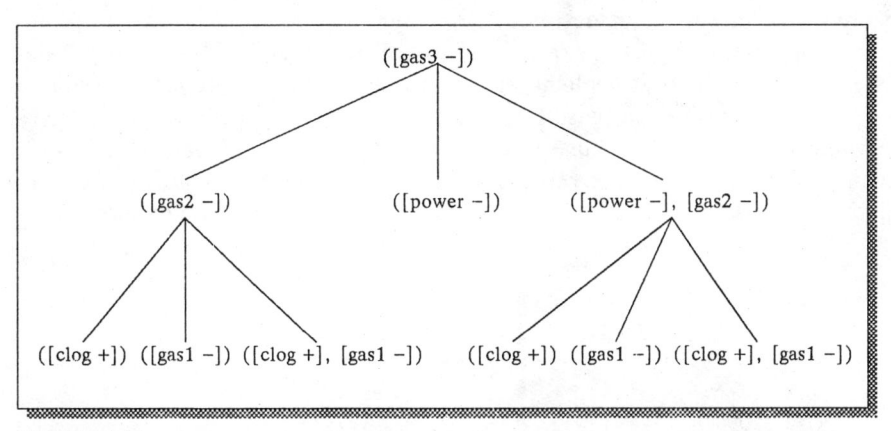

FIGURE 4-10
Tree resulting from BUILD-CAUSES-TREE on the initial state ([gas3 −]).

exist, then the expert's input does not agree with the current domain model and the expert may either retract the statement or ask to augment the knowledge base to contain the knowledge of the statement.

Assuming an expert's claim is justifiable, we may also be able to sharpen her or his description of the conditions required for the causal connection to be realized. For example, consider the component in Fig. 4-11. It is a model for a tube with two inputs and one output. Suppose an expert claims that a decrease in fluid1 causes a decrease in fluid3. We can justify this since there is a state of the domain ([fluid1 $-$]) that causes a state of the domain where fluid3 is $-$, namely, ([fluid1 $-$], [fluid3 $-$]). However, if fluid2 increases, the statement is untrue or uncertain at best. We therefore complete the expert's input by adding the precondition of fluid2 not having a value of $+$.

Experts can augment the knowledge base whenever omissions or mistakes are encountered. This might entail adding components to an existing knowledge base as well as the further specification of the causal equations. Initially the causal default equations are all that is available to a component, and the component is limited by the scope and accuracy of these equations. The expert may encounter two types of behavior that need correction. The first is the calculation of an incorrect causal value for some component in a certain state. The second is the inability to calculate a causal value for some component state. In both cases to correct the problem, the expert must specify what the value should be. The system then stores this value as an annotation to the default equations. Additional defaults may be inferred from the annotation, and theses may be stored for future use (or their evaluation may be deferred until needed).

We demonstrate this by an example using the domain model shown in Fig. 4-9. The domain model in this figure contains only default causal equations, so there exist input states for the two components for which their equations cannot produce an output value. For example, if power is $+$ and gas2 is $-$, we are confronted with adding a $+$ and a $-$. According to Table 4-4, this results in the unknown value of ?. How might this be discovered and remedied? Suppose the expert has requested to see the causes of gas1 being $-$ and power being $+$. The first application of causal equations will reveal that in one step gas2 becomes $-$ and gas3 becomes $+$. The next step, however, revises the value of gas3 to ? because gas3 is equal to the sum of power and gas2, which are $+$ and $-$, respectively. At this point the system, realizing it has an indeterminate value, requests assistance from the expert. If the expert knows

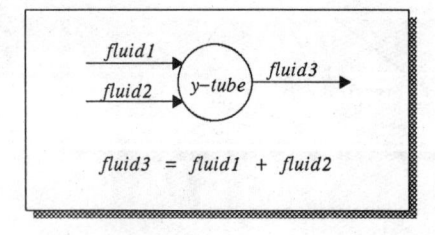

FIGURE 4-11
A Y-tube component.

the solution of the problem, he or she may enter it (i.e., the correct causal value for the situation), and the system will save it as an annotation to the equations of the gas pump.

Suppose at some future point another expert is checking out causes of a decrease in gas flow from the gas pump (gas3). According to the default causal equations, this can happen directly (in one step) in only one of three ways: both power and gas2 may be − or either power or gas2 is − and the other is 0. The existence of the annotation allows for one more possible cause; the power may be + and gas2 may be −. This means that prior to adding the annotation, the system would have failed had it, while trying to determine the cause of the gas3s decrease, asked for the value of power and been told +. The annotation being told that + is the value of power narrows the direct cause of the decrease in gas3 to one possibility—a decrease in gas2.

CONCLUSIONS

In the beginning of this chapter we listed five issues which make knowledge acquisition—and, by implication, its automation—a formidable task: vocabulary, completeness, integration, analysis, and transparency. The various model-based knowledge acquisition approaches we have discussed differ in terms of the extent to which they address these issues. Although an attempt to classify each of the modeling techniques discussed is bound to be subjective, and relative to the example systems used in the discussion, it is nevertheless useful to characterize each approach in terms of the five issues. Table 4-5 summarizes this characterization.

The major strength of similarity-based learning lies in its power to integrate information. New information (e.g., new examples) can simply be added to the existing pool, and the knowledge base is updated simply by rerunning the induction algorithm. Modeling the domain, be it in terms of syntactical structures, semantics, or causal properties, provides ways to store and retrieve the terminology of the domain. It would appear, on one hand, that syntactic models are likely to be more transparent because they are based on existing knowledge structures. Semantic modeling, on the other hand, provides better means for explaining the meaning of a rule and its constituent parts. Neither syntactic nor semantic models are sufficiently strong to determine the

TABLE 4-5
Summary of characteristics of knowledge acquisition models

	Vocabulary	Completeness	Integration	Analysis	Transparency
Similarity-based learning			●		
Syntactic models	●		●		●
Semantic models	●		●	●	
Causal models	●	●	●	●	●
Explanation-based learning	●		●	●	●

completeness of a knowledge base, however. This is also a weak point in explanation-based learning systems. Provided that the domain of application is suitable to the approach, causal models may provide an answer in this regard.

Our presentation has stressed the use of causal models for three main reasons. First, causal models can be conveniently represented in a graphical form, providing a means to focus the dialogue between the expert and the acquisition system. Second, causal modeling is cumulative. Once domain models have been developed, it is highly likely that the definition of standard components can be transferred to similar domains. Third, a causal model provides the acquisition system with a means to react to the experts' input in an intuitive fashion. The system can trace the consequences of adding information step by step, allowing experts to see whether their additions and refinements indeed produce the desired effects.

The integration of these three strong advantages with those of the other modeling techniques will provide an important step toward realization of the **expert** → **acquisition system** → *knowledge base* paradigm.

REFERENCES

Bachant, J., and J. McDermott (1984). R1 revisited: Four years in the trenches. *AI magazine* 5(3):21–32.

Boose J. (1984). Personal construct theory and the transfer of human expertise. In: *Proceedings of the Fourth National Conference on Artificial Intelligence.* Austin, Texas.

Buchanan, B. G., and E. H. Shortliffe (Eds.) (1984). *Rule-Based Expert Systems.* Reading, Mass.: Addison-Wesley.

Buchner, B. A. (1988). A domain theory for knowledge acquisition. Report No. UIUCDCS-R-88-1424, Department of Computer Science, University of Illinois at Urbana-Champaign, April.

Clancey, W. J. (1983). The epistemology of a rule-based expert system: A framework for explanation. *Artificial Intelligence* 20:215–251.

Davis, R. (1984). Interactive transfer of expertise. In: B. G. Buchanan and E. H. Shortliffe (Eds.), *Rule-Based Expert Systems.* Reading, Mass.: Addison-Wesley, pp. 171–205.

Davis, R., and D. B. Lenat (1982). *Knowledge Based Systems in Artificial Intelligence.* New York: McGraw-Hill.

DeJong, G. F. (1981). Generalizations based on explanations. *Proceedings of the Seventh International Joint Conference on Artificial Intelligence.* Vancouver, British Columbia, pp. 67–70.

DeJong, G., and R. Mooney (1986). Explanation based learning: An alternative view. *Machine Learning* 1(2):145–176.

de Kleer, J., and J. Brown (1984). A qualitative physics based on confluences. *Artificial Intelligence* 24:7–84.

Duda, R. O., J. G. Gachnig, and P. E. Hart (1979). Model design in the PROSPECTOR Consultant system for mineral exploration. In: D. Michie (Ed.), *Expert Systems in the Micro-Electronic Age.* Edinburgh: Edinburgh University Press, pp. 153–167.

Emde, W., and K. Morik (1986). Consultation-independent learning in BLIP. In: Y. Kodratoff (Ed.), *Machine and Human Learning.* Chichester, England: Horwood Publishing.

Eshelman, L. D., D. Ehret, J. McDermott, and M. Tan (1987). MOLE: A tenacious knowledge acquisition tool. *International Journal of Man-Machine Studies* 26(1):41–54.

Feigenbaum, E. A. (1978). The art of artificial intelligence: Themes and case studies of knowledge engineering. In: *AFIPS Conference Proceedings of the 1978 National Computer Conference,* vol. 47. Anaheim, Calif., pp. 227–240.

Fikes, R. E., P. E. Hart, and N. J. Nilsson (1972). Learning and executing generalized robot plans. *Artificial Intelligence* 3:251–288.

Forbus, K. (1984). Qualitative process theory. *Artificial Intelligence* 24:85–168.
Forgy, C. (1981). The OPS5 users manual. Technical Report, Carnegie-Mellon University, Department of Computer Science, Pittsburg.
Hayes-Roth, F., D. A. Waterman, and D. B. Lenat (1983). *Building Expert Systems*. Reading, Mass.: Addison-Wesley.
Hunt, E. B. (1962). *Concept Learning: An Information Processing Approach*. New York: Wiley.
Lange, R., and M. T. Harandi (1986). The elements of a distributed knowledge acquisition system. *Proceedings of the Sixth International Workshop on Expert Systems and Their Applications*, Avignon, France.
Lange, R., L. Hearn, and F. W. Kearney (1986). The use of knowledge engineering teams as a method for the development of expert systems. In: D. Sriram and R. Adey (Eds.), *Applications of Artificial Intelligence in Engineering Problems*, Berlin: Springer-Verlag, pp. 45–54.
McDermott, J. (1982). R1: A rule based configurer of computer systems. *Artificial Intelligence* 19(1):39–88.
Mescheder, B. (1985). Funktionen un Arbeitsweise der Expertensystem-Shell TWAICE. In: S. Savory (Ed.), *Kuenstliche Intelligenz und Expertensysteme*, Oldenbourg Verlag, pp. 57–90.
Michalski, R. S., J. G. Carbonell, and T. M. Mitchell (Eds.) (1983). *Machine Learning: An Artificial Intelligence Approach*. Palo Alto, Calif.: Tioga Publishing.
Michalski, R. S., J. G. Carbonell, and T. M. Mitchell (Eds.) (1986). *Machine Learning*, vol. 2. Los Altos, Calif.: Morgan Kaufmann.
Michalski, R. S., and R. L. Chilausky (1980). Learning by being told and learning from examples: An experimental comparison of two methods of knowledge acquisition in the context of developing an expert system for soybean diagnosis. *Policy Analysis an Information System*, vol. 4(2).
Mitchell, T. M. (1983). Learning and problem solving. In: *Proceedings of the Eighth International Joint Conference on Artificial Intelligence*. Karlsruhe, West Germany: Morgan Kaufmann, pp. 1139–1151.
Mitchell, T. M., R. Keller, and S. Kedar-Cabelli (1986). Explanation based generalization: A unifying view. *Machine Learning* 1:47–80.
Mitchell, T. M., S. Mahedevan, and L. I. Steinberg (1985). LEAP: A learning apprentice for VLSI design. In: *Proceedings of the Ninth Joint Conference on Artificial Intelligence*. Los Angeles: Morgan Kaufmann, pp. 573–580.
Morik, K. (1986). Anything you can do I can do meta. *Technische Universitaet Berlin*, KIT-Report 40, November.
Quinlan, J. R. (1983). Learning efficient classification procedures and their application to chess endgames. In: R. S. Michalski, J. G. Carbonell, and T. M. Mitchell (Eds.), *Machine Learning: An Artificial Intelligence Approach*. Palo Alto, Calif.: Tioga Publishing, pp. 463–482.
Steier, D., and A. Newell (1988). Integrating multiple sources of knowledge into designer-soar, an automatic algorithm designer. In: *Proceedings of the Seventh National Conference on Artificial Intelligence*, pp. 8–13.
Swartout, W. R. (1983). XPLAIN: A system for creating and explaining expert consulting programs. *Artificial Intelligence* 21:285–325.
van Melle, W., E. H. Shortliffe, and B. Buchanan (1984). EMYCIN: A knowledge engineer's tool for constructing rule-based expert systems. In: B. G. Buchanan and E. H. Shortliffe (Eds.), *Rule-Based Expert Systems*. Reading, Mass.: Addison-Wesley, pp. 302–313.
Waldinger, R. (1977). Achieving several goals simultaneously. In: E. Elcock and D. Michie (Eds.), *Machine Intelligence*, vol. 8.
Waterman, D. (1985). Higher-order concepts in a tractable knowledge representation. In: *Proceedings of GWAI-87*. Berlin: Springer-Verlag.
Weiss, S. C., C. Kulikowski, S. Amarel, and A. Safir (1978). A model based method for computer aided medical decision making. *Artificial Intelligence* 11(1, 2):145–172.
Wrobel, S. (1987). Higher-order concepts in a tractable knowledge representation. In: *Proceedings of GWAI-87*. Berlin: Springer-Verlag.

CHAPTER 5

MODELS OF EXPERTISE IN KNOWLEDGE ENGINEERING

PHILIP E. SLATTER

1 INTRODUCTION

What is the nature of human expertise? Researchers and practitioners in knowledge engineering (KE) give many different answers to this question. The answer is not merely one of scientific interest, however—it can have profound implications for KE practice. For example, a view of the nature of expertise may suggest

- Fundamental limits of what can be achieved by expert systems
- The appropriate role of such systems
- What type of knowledge can be elicited from an expert
- How that knowledge should be represented

Each answer, whether a formal theory or an informal description, whether explicitly stated or implicitly invoked, can be viewed as a model of human expertise. This is the sense in which the term *model of expertise* is used in this chapter. A model focuses attention on certain aspects of a phenomenon and deemphasizes others. Which aspects are identified as important depends, in part, on the purpose of the model. In KE it is a question of which aspects of human expert thinking and behavior are considered relevant to expert system construction.

The aim of this chapter is to

- Draw attention to the range and variety of models of expertise presented or invoked by workers in KE (Section 3)
- Present a comparative review of the major classes of model in the field (Section 4)

- Briefly consider other models that fall outside this classification, but which are nonetheless of some theoretical interest (Section 5)

Prior to this, Section 2 examines related uses of the term *model* by workers in KE. The aim is twofold: to show that the different usages are conceptually distinct and to demonstrate that in practice these usages tend to overlap.

2 RELATED USES OF THE TERM *MODEL*

One definition of a model (of expertise) has already been given. This can be contrasted with four common uses of the term in the KE literature:

(1). *The sense in which all expert systems are based on models.* Whether or not the designer is explicitly aware of it, an expert system can be regarded as a model of expertise in some domain (e.g., Breuker et al., 1987; Smith, 1985). That is, even the most rudimentary system contains a knowledge base in which selected domain entities and relations between them are represented in codified form. The system architecture may model other aspects of the expert domain. A model of this sort is limited in scope to the application domain of a particular expert system, whereas the models of human expertise reviewed in this chapter are domain-general characterizations of human expert functioning.

(2). *Formal models of expertise.* There is a growing tendency among some workers in KE to view knowledge acquisition as a modeling process (e.g., Addis, 1987; Breuker et al., 1987; Gaines, 1987a). A modeling metaphor is considered more appropriate than the conventional view, which sees knowledge acquisition as a process of extracting, or "mining," chunks of knowledge from a human expert and codifying it within an expert system (e.g., Feigenbaum and McCorduck, 1983). The modeling approach, however, regards the results of knowledge elicitation as "data" that can be used to construct and test a model of expertise in some domain. This process usually employs formal techniques derived from model theory. However, a model of expertise of this sort is no different in kind from a formal model of any other observable phenomenon—no commitment to any particular model of the nature of human expertise is necessarily implied.

(3). *Cognitive models of expertise.* Cognitive scientists investigate human expertise by constructing theories which are tested by attempting to express them as computational models and by carrying out empirical studies comparing expert and novice performance [see, e.g., Anderson (1985) and Slatter (1987) for reviews of this work]. The principal aim of cognitive research on expertise is to further scientific understanding of expert cognition. In contrast, models of expertise in KE aim to define and support the construction of knowledge-based expert systems.

(4). *The view that human expertise is model-based.* One perspective on human expertise in KE holds that an expert's ability to solve difficult problems and provide convincing explanations of their solutions depends on an ability to utilize mental models of entities in the task domain (see Section 4.2). A further complication is that researchers in this field are investigating ways of representing an expert's mental models in knowledge-based systems. Clearly, though, both the expert's mental model and the researcher's conceptual model of it (cf. Norman, 1983) are quite different from a model about the general nature of human expertise.

In summary, it is possible to distinguish between, on the one hand, general models of human expertise and, on the other, a variety of other usages of the term *model* in KE. However, while these conceptual distinctions are valid, in practice the different usages are less clearly differentiated than the above presentation might suggest. So we return to the set of distinctions made above:

(1). Even where an expert system builder is not consciously aware of employing any explicit model, it can often be inferred that a certain model of expertise is being implicitly invoked. For example, representing knowledge as a set of if-then rules (as is still the norm in commercial applications based on simple expert system shells) may imply a commitment to a heuristic rule-based conception of human expertise (described in Section 4.1).
(2). Although formal modeling techniques contain no explicit presuppositions about the nature of human expertise, the modeling approach to knowledge acquisition is, in fact, more closely associated with certain classes of model of human expertise than with others (see Section 4). Moreover, some of the models to be considered (e.g., Breuker et al., 1987; Gaines, 1987a) provide their own modeling language for specifying expertise in a given task domain.
(3). The proponents of models of expertise in KE can invariably point to theoretical and empirical work in cognitive science which lends support to their own formulations (see Sections 4 and 5).
(4). The view that human experts rely on mental models is itself one of the five major models of human expertise considered in this chapter (see Section 4.2).

3 A SURVEY AND CLASSIFICATION OF MODELS

Books and articles on KE tend to adopt one of four basic positions regarding models of expertise:

(1). *No model is required.* The author may hold that building an expert system requires no special assumptions about the nature of human expertise—or may imply this attitude by making no reference to the subject. Proponents of formal logic as a principled basis for expert system construction often appear to be adopting such a position.

(2). *A single model is implied.* Many publications contain no explicit statement concerning the nature of human expertise, yet it is apparent from the occasional remark or general tone that certain tacit assumptions about human expertise are being made.

(3). *A single model is explicitly stated.* A minority of publications contain a section on the nature of human expertise. The implication here, as in item 2 above, is that there is an agreed consensus within the KE community about at least the general nature of human expertise. This is not, in fact, the case, as Sections 4 and 5 of this chapter seek to demonstrate.

(4). *More than one model is presented.* Occasionally, more than one model of expertise is distinguished (usually two): an existing model, which is shown to be seriously flawed, and a new model, which is claimed to remedy the deficiencies of the earlier one. The heuristic rule-based conception of expertise (see Section 4.1) is often cast in the former role.

A survey of the KE literature reveals that a wide range and variety of specific models of expertise are in use. Unfortunately, no consensus exists as to which are the major models or how the full range of models might be classified (perhaps because discussion of more than a small subset of models is so rare). However, an analysis of the models surveyed identified five major classes of model:

- Heuristic models
- Deep models
- Implicit models
- Competence models
- Distributed models

The rationale for this tentative classification should become clear in Section 4, which makes a comparative presentation of these five generic models. Each class of model is shown to selectively emphasize different aspects of human expertise, to be exemplified by a number of specific models and application systems, to appeal to different theoretical and empirical research for justification, and to be associated with different KE practices and limitations.

There are obvious pitfalls in attempting a classification of this sort, such as identifying either too many or too few categories or misclassifying models. There is also a danger of pigeonholing specific models into a single class when, in fact, the model contains elements characteristic of two or more classes. The more obvious examples of such hybrid models are thus dealt with separately in Section 5, together with a range of other models that cannot be naturally accommodated within the classificatory scheme presented.

4 FIVE MAJOR CLASSES OF MODEL

To facilitate comparisons between the five classes of models, a standard set of headings is employed:

A GENERAL DESCRIPTION

Level of description. Models exist at different levels of description. For present purposes it is sufficient to distinguish just two levels: a knowledge level and a symbolic or representational level (cf. Newell, 1982). Explanations at the symbolic level are in terms of knowledge representations and the processes that operate them. The knowledge level is a competence-type notion: it denotes a capacity for generating action and serves as a specification for what some symbolic systems should perform.

Knowledge types. Each class of model tends to categorize the knowledge involved in expert performance in different ways.

Representative view of expertise. Each class of model selectively emphasizes different aspects of human expertise. Representative features of each class of model are outlined.

> **Examples.** Some influential examples of each class are discussed and important differences in emphasis between specific models highlighted.

B RATIONALE

Theoretical foundations and empirical support. These refer to the key theoretical and empirical research (mostly in cognitive science) invoked by proponents of different classes of models.

KE rationale. Refers to the arguments offered by proponents of models in terms of a model's beneficial consequences for the practice of KE.

C ASSOCIATED KE PRACTICE

Role of the system. The "proper" role of an expert system tends to vary between classes of model.

Tools and techniques. A different set of expert system development tools and techniques is often advocated, depending on the class of model.

> **Example applications.** Some well-known expert systems associated with each class of model are cited.

D LIMITATIONS

As a cognitive model. Viewed as a cognitive model of expertise, what are the main limitations?

KE limitations. What are the main disadvantages from a practical KE point of view?

4.1 Heuristic Models

A GENERAL DESCRIPTION

Level of description. Heuristic models offer a symbolic, or representational, level account.

Knowledge types. The principal distinction is between facts and heuristic knowledge.

Representative view of expertise

- Expertise involves acquiring lots of facts about a domain. This is widely shared knowledge available in textbooks and other public sources.
- However, it is the possession of heuristic knowledge which is the hallmark of expertise. This is the compiled know-how—rules of thumb, rules of good practice, etc.—which is accumulated from many years' experience in a specialist domain. Such knowledge tends to be inexact, ill-specified, uncertain, incomplete, and inaccessible.
- Experts store their heuristic knowledge in a rulelike form, with each rule approximating an independent "chunk" of know-how.
- Expertise is acquired incrementally, as new domain facts and rules are learned. Domain experts are people who have acquired tens of thousands of specialist rules.

> **Examples.** Edward Feigenbaum has done as much as anyone to advance this conception of expertise, through both his published work (e.g., Feigenbaum and McCorduck, 1983) and his involvement in the Stanford University heuristic programming project. A heuristic model of expertise is implicit in much of the KE literature: it constitutes a kind of default model for workers in this area. Where it is explicitly dealt with, the rule-based nature of heuristic knowledge is usually emphasized (e.g., Feigenbaum and McCorduck, 1983; Hayes-Roth, 1985). Others (e.g., d'Agapeyeff and Barrett, 1986), however, sidestep the issue of how expertise is represented in the human mind, claiming only that experts tend to *express* their knowledge as rules of thumb.

B RATIONALE

Theoretical foundations. The heuristic rule-based conception of expertise is rooted in what Partridge (1987) terms the *symbolic search space paradigm* of artificial intelligence (AI). As formalized by Newell and Simon (1976), problem solving entails finding a path through a search space comprising a set of discrete, independently defined states. However, because the search space is often too large to search exhaustively, problem solvers are forced to adopt heuristic shortcuts to solve problems in a reasonable time. A heuristic rule-based model also assumes that (1) knowledge can be represented as a collection of more or less context-free rules, (2) intelligent decision making can be implemented as a logical truth derivation mechanism, and (3) human expertise can be represented in terms of our normal symbolic concepts (Partridge, 1987). The utility of this paradigm is supported by the vigorous use of rule-based approaches in the computational modeling of human cognition (e.g., Anderson, 1983; Newell and Simon, 1972).

Empirical support. In many psychological studies of expert thinking, the experimental findings are often interpreted or modeled in terms of rules (see, e.g., Slatter, 1987, for a review). Thus, for example, skill acquisition may be viewed as "knowledge compilation"—a process whereby declaratively represented factual knowledge is compiled into a procedural form, represented by production rules (e.g., Anderson, 1983; Larkin, 1981).

KE rationale. Rule-based techniques are well understood and widely used in KE. A heuristic model of expertise accords well with this technology, offering a strong rationale for the existing practice of acquiring, representing, storing, and applying expert knowledge via rule-based methods. In this view, each rule in a rule-based expert system approximates an independent chunk of human know-how, opening the possibility of automating expertise in specialized tasks with a few hundred to a few thousand heuristic rules (Hayes-Roth, 1985). The modularity and the understandability of individual rules are a further benefit.

C ASSOCIATED KE PRACTICE

Role of the system. A heuristic model of human expertise, allied to the existing rule-based technology, implies that expert systems should be able to act as autonomous problem solvers or powerful advisory systems (Feigenbaum and McCorduck, 1983; Hayes-Roth, 1985).

Tools and techniques. A heuristic model is associated with the view that an intelligent agent's problem-solving ability is the consequence of primarily its knowledge base and only secondarily the inferencing method employed. That is, "in the knowledge lies the power" (Davis, 1982). Systems based on a heuristic model typically have "shallow" knowledge bases, consisting of a uniform set of rules. Simple forward and backward inference control structures are often considered adequate. The uncertain nature of heuristic knowledge is modeled by attaching uncertainty factors, Bayesian probabilities, etc., to rules and propagating these uncertainties through the system at runtime.

There are also implications for knowledge acquisition. If expert knowledge consists of chunks of know-how, then it should be possible to extract these one at a time from a specialist. This "mining" metaphor (Feigenbaum and McCorduck, 1983) also justifies the development of (semi-) automated tools for the direct transfer of bits of knowledge from human expert to expert system, with little or no intervention from a knowledge engineer. Protocol analysis is a favored elicitation technique because "rules express situation-action heuristics in the problem-solving protocols of experts" (Hayes-Roth, 1985, p. 927).

Examples applications. Famous first-generation expert systems such as MYCIN (Shortliffe, 1976) and XCON/R1 (McDermott, 1982) reflect a heuristic conception of expertise. Although neither system is intended as a cognitive model, both represent expert knowledge as a uniform set of rules, interpreted by using (relatively) simple backward and forward chaining inference mechanisms.

D LIMITATIONS

Limitations as a cognitive model. While cognitive research confirms that heuristic knowledge is important for expert performance, it also emphasizes the role of other cognitive factors not dealt with by the heuristic model, in particular:

- The use of specialist representations and causal models of domain entities in expert problem solving (see Section 4.2)
- The dependence of skilled human performance on implicit mental processes and knowledge (see Section 4.3)
- The variety and complexity of experts' problem-solving strategies (see Section 4.4)
- The cooperative nature of much expert problem-solving behavior (see Section 4.5)

KE limitations. The deficiencies of expert systems based on a simple heuristic model of expertise are well known (e.g., Clancey, 1983; Davis, 1982). They include difficulties in solving nonstandard problems, weak explanation, brittleness, lack of learning capability, and inability to support roles other than problem solving (e.g., tuition, which requires the principles underlying heuristic associations to be made explicit). These deficiencies are elaborated in Sections 4.2 to 4.5 since they provide the KE rationale for the four other classes of models identified.

4.2 Deep Models

A GENERAL DESCRIPTION

Level of description. Symbolic level. Deep models suggest that experts use a variety of deep knowledge structures represented differently from heuristic rules.

Knowledge types. Principally, shallow (i.e., heuristic) and deep knowledge.

Representative view of expertise

- Experts do employ heuristic rules, but these rules form only the top layer of an expert's knowledge.
- Complementing this shallow layer of heuristic knowledge are other knowledge structures representing an expert's deep knowledge about the area of expertise.

- Deep knowledge is the background knowledge an expert has for example, about:

  1. domain objects (e.g., symptoms, diseases, patients)
  2. hierarchical relationships between domain objects (e.g., disease taxonomies)
  3. causal models about how domain objects are related (e.g., how Disease A causes Symptom B through X, Y, and Z intermediate states)
  4. specialist representations of domain objects, capturing temporal, spatial, or other analogical properties (e.g., a structural model of a "device" such as the gastrointestinal tract)
  5. how to reason from first principles (e.g., applying standard diagnostic strategies)

- Heuristic rules are the shortcuts in applying deep knowledge that an expert learns through experience. Routine problems are thereafter solved by using this synthesized heuristic knowledge. However, when a problem is too complex or unusual to yield to heuristic reasoning, the expert falls back on reasoning from first principles by using deep knowledge.

- Published material such as textbooks are a major source of taxonomies, etc. Practical experience may lead the expert to refine such textbook models and to create new ones which better accord with the problem-solving structure of the domain.

- Specialists are usually expected to be able to justify their decisions in terms of the deeper principles of a domain.

> **Examples.** Davis (1982) was among the first to distinguish between deep and shallow knowledge, although preferring the phrases *causal models* and *empirical associations* because of the negative overtones of the term *shallow*. Not all researchers have accepted a simple distinction between deep and shallow knowledge, however. Instead, they are more attracted by the idea of a continuous scale of shallow and deeper knowledge (e.g., Chandrasekaren and Mittal, 1984; Leary and Rodriguez-Bachiller, 1988). Knowledge at the shallower end might be represented as simple facts or look-up tables while at the other end are causal models and other deep structures. Compiled, heuristic knowledge falls toward the middle of the range. Leary and Rodriguez-Bachiller (1988) also suggest that many experts are no longer conscious of the causal model that underlies their operational (heuristic) knowledge. Davis (1984) uses a deep model to explain how experienced electrical engineers troubleshoot a device by reasoning from first principles, i.e., from a knowledge of the structure and function of the device they are examining.

B RATIONALE

Theoretical foundations. Deep models make claims about how expert knowledge is represented (as symbolic representations). As in heuristic rule-based models, they are therefore founded on the symbolic search space paradigm of AI (see Section 4.1). The more immediate theoretical background, however, is AI research on qualitative reasoning. Workers in this field are investigating ways to represent the structure and function of a system in order to reason about it. The methods developed vary according to the type of application and domain, but perhaps the main concept to emerge so far is that of qualitative simulation, or *envisionment*. As developed by de Kleer (1979) in relation to problems in Newtonian mechanics, this can be thought of as a series of "snapshots" of an object, which describe its change in position over time. By organizing these snapshots into a treelike data structure, it is possible to represent an entire event, with forks in the tree standing for qualitative ambiguities and descendant nodes representing alternative futures based on incomplete information. de Kleer's (1979) work exemplifies various features typical of research in this field, including using a qualitative description of a system to produce a nonnumerical account of what will happen over time: only resorting to quantitative methods (e.g., use of equations in physics) where qualitative reasoning proves insufficient; and using commonsense reasoning to detect nonsensical problems. A key reference to work on qualitative reasoning is Bobrow (1985).

Empirical support. Empirical evidence is accumulating for the widespread use of deep knowledge in everyday life and in expert reasoning. The volume of papers entitled *Mental Models* (Gentner and Stevens, 1983) is a major source. It includes chapters on such topics as the models people have of their electronic calculators (Young) and the model of position and movement that Micronesians use to accurately navigate between remote South Pacific islands (McCloskey). Other chapters discuss the changes that occur in models with the development of expertise in, for example, physics problem solving (Larkin).

KE rationale. Deep models are usually justified in terms of the KE benefits they confer relative to a simple heuristic model. The principal benefits are

- *More powerful problem solving.* In cases not covered by heuristic rules, it is possible to fall back on reasoning from first principles, using a deep representation of the domain.
- *Better explanation.* System behavior can be justified in terms of domain principles rather than by simply tracing the firing of rules.
- *Tuition.* A system incorporating a deep model has the potential to serve a tutorial role, explaining the application of domain principles.
- *Learning capability.* There is a capability for automatic knowledge acquisition (see below).

C ASSOCIATED KE PRACTICE

Role of the system. Systems based on a deep model of expert reasoning can perform as autonomous problem solvers or support a tutorial role.

Tools and techniques. The technology for eliciting, representing, and utilizing experts' deep knowledge is less well developed than the rule-based technology associated with heuristic models. Elicitation techniques derived from cognitive psychology, such as conceptual sorting and multidimensional scaling, are suggested as appropriate for capturing an expert's conceptual model of a domain (Gammack and Young, 1985). At its simplest, conceptual sorting can involve the expert's sorting a set of domain concepts into several groups, identifying what each group has in common, and iteratively combining these groups to form a hierarchy. Multidimensional scaling techniques typically require making intuitive judgments about the strength of relationship between pairs of concepts in a sample. With both techniques the data may require statistical analysis to yield useful results (see, e.g., Gammack, 1987, for more discussion of these techniques). It is usually recommended that experts' domain models—however elicited—be mapped into some form of intermediate (i.e., diagrammatic) representation prior to implementation (e.g., Gammack, 1987; Steels, 1987). Workers in this area thus tend to view knowledge acquisition as a modeling process.

Regarding representation, networks of structured descriptions are typically favored (Steels, 1987). The formalisms offered by frame systems, semantic networks, or object-oriented techniques are considered suitable for expressing taxonomic, causal, or other relationships between entities. Other sophisticated techniques, such as temporal logic, may be required to model an expert's ability to reason analogically about time and space.

> **Example applications.** Expert systems which embody some form of deep model of domain expertise include CASNET (Weiss et al., 1978), CADUCEUS (Pople, 1982), and MDX (Chandrasekaren and Mittal, 1984). All these programs are concerned with medical reasoning: CASNET and CADUCEUS contain knowledge represented as a set of cause-and-effect relations, whereas MDX is designed to show that deep knowledge can be compiled into a form permitting more efficient runtime performance.

D LIMITATIONS

As a cognitive model. Deep models acknowledge the role of both heuristic and model-based reasoning in expert cognition. However, they do not reflect cognitive research findings which indicate the importance of implicit cognitive processes (see Section 4.3), domain-independent reasoning strategies (see Section 4.4), or the cooperative nature of much expert problem solving (see Section 4.5).

KE limitations

- Relative to a heuristic model of a domain, a deep model will tend to have a much larger search space and will take correspondingly longer to produce a solution.

- Research on deep knowledge-based systems (DKBS) is still at an early stage; a mature technology for implementing deep models of expertise is not yet available. Techniques for representing domain objects and the relationships between them are more advanced in expert system development tools than are techniques for running qualitative simulations or for reasoning about time and space.
- The focus of DKBS research has tended to be the expert's models of domain entities and processes. As such, this approach has had little to say about the construction of user-system dialogues or user models, now recognized as crucial aspects of expert system design (see Section 4.5).

4.3 Implicit Models

A GENERAL DESCRIPTION

Level of description. Implicit models offer a representational level account, although the precise nature of this representation is left undefined.

Knowledge types. Terminology differs between models, but the key distinction is between implicit and explicit knowledge.

Representative view of expertise. Implicit models are varied, but a common core might include these:

- Experts represent some of the knowledge explicitly, i.e., in a declarative or symbolic form. This includes facts, theories, and models acquired from textbooks and rules of thumb acquired through experience. While an expert may not always have conscious access to such explicit knowledge, it can—in principle at least—be elicited and represented explicitly in a declarative formalism such as production rules or semantic networks.
- Implicit knowledge is represented in some complex, poorly understood form. It is the nonarticulable experience-based knowledge that enables a skilled expert to perform a task or solve a problem in an effortless, seemingly intuitive fashion. Such knowledge eludes capture by existing knowledge elicitation and representation techniques. Some explicit knowledge, through repeated application in specific situations, may get "compiled" into some implicit (e.g., procedural) form. Other implicit knowledge (e.g., perceptual, motor, cultural) may be acquired directly through an implicit learning process.
- Context and background play an important role in the learning and subsequent application of expert skills.

> **Examples.** An often-cited model in this class is the five-stage model of expertise acquisition developed by Dreyfus and Dreyfus (1986). The acquisition of expertise is characterized by a shift from a reliance on explicit knowledge (e.g., context-free rules) to an almost complete reliance on intuitive processes. Mature

experts have experienced a vast number of problem situations. This enables them to deal with new situations intuitively (i.e., nonanalytically), by holistically matching the new situation to prototypical situations stored in memory and automatically applying the associated action, strategy, or decision that was found to work in earlier similar situations. The explicit use of rules, plans, and strategies is characteristic of subexpert stages of development. Partridge (1987) evaluates the possibility that intuitive expertise of this sort can be accounted for in terms of a connectionist model. Collins et al. (1986), on the other hand, liken expertise to chicken soup and dumplings. The "dumplings" are readily explicable bits of knowledge such as factual information and articulable heuristics; the "soup" is the context of meaning and nonarticulated but taken-for-granted working practices, in short, tacit knowledge.

B RATIONALE

Theoretical foundations. The Dreyfus model is rooted in "the somewhat intangible arguments from phenomenology" (Partridge, 1987, p. 44). However, the nonsymbolic framework provided by research on parallel distributed processing might provide a more solid foundation for implicit models of this kind. The Collins et al. (1986) model owes much to sociological and philosophical accounts of "tacit knowledge."

Empirical support. Dreyfus and Dreyfus (1986) suggest that research in cognitive science (e.g., prototype theory) is consistent with their model. They also point to an empirical study of how expertise in nursing is acquired, where the five-stage model offered a good fit to the data. Sociological research findings, including studies conducted from an ethnomethodological perspective, are relevant to the Collins et al. (1986) model. However, probably the most clear-cut evidence for the role of implicit knowledge and processes in skilled task performance comes from a series of highly controlled experimental demonstrations conducted by cognitive psychologists (see Berry, 1987).

KE rationale. Models of expertise based on explicit knowledge (heuristic or deep) are said to overlook the implicit nature of much expert thinking. For example, Dreyfus and Dreyfus (1986) argue that there are far fewer successful expert systems than might be assumed from examining the published literature. They claim this is because intuitive expertise eludes capture—certainly by using existing techniques. Implicit models thus suggest severe limitations on what expert system developers can realistically hope to achieve. On the other hand, an implicit model of expertise has sometimes been invoked as a rationale for the use of machine induction techniques in KE (e.g., Michie, 1982).

C ASSOCIATED KE PRACTICE

Role of the system. According to Dreyfus and Dreyfus (1986), unless the nature of the domain inhibits the development of intuitive expertise, as in the case of the successful XCON configuration system, expert systems can only be expected to attain a subexpert level of performance. This would enable the building of "competent systems" as intelligent aids but rule out the possibility of autonomous machine experts. Similarly, Collins et al. (1986) argue that expert systems can only embody explicit knowledge, devoid of its interpretive context. The users of a system must be sufficiently skilled to supply the tacit understanding needed to interpret the system's behavior. In the short term, this restricts the role of an expert system to being a productivity aid for the already skilled or to replacing the relatively unskilled. A further implication is that expert systems applications should concentrate on areas where the proportion of explicit knowledge is high, such as legislation-based domains (Berry, 1987).

Tools and techniques. It is in the nature of implicit knowledge that it is extremely difficult to capture from human experts. Nevertheless, several techniques have been advanced as relevant to this type of knowledge:

- Protocol analysis, considered a means of inferring procedural problem-solving knowledge from the talking aloud protocols of experts as they perform a task
- Participant observation (advocated by Collins et al., 1986) in which the knowledge engineer acquires some of the tacit knowledge through a period of practical domain experience
- Machine induction, considered useful because experts are usually able to supply examples of their decision making, even if they cannot articulate the underlying knowledge (e.g., Michie, 1982)

Unfortunately, there are serious practical problems in employing all these techniques (see Berry, 1987; Bloomfield, 1987). The problems of acquiring and representing implicit knowledge suggest that the modeling metaphor for knowledge acquisition is more profitable than the knowledge transfer perspective, outlined in Section. 4.1(C).

> **Example applications.** Successful applications of machine induction methods might be considered relevant here. For example, in one reported study (Michalski and Chilausky, 1980), computer-induced classification rules were compared with those obtained from a plant pathologist for diagnosing diseases in soybean plants. The induction algorithm was trained with several hundred examples linked to confirmed diseases. After training, the computer-derived rules were found to more accurately diagnose new cases than the expert's explicit rules.

D LIMITATIONS

As a cognitive model. Perhaps because of its intangible nature, there is a lack of consensus as to how intuitive expertise might be represented in human cognition. Connectionism may offer the best modeling approach at present (Partridge, 1987). Some of the detailed claims made by proponents of implicit models are also open to question. For example, the claim by Dreyfus and Dreyfus that experts process situations in an unanalyzed holistically fashion is not altogether consistent with reviews of the relevant cognitive research literature (see, e.g., the papers in the collection edited by Neisser, 1987). In general, implicit models tend to deemphasize the role of deep knowledge, strategic reasoning, and cooperative problem solving in human expert performance.

KE limitations. As already noted, implicit models are rather pessimistic about what can be achieved in KE. The lack of existing techniques for capturing implicit knowledge is placing major constraints on expert system development at present. Based on this view, only major advances in AI, cognitive psychology, and other foundation sciences will enable these constraints to be lifted.

4.4 Competence Models

A GENERAL DESCRIPTION

Level of description. These models are intended as representation-independent, or knowledge-level (Newell, 1982), descriptions of expertise.

Knowledge types. The principal distinction made is typically between domain and task knowledge.

Representative view of expertise
- Expertise is a competence-level term denoting the potential for doing something. It refers to the capacity for generating action (i.e., expert behavior) and serves as a specification for what some symbol-processing systems (i.e., a human or machine expert) should be able to do. Expertise is a kind of operative knowledge and does not refer to the specific mental operations and representations used in computing the answer to a particular problem-solving task (Johnson, 1987).
- Experts know a great deal about a specialist domain: facts, theories, taxonomies, models, etc. Domain knowledge such as this is relatively static. Experts also need to know how to use their domain knowledge effectively in solving problems. For this, task knowledge is required, i.e., knowledge about the task structure of the domain, possible inferences between domain entities, useful strategies, and so forth.
- While in practice an expert's domain knowledge and task knowledge tend to get compiled in the search for heuristic shortcuts, in principle, this heuristic knowledge can be "decompiled" to generate an explicit, knowledge-level description of a specialist's expertise.
- Different domains (e.g., electronic fault finding, medicine) may share a similar task strategy (e.g., diagnostic reasoning).

> **Examples.** A good example of this class of model is the four-layer model of expertise developed by Breuker et al. (1987; Wielinga and Breuker, 1986). The first level contains the expert's static domain knowledge (domain concepts, relations,

model of processes, and devices). The second layer describes inferences that can be made by using the static knowledge. The basic objects at the third layer are tasks and goals. The fourth layer specifies strategies for controlling and monitoring the execution of tasks and for diagnosing failures in reasoning, and finds repairs for these impasses. Alternative competence-level ways of characterizing expertise in a particular domain are described by, among others, Keravnou and Johnson (1986), who speak of *models of competence,* and Reichgelt and van Harmelin (1986), who use the term *models of rationality.* Clancey (1985a) outlines an approach to building a competence model of expert reasoning which separates domain and task knowledge. Johnson's (e.g., Johnson et al., 1987b) specification of expertise enables knowledge-level models of expertise to be constructed by using the semantic categories of actions, abilities, goals, conditions, strategies, and solutions.

B RATIONALE

Theoretical foundations. Competence models of expertise assume the existence of a knowledge level, or competence level, at which intelligent behavior can validly be described. Newell's (1982) distinction between symbolic-level and knowledge-level descriptions has already been noted. It loosely parallels Chomsky's (1965) earlier—and equally influential—distinction between *performance* and *competence* levels of explanation [indeed, Newell (1982, p. 100) notes that the terms *knowledge level* and *competence level* are almost interchangeable].

Empirical support. Psychometric and other psychological studies of human intelligence provide some support for a separation of domain knowledge from task/strategic reasoning. It mirrors the distinction drawn between metacomponents and performance components in some models of human intelligence and problem solving: the metacomponents plan, monitor, and revise strategies of information processing (see Rabbitt, 1988). They form hierarchical structures in which metacomponents organize lower-level components. Expert and novice problem solving alike can be modeled in this kind of fashion (Janswiejer et al., 1986; Wielinga and Breuker, 1986).

KE rationale. Expert systems embodying a shallow (heuristic) model of expertise are unable to provide strategic explanations or support other than a narrow problem-solving role. This was most clearly revealed in Clancey's (1983) attempt to adapt the knowledge base of MYCIN for tutorial purposes. He drew attention to how much "deep" and strategic knowledge was effectively compiled into MYCIN's heuristic rules. A knowledge-level (or *epistemological*) analysis made explicit a large amount of unstated knowledge about the strategies, justifications, taxonomies, etc., used by domain experts. It has been subsequently argued (e.g., Breuker et al., 1987) that knowledge-level descriptions of expertise also

- Serve as a specification for the design of an expert system
- Offer an implementation-independent way of representing elicited knowledge

- Support model-driven knowledge acquisition (see below)
- Can be implemented as either deep or shallow systems
- Provide a valuable intermediate step between elicited verbal data and knowledge representation

C ASSOCIATED KE PRACTICE

Role of the system. The explicit, decompiled nature of competence-level descriptions of expertise provides the flexibility needed to adequately support system functions other than dedicated problem solving; in particular, deeper explanations and tuition (Clancey, 1983).

Tools and techniques. The models of Wielinga and Breuker (1986), Johnson et al. (1987b), and others are intended as tools to facilitate expert system development—especially the knowledge acquisition phase. Each provides a different set of descriptors for modeling expertise, but their overall purpose is similar. Protocol analysis is the favored elicitation technique: it is considered the best suited for capturing an expert's problem-solving strategies (task knowledge). A knowledge-level analysis is performed on the elicited verbal data. Put simply, this involves classifying domain concepts and entities into types (e.g., symptom, hypothesis, cause), and identifying the role of each type in the reasoning process. The KE rationale for constructing such a knowledge-level description has already been mentioned. This description, stripped of its domain knowledge, can be used to guide knowledge acquisition in other domains considered to share a similar task structure (e.g., diagnosis). Such a model-driven approach to knowledge acquisition is at the center of the KADS methodology for knowledge-based system construction (Breuker et al., 1987). A model-driven approach contrasts strongly with the more usual rapid prototyping approach.

The distinction between domain and task knowledge may also be reflected in the system architecture of expert systems by introducing a control layer of metaknowledge, which can be implemented as metarules (e.g., Clancey, 1985a), task frames (Keravnou and Johnson, 1986), etc.

> **Example applications.** Examples of expert systems based on a competence-level analysis of domain expertise include NEOMYCIN (Clancey, 1985a), a successor to the MYCIN medical diagnosis system in which the principles underlying MYCIN's heuristic rules are represented explicitly; neo-CRIB (Keravnou and Johnson, 1986), a system for diagnosing faults in computer hardware; and ROGET (Bennett, 1985), an expert system development tool for creating MYCIN-like diagnostic systems.

D LIMITATIONS

As a cognitive model

- Most models of expert thinking in cognitive science are at a representational, rather than a competence, level.

- Competence models give rise to highly explicit, decompiled descriptions of domain reasoning. However, as made clear by heuristic models (Section 4.1) and implicit models (Section 4.3), expert thinking is often the reverse of explicit or decompiled. In many respects, competence-level models are more akin to novice reasoning or to an expert reasoning in "careful mode" (Clancey, 1985a).
- The cooperative nature of much expert problem solving is deemphasized.

KE limitations

- Existing KE technology may not be powerful enough to implement all formalizable competence-level domain descriptions.
- It may prove disadvantageous to implement an explicit competence model of domain expertise in applications where cognitive compatibility with expert reasoning is sought.
- The techniques involved (e.g., epistemological analysis) are relatively difficult to understand.
- The advantages of rapid prototyping are lost with a model-driven approach.

4.5 Distributed Models

A GENERAL DESCRIPTION

Level of description. This is not relevant, strictly speaking, because these are multiagent models. However, concern is more often with *what* an expert must know (knowledge level), in order to cooperate with others, rather than with *how* that knowledge is represented (symbolic level).

Knowledge types. In addition to task and domain knowledge, emphasis is placed on the expert's knowledge of the other person in a cooperative dialogue and on the conventions underlying their dialogue.

Representative view of expertise

- In a complex technological society, the expertise needed to solve a problem may be distributed among many individuals. Consequently, expert problem solving often involves some form of cooperative interaction between people.
- Users tend to be active participants in the problem-solving process, negotiating with the expert to agree on a formulation of the problem, offering constraints, critiquing initial solutions, etc.
- To engage in a cooperative interaction with a colleague or client, an expert requires knowledge (or a model) about the other person's goals, beliefs, preferences, level of expertise, and formulation of the problem. General knowledge about the rules governing cooperative communication is also required.
- In a cooperative dialogue different agents contribute different types of knowledge: functions and subtasks are apportioned among the participants according to their relative abilities.

Examples. A number of (probably compatible) distributed models have been proposed:

- Several researchers have modeled the interactions between experts and their clients as a negotiation process (e.g., Hawkins, 1983; Kidd, 1986; Pollack et al, 1982). The expert and the client try to agree on a mutually acceptable model of the problem and the constraints that any solution must satisfy. Throughout the dialogue, the expert's remarks are tailored to the perceived goals, expectations, level of understanding, etc., of the client.
- An expert may function as a "knowledge broker" between other specialists (e.g., Hawkins, 1983). A knowledge broker is able to build a model of the problem expressed in the domain terminology of one specialist and to re-represent it in a form understandable to a specialist in a related field (and vice versa).
- Interactions between experts in related domains have also been characterized as a form of guided discovery learning (Coombs and Alty, 1984). That is, experts often appear to call upon each other, not to answer well-defined problems, but rather to extend and refine their understanding of a problem area at the junction of their two fields of knowledge.
- Gaines (1987a) proposes a distributed model of expertise which suggests that the basic cognitive system that should be considered is a social organization rather than the individual. Elsewhere (Gaines, 1987b) this model is elaborated to describe the role of social processes in acquiring expertise. Experts are seen as trading the transfer of their existing knowledge for access to new problems which will enable them to acquire more.

B RATIONALE

Theoretical foundations. Conversation theory (e.g., Pask, 1975), a cybernetic approach to human cognition, is a basis for more than one distributed model of human expertise (e.g., Coombs and Alty, 1984; Gaines, 1987a). It defines the concept of a P individual as the basic unit of psychosocial processes, enabling roles, people, groups, and societies to be handled within a uniform framework. Gaines' (1987a) "distributed anticipatory formulation" of human expertise also has roots in mathematical modeling theory and personal construct theory. Research in natural language understanding and philosophy (e.g., speech act theory) has helped reveal the conventions underlying human communication. Such research offers a principled basis for modeling cooperative dialogues involving experts and, by extension, for building cooperative interfaces to knowledge-based systems (Stenton, 1987).

Empirical support. This support is provided by a number of observational studies of naturally-occurring dialogues between experts and their clients and colleagues (Coombs and Alty, 1984; Kidd, 1986; Pollack et al., 1982; see also Hawkins, 1983).

KE rationale. The first generation of expert systems tended to perform the role of autonomous problem solvers. In such a system the user may be required to do no more than passively respond to requests for information from the system and accept or reject the advice generated. This mode of system operation is linked to conceptions of expertise (e.g., heuristic and deep models) which emphasize the cognitive processes and knowledge organization of individual experts and deemphasize the cooperative nature of expertise. Proponents of distributed models suggest that many of the problems of user acceptance encountered with first-generation systems may be attributed to the unnatural passive role ascribed to the user (e.g., Coombs and Alty, 1984; Kidd, 1986). Modeling system operation on the cooperative patterns of expert-expert and user-expert interaction noted in observational studies is seen as a more promising strategy.

C ASSOCIATED KE PRACTICE

Role of the system. Systems based on a distributed model of human expertise tend to be cooperative problem solvers of one sort or another. Favored cooperative roles include negotiation (e.g., Pollack et al., 1982), critiquing (e.g. Miller, 1984), guidance (Coombs and Alty, 1984), and the role of an intelligent assistant to an expert (Worden et al., 1986).

Tools and techniques

- Observational studies of the expert at work are a frequently used technique (e.g., Coombs and Alty, 1984; Kidd, 1986; Pollack et al., 1982) to establish what actually takes place between experts and users in naturally occurring situations. This might form part of a detailed task analysis of the expert's problem domain or, alternatively, be part of a thorough user requirements analysis.
- Supporting a cooperative problem-solving role usually demands more complex and sophisticated approaches to knowledge representation and system architectures than are associated with first-generation expert systems (e.g., Kidd, 1986; Worden et al., 1986).
- *User modeling.* To engage in a cooperative dialogue with a user, a system requires a dynamic model of the user's goals, expectations, view of the problem, etc. (Pollack et al., 1982).
- *Dialogue design.* Techniques derived from natural language research and related fields are being applied to the design of cooperative interfaces between users and knowledge-based systems (Stenton, 1987). The conventions of human communication provide the underlying model.

> **Example applications.** Applications cited in the literature include: ATTENDING (Miller, 1984), a program for critiquing anaesthetia management; MINPAD (Coombs and Alty, 1984), a guidance system which helps users debug simple Prolog programs; and NEOMYCIN (Clancey, 1985a), in which the strategic and other principles underlying MYCIN's knowledge base are represented explicitly to support a tutoring role.

D LIMITATIONS

As a cognitive model. Experts are not continually engaged in cooperative interactions with others. Much of their time may be spent working alone, classifying, planning, designing, and so forth. Distributed models of expertise do little to illuminate the nature of the cognitive processes and knowledge organization underlying such dedicated problem-solving activity.

KE limitations

- Increasingly, expert system modules are being embedded into conventional systems. The utility of a distributed model seems limited where the expert component is fully hidden from the user.
- Distributed models focus on the communication aspects of expert performance; they provide minimal assistance in handling the problem-solving component of the expert's task.
- The main practical limitation on distributed models is the absence of a mature technology for implementing cooperative systems. Indeed, this is still very much a research area, with workers in the field acknowledging that it will be many years before the software techniques are developed for modeling the full complexity and subtlety of cooperative interactions involving human experts (e.g., Kidd, 1986).

5 HYBRID AND OTHER MODELS

Section 4 distinguished five major classes of models of expertise found in the KE literature. While these are arguably the most influential conceptions of human expertise in terms of KE practice, there are many others. It is possible to view some of these as hybrids of the classes of models already discussed; others introduce completely new principles:

- The heuristic classification problem-solving model (Clancey, 1985b) is a knowledge-level description of the inference structure underlying heuristic reasoning in human experts and expert systems. It thus combines elements of both the representation-level heuristic models (Section 4.1) and the knowledge-level competence models (Section 4.4).
- Steels (1987) offers an amount of deep knowledge which incorporates both domain theories and model-based reasoning (as emphasized by most deep models in Section 4.2) as well as problem-solving knowledge and strategic reasoning (as emphasized by the competence models outlined in Section 4.4).
- One aspect of Hawkins' (1983) model of expertise has already been described in relation to distributed models in Section 4.5. However, in addition to characterizing interactions between experts and users as a negotiation process, he also proposes a theory of human expert thinking. The theory of-

fers a "deep" account of how experts learn, understand, and apply their knowledge.
- The models of Kolodner (1984) and Riesbeck (1984) are case-based rather than rule-based accounts of expert thinking. These models focus on the role of experience in reorganizing an expert's knowledge in long-term memory and the incremental learning and refinement of both reasoning processes and domain knowledge.
- The reformulation model (Johnson et al., 1987a) is based on the perceptual cycle notion developed in cognitive psychology. It consists of a three-stage process in which the environment supplies information that is used to create or modify a representation which in turn directs the search for a problem's solution. This process repeats until the representation is stable or a solution is obtained.
- A research team at the University of Connecticut is attempting to build "cognitive expert systems" based on cognitive models of human reasoning and learning, including deep models (Selfridge et al., 1987). One computer system reported as embodying this approach is CMACS, a system for learning causal models of physical mechanisms by understanding real-world natural language explanations of those mechanisms.
- Several researchers (e.g., Carbonell, 1986; Eliot, 1987) have proposed models which center on the role of reasoning by analogy in expert problem solving, although they offer very different theoretical accounts.

6 CONCLUSIONS

(1) MODELS OF EXPERTISE AS REPRESENTING ALTERNATIVE KE PARADIGMS. This chapter has reviewed a wide range of models of expertise in KE. Five major classes of models were identified. Each class selectively emphasizes a different aspect of human expertise, appeals to different theoretical and empirical work for justification, has a distinctive KE rationale, and is associated with different KE practices. To a significant extent, then, each type of model both exemplifies and supports an alternative paradigm for workers in the field.

(2) RELATIONSHIPS BETWEEN MODELS. Many overlaps, dependencies, and patterns of similarity were noted between classes of model. For example:

- One form of overlap is exemplified by hybrid models such as those cited in Section 5.
- A special case of overlap is subsumption. Later models often attempt to subsume earlier ones. This pattern of subsumption is represented in simplified form in Fig. 5-1, with reference to the principal knowledge types distinguished by each class of model.
- Various dependencies between classes of models may emerge when models are applied to building expert systems. For example, a deep model of a do-

```
1. Heuristic models
   Expert knowledge      =   Factual knowledge + Heuristic knowledge
2. Deep models
   Shallow knowledge     =   Factual knowledge + Heuristic knowledge
   Expert knowledge      =   Shallow knowledge + Deep knowledge
3. Implicit models
   Explicit knowledge    =   Shallow knowledge + Deep knowledge
   Expert knowledge      =   Explicit knowledge + Implicit knowledge
4. Competence models
   Domain knowledge      =   Shallow knowledge + Deep knowledge
   Expert knowledge      =   Domain knowledge + Task knowledge
5. Distributed models
   Expert knowledge      =   Domain knowledge + Task
                             knowledge + Cooperative knowledge
```

FIGURE 5-1
Patterns of subsumption typically found among the five classes of models, with reference to the principal knowledge types that distinguishes each (see Section 4 for details).

main may be required to support the cooperative system behavior specified by a distributed model. In turn, implementing a deep model may necessitate that a competence-level description of the domain be formulated first.
• Patterns of similarity and contrast include the following:

1. Heuristic, deep, and implicit models are all concerned with how expert knowledge is represented, whereas competence and distributed models tend to be pitched at a higher (knowledge) level.
2. Only distributed models offer truly multiagent explanations; the others are essentially concerned with the capabilities of a single expert—even where the knowledge of several specialists is pooled (see Fig. 5-2).
3. While heuristic models are the most closely associated with the transfer approach to knowledge acquisition, the four other classes of model tend to be more closely aligned to a modeling approach.

(3) IMPLICATIONS OF HAVING A MULTIPLICITY OF MODELS. The diversity of models of expertise in KE seems primarily the consequence of the different KE objectives that different models are intended to support and the absence of a definitive cognitive science account of the nature of human expertise. In the absence of such an account, existing models should be viewed with caution as comprehensive descriptions of the nature of expertise. To be fair, few proponents of models would claim comprehensiveness, and many are explicit about the limited scope of their formulations (e.g., Wielinga and Breuker, 1986; Johnson et al., 1987b). In any case, for the KE practitioner, it is more important to select models pragmatically, i.e., on the basis of their perceived utility rather than their cognitive status (although the two are clearly not

Orientation	Representational level	Knowledge level
Single-agent	Heuristic models Deep models Implicit models	Competence models
Multiagent		Distributed models

FIGURE 5-2
Five classes of models grouped according to level of description and single- or multiagent orientation.

unrelated). This implies that the primary criterion of evaluation of a model of human expertise in KE is how well it serves its intended function, with validity as a cognitive model accorded only secondary importance.

(4) ONE MODEL OR MANY MODELS? Human expertise is clearly a complex, multifaceted phenomenon. It is therefore open to question how useful a single, all-embracing theory of expertise would be—even if one could be developed. Many researchers share a similar view. For example, Johnson et al. (1987a) point to the benefit of developing both a knowledge-level specification and a process model of domain expertise. Indeed, there is now a clear trend toward combining several models and representations within a single KE methodology or framework (e.g., Breuker et al., 1987; Gaines, 1987a; Gammack, 1987). Based on this view—and based on the discussion in Section 4—the five major classes of model might most appropriately be invoked at different stages of expert system development:

Stage	Most relevant model
Application selection	Implicit
Role or requirements analysis	Distributed
Knowledge acquisition	Competence
Implementation	Heuristic or deep

(5) MODELS OF THE EXPERT AND THE USER. First-generation expert systems were often intended as autonomous problem solvers. The underlying model of expertise invoked (i.e., heuristic) reflected this ambition. One response to the problems of these earlier systems (in particular, acceptance problems) is a move toward systems which offer some form of intelligent assistance to users in carrying out their normal tasks, while allowing the user greater control and initiative in dealing with the system. This is partly reflected in a shift in the underlying model of expertise invoked—away from dedicated problem-solving models and toward those emphasizing cooperative problem solving (i.e., distributed models). But it is also reflected in a greater concern with *user* modeling per se in expert systems (e.g., Sparck Jones, 1985). The recent advent of user-centered KBS design methodologies (e.g., Eason et al., 1987), in which

end users actively participate in the design process, is another manifestation of this concern. These developments notwithstanding, providing that expert systems remain expert in some real sense, models of expertise should retain their central importance in KE.

7 ACKNOWLEDGMENT

The author wishes to acknowledge the helpful comments of Dianne Berry (Oxford University, UK) and Tony Priest (Oxford Polytechnic, UK) on earlier drafts of this manuscript.

8 REFERENCES

Addis, T. R. (1987) A framework for knowledge elicitation. In: *Proceedings of the First European Conference on Knowledge Acquisition for Knowledge Based Systems* (Section A1).
Anderson, J. R. (1983) *The Architecture of Cognition.* Cambridge, Mass.: Harvard University Press.
Anderson, J. R. (1985) *Cognitive Psychology and Its Implications.* New York: Freeman.
Bennett, J. S. (1985) ROGET: A knowledge-based system for acquiring the conceptual structure of an expert system. *Journal of Automated Reasoning* 1:49–74.
Berry, D. C. (1987) The problem of implicit knowledge. *Expert Systems* 4(3):144–151.
Bloomfield, B. P. (1987) Capturing expertise by rule induction. *Knowledge Engineering Review* 2(1):55–63.
Bobrow, D. G. (Ed.) (1985) *Qualitative Reasoning about Physical Systems.* Cambridge, Mass.: M.I.T. Press.
Breuker, J., B. Wielinga, M. van Someren, R. de Hoog, G. Schreiber, B. Bredeweg, J. Wielemaker, J-P. Billault, M. Davoodi, and S. Hayward (1987) Model-driven knowledge acquisition: Interpretation models. Deliverable Task A1, Esprit Project 1098 Memo 87.
Carbonell, J. G. (1986) Derivational analogy: A theory of reconstructive problem solving and expertise acquisition. In: R. S. Michalski, J. G. Carbonell, and T. M. Mitchell (Eds.), *Machine Learning, An Artificial Intelligence Approach,* vol. 2. Los Altos, Calif.: Kaufman, pp. 371–392.
Chandrasekaren, B., and S. Mittal (1984) Deep versus compiled knowledge approaches to diagnostic problem solving. In: M. J. Coombs (Ed.), *Developments in Expert Systems.* London: Academic Press, pp. 23–34.
Chomsky, N. (1965) *Aspects of the Theory of Syntax.* Cambridge, Mass.: M.I.T. Press.
Clancey, W. J. (1983) The Epistemology of a rule-based expert system: A framework for explanation. *Artificial Intelligence* 20:215–251.
Clancey, W. J. (1985a) Acquiring, representing, and evaluating a competence model of diagnostic strategy. Stanford Knowledge Systems Laboratory, Report no. KSL-84-2.
Clancey, W. J. (1985b) Heuristic classification. *Artificial Intelligence* 27:215–251.
Collins, H. M., R. H. Green, and R. C. Draper (1986) Where's the expertise? Expert systems as a medium of knowledge transfer. In: M. Merry (Ed.), *Expert Systems 85.* London: Cambridge University Press, pp. 323–334.
Coombs, M. J., and J. Alty (1984) Expert systems: An alternative paradigm. In: M. J. Coombs (Ed.), *Developments in Expert Systems.* London: Academic Press, pp. 135–157.
d'Agapeyeff, A., and M. Barrett (1986) Tools for exercising know-how. In: *Proceedings of the Second International Expert System Conference.* Oxford: Learned Information, pp. 351–361.
Davis, R. (1982) Expert systems: Where are we? And where do we go from here? *AI Magazine* 3(2):3–22.
Davis, R. (1984) Reasoning from first principles in electronic trouble-shooting. In: M. J. Coombs (Ed.), *Developments in Expert Systems.* London: Academic Press, pp. 1–21.

de Kleer, J. (1979) Qualitative and quantitative reasoning in classical mechanics. In: P. H. Winston and R. H. Brown (Eds.), *Artificial Intelligence: An M.I.T. Perspective,* vol. 1. Cambridge, Mass.: M.I.T. Press, pp. 11–30.

Dreyfus, H. L., and S. E. Dreyfus (1986) *Mind over Machine.* Oxford: Basil Blackwell (first published in New York: Macmillan/Free Press).

Eason, K. D., S. D. Harber, P. F. Raven, J. R. Brailsford, and A. D. Cross (1987) A user-centered approach to the design of a knowledge based system. In: H. J. Bullinger and B. Schackel (Eds.), *Human-Computer Interaction—INTERACT '87.* Amsterdam: North-Holland, Elsevier, pp. 341–346.

Eliot, L. B. (1987) Investigating the nature of expertise: Analogical thinking, expert systems, and ThinkBack. *Expert Systems* 4(3): 190–195.

Feigenbaum, E. A., and P. McCorduck (1983) *The Fifth Generation.* Reading, Mass.: Addison-Wesley.

Gaines, B. R. (1987a) An overview of knowledge acquisition and transfer. *International Journal of Man-Machine Studies* 26:453–472.

Gaines, B. R. (1987b) How do experts acquire expertise? In: *Proceedings of the First European Workshop on Knowledge Acquisition for Knowledge Based System* (Section B2).

Gammack, J. G. (1987) Modeling expert knowledge using cognitively compatible structures. In: *Proceedings of the Third International Expert Systems Conference.* Oxford: Learned Information, pp. 191–200.

Gammack, J. G., and R. M. Young (1985) Psychological techniques for eliciting knowledge. In: M. A. Bramer (Ed.), *Research and Development in Expert Systems.* Cambridge: Cambridge University Press, pp. 105–112.

Gentner, D., and A. S. Stevens (Eds.) (1983) *Mental Models.* Hillsdale, N.J.: Erlbaum.

Hawkins, D. (1983) An analysis of expert thinking. *International Journal of Man-Machine Studies* 18:1–47.

Hayes-Roth, F. (1985) Rule-based systems. *Communications of the ACM* 28(9):921–932.

Janswieger, W. H. N., J. J. Elshout, and B. J. Wielinga (1986) The expertise of novice problem solvers. In: *Proceedings of ECAI '86,* vol. 1, pp. 576–585.

Johnson, P. E. (1987) Introduction to the theme issue on the nature of expertise. *Expert Systems* 4(3): 140–141.

Johnson, P. E., C. J. Nachtsheim, and I. A. Zualkernan (1987a) Consultant expertise. *Expert Systems* 4(3): 180–188.

Johnson, P. E., I. Zualkernan, and S. Garber (1987b) Specification of expertise. *International Journal of Man-Machine Studies* 26:161–181.

Keravnou, E. T., and L. Johnson (1986) *Competent Expert Systems: A Case Study in Fault Diagnosis.* London: Kogan Page.

Kidd, A. L. (1986) What do users ask?—Some thoughts on diagnostic advice. In: M. Merry (Ed.), *Expert Systems 85.* Cambridge: Cambridge University Press, pp. 9–19.

Kolodner, J. L. (1984) Towards an understanding of the role of experience in the evolution from novice to expert. In: M. J. Coombs (Ed.), *Developments in Expert Systems.* London: Academic Press, pp. 95–116.

Larkin, J. H. (1981) Enriching formal knowledge: A model for learning to solve textbook problems. In: J. R. Anderson (Ed.), *Cognitive Skills and Their Acquisition.* Hillsdale, N. J.: Erlbaum, pp. 311–334.

Leary, M., and A. Rodriguez-Bachiller (1988) The potential of expert systems for development control in British town planning. In: S. D. Moralee (Ed.), *Research and Development in Expert Systems IV.* Cambridge: Cambridge University Press, pp. 198–210.

McDermott, J. (1982) R1: A rule-based configurer of computer systems. *Artificial Intelligence* 19:39–88.

Michalski, R. S., and R. L. Chilausky (1980) Knowledge acquisition by encoding expert rules versus computer by induction from examples: A case study involving soybean pathology. *International Journal of Man-Machine Studies* 12:63–87.

Michie, D. (1982) The state of the art in machine learning. In: D. Michie (Ed.), *Introductory Readings in Expert Systems.* London: Gordon and Breach, pp. 208–229.

Miller, P. L. (1984) *A Critiquing Approach to Expert Computer Advice: ATTENDING*. New York: Pitman.

Neisser, U. (1987) *Concepts and Conceptual Development: Ecological and Intellectual Factors in Categorization*. Cambridge: Cambridge University Press.

Newell, A. (1982) The knowledge level. *Artificial Intelligence* 18:87–127.

Newell, A., and H. A. Simon (1972) *Human Problem Solving*. Englewood Cliffs, N. J.: Prentice-Hall.

Newell, A., and H. A. Simon (1976) Computer science as empirical enquiry: Symbols and search. (1975 ACM Turing Lecture) *Communications of the ACM* 19(3):113–126.

Norman, D. A. (1983) Some observations on mental models. In: D. Gentner and A. L. Stevens (Eds.), *Mental Models*. Hillsdale, N. J.: Erlbaum, pp. 7–14.

Partridge, D. (1987) Is intuitive expertise rule based? In: *Proceedings of the Third International Expert Systems Conference*. Oxford: Learned Information, pp. 343–353.

Pask, G. (1975) *Conversation, Cognition and Learning*. Amsterdam: Elsevier.

Pollack, M. E., J. Hirschberg, and B. Webber (1982) User participation in the reasoning processes of expert systems. In: *Proceedings of AAAI '82*, Los Altos, Calif.: Kaufmann, pp. 358–361.

Pople, H. E. (1982) Heuristic methods for imposing structure on ill-structured problems: The structuring of medical diagnostics. In: P. Szolovits (Ed.), *Artificial Intelligence in Medicine*. Boulder, Colo.: Westview, pp. 119–190.

Rabbitt, P. (1988) Human intelligence. *Quarterly Journal of Experimental Psychology* 40A(1):167–185.

Reichgelt, H., and F. van Harmelin (1986) Criteria for choosing representational languages and control regimes for expert systems. *Knowledge Engineering Review* 1(4):2–17.

Riesbeck, C. K. (1984) Knowledge reorganization and reasoning style. In: M. J. Coombs (Ed.), *Developments in Expert Systems*. London: Academic Press, pp. 159–175.

Selfridge, M., D. J. Dickerson, and S. F. Biggs (1987) Cognitive expert systems and machine learning: Artificial intelligence research at the University of Connecticut. *AI Magazine*, Spring, pp. 75–79.

Shortliffe, E. H. (1976) *Computer-Based Medical Consultations: MYCIN*. New York: American Elsevier.

Slatter, P. E. (1987) *Building Expert Systems: Cognitive Emulation*. Chichester: Ellis Horwood.

Smith, B. (1985) Models in expert systems. In: *Proceedings of IJCAI '85*, Los Altos, Calif.: Kaufmann, p. 1308.

Sparck Jones, K. (1985) Issues in user modeling for expert systems. In: *Proceedings of AISB '85*, pp. 177–186.

Steels, L. (1987) The deepening of expert systems. *Artificial Intelligence Communications* 0(1):9–16.

Stenton, S. P. (1987) Dialogue management for cooperative knowledge based systems. *Knowledge Engineering Review* 2(2):99–122.

Weiss, S. M., C. A. Kulikowski, S. Amaral, and A. Safir (1978) A model-based method for computer-aided medical decision making. *Artificial Intelligence* 11:145–172.

Wielinga, B. J., and J. A. Breuker (1986) Models of expertise. In: *Proceedings of ECAI '86*, vol. 1, pp. 306–318.

Worden, R. P., M. H. Foote, J. A. Knight, and S. K. Andersen (1986) Co-operative expert systems. In: *Proceedings of ECAI '86*, pp. 319–334.

CHAPTER 6

AI PLANNING

M. DRUMMOND
A. TATE

1 OVERVIEW

This chapter attempts to put some structure on the field of artificial intelligence planning. It assumes that the reader is familiar with AI and does not cover such general issues as pattern matching and inference. The goal of this chapter is to have the reader go away with a feeling of organization: what was once an apparently unorganized collection of research reports, conference papers, journal articles, and workshop notes will be seen as a library of documents, each one an account of the creation of a critical new idea.

This chapter is organized into sections as follows. Section 2 provides a brief account of what an AI planning system is expected to do. Section 3 talks about the different possible plan representations that have been considered in the literature. Following this, Section 4 considers the business of constructing plans through search. These first three sections set the stage for an historical account of planners through the ages, given in Section 5. An attempt is made to show how successive systems have built on each other, exploiting successes and overcoming failures. Section 6 takes an alternative path through the planning literature by considering recurrent planning themes. These themes are independent of particular systems and so merit independent discussion. Section 7 concludes. At the end there is a two-part bibliography; the first part is concerned with planning systems specifically, and the second part is concerned with planning system implementation details.

2 THE BASIC IDEA OF AI PLANNING

An AI planning system is charged with generating a *plan* which is one possible *solution* to a specified *problem*. The plan generated will be composed of *oper-*

ator schemata, provided to the system for each domain of application. This section briefly considers the meaning of these terms and how they relate to one another.

A problem is characterized by an *initial state* and a *goal state description*. The initial state description tells the planning system the way the world is "right now." The goal state description tells the planning system the way we want the world to look when the plan has been executed. The world in which planning takes place is often called the *application domain*. We sometimes refer to the goal state description as simply the *goal*.

For example, imagine that you are in New York wearing your favorite blue suit, that you have $100 in cash in your jacket pocket and a charge card in your wallet, and that it is 1:30 p.m. This is the problem's initial state. The goal is to be in San Francisco no later than 11:30 p.m. with at least $50 in cash in pocket. Notice that the goal does not say anything about what you are wearing. The blue suit is acceptable according to the goal, as is arriving naked, having sold the suit to pay for the airfare. The goal only gives relevant aspects of a situation and is not typically required to give all details. Any state that satisfies the goal is acceptable.

Operator schemata characterize *events*. The terms *action* and *event* are often used interchangeably in the AI planning literature and certainly are so used here. Schemata describe events, or actions, in terms of their *preconditions* and *effects*. Plans are built out of these operator schemata. Operator schemata can be organized into plans in various ways; this is the topic of Section 3. Each operator schema actually characterizes a *class* of possible events. The term *schema* means that each event description contains variables. These variables can be replaced by constants to derive operator *instances* that describe specific, individual events. When the distinction does not matter, we use the term *operator* to stand for both operator schemata and operator instances.

Throughout the AI planning literature, and throughout this chapter, STRIPS operators are used (Fikes, Hart, and Nilsson, 1972a, 1972b), if not exclusively, then certainly extensively. STRIPS operators describe an action by three parts: a *precondition formula,* an *add list,* and a *delete list.* An operator's precondition formula (or simply, the operator's preconditions) gives facts that must hold before the operator can be applied. The add list and delete list are used in concert to simulate action occurrence. If an operator's preconditions hold in a state, then the operator can be applied. Applying the operator means acting on its add list and delete list to produce a new state. The new state is produced by first deleting all formulas given in the delete list and then adding all formulas in the add list. While this approach is somewhat limited, it *is* the industry standard, and we stick with it in this chapter.

A *plan* is an organized collection of operators. A plan is said to be a *solution* to a given problem if the plan is applicable in the problem's initial state and if, after plan execution, the goal is true. What does it mean for a plan to be "applicable"? Assume that there is some operator in the plan that must be executed first. Then the plan is *applicable* if all the preconditions for the execution of this first operator hold in the initial state. Repeated analysis can determine whether all operators can be applied in the order specified by the plan. This analysis is referred to as *temporal projection*. The first state considered in the projection is the problem's initial state. Repeated operator applications produce intermediate state descriptions. If the goal is true at the end of this projection, then the plan is a solution to the specified problem.

So the inputs to a typical AI planning system are a set of operator schemata and a problem that is characterized by an initial state description and goal. The output from the planner is a plan which under projection satisfies the goal. The process connecting the input and output is known by various names. Common names are *plan generation, synthesis*, and *construction*. A planner is called *domain independent* in the sense that the plan representation language and plan generation (or synthesis, or construction) algorithms are expected to work for a reasonably large variety of application domains. While this view of planning is slightly restrictive, it will suffice for the purposes of this chapter.

3 PLAN REPRESENTATION

To generate plans, we must first be able to represent plans. Thus, the topic of this section is plan representation. Two main types of plan representation can be found in the planning literature: state space and action ordering.

3.1 State-Space Plans

State-space problem representations form the backbone of much work in AI. A state-space representation is composed of two sorts of objects:

> ...*states,* which are data structures giving "snapshots" of the condition of the problem at each stage of its solution, and *operators,* which are means for transforming the problem from one state to another. (Barr and Feigenbaum, 1981, p. 32)

State-space representations are not unique to AI.

> The state-space representation is borrowed from the classical representations of physics and other domains of applied mathematics. In these domains, a set of basis variables is selected (position and velocity in the case of classical dynamics), and each space-time point is characterized by a vector of the values of these variables. The laws of the system, typically in the form of differential equations, are the "move operators." (Simon, 1983, p. 15)

State-space representations are used in many domains, e.g., in applied mathematics, economics, and automata theory and in providing operational semantics for programming languages. For an excellent review of the use of state-space representations, see Simon (1983). In AI in general and planning in particular, state-space representations are used extensively. If a state-space representation is used *for a plan,* then the "vectors" (to use Simon's term) describe states of the application domain and the "move operators" describe the planned actions.

See Fig. 6-1 for an example of a state-space plan. The application domain is the blocks world. And S_0 is the plan's initial state, where blocks *A, B, C,* and *D* are clear and on the table. The first operator describes the action of moving block *A* from the table to the top of block *B*. The results of this action are given in state S_1. In S_1, the operator *move(C, table, D)* is applied, producing S_2. In S_2, block *A* is on block *B* and block *C* is on block *D*.

Why use a state-space representation for a plan? The reason is that it is easy to *recognize* a state-space solution to a given problem. Any state space can be easily represented in a computer as a directed graph structure, where nodes describe states and arcs describe the application of operators which transform one state to another. Encoded this way, a solution to a problem can be derived from a path through the graph which starts at one of the initial nodes and terminates at one of the nodes satisfying the goal.

3.2 Action-Ordering Plan Representations

Another common plan representation is a set of actions with constraints on the order in which the actions can be executed. For example, it is often convenient to represent a plan as a simple *list* of actions. The actions are to be executed in the order in which they occur in the list. This representation is *not* a state-space. It consists of actions and ordering relations on those actions. We refer to the general class of representations for plans which use actions and ordering relations as *action-ordering representations*. See Fig. 6-2 for an example in which the order is total, i.e., the actions are drawn out in a sequence. Lansky (1985) has called action-ordering representations *behavioral,* in the sense that they focus more on the actions to be performed and less on the conditions that each action affects.

Sacerdoti's (1975a, 1975b) planning system, NOAH, used an action-ordering representation for its plans. NOAH's representation was called the *procedural net*. A procedural net has been defined as "a network of actions at varying levels of detail, structured into a hierarchy of partially ordered time sequences" (Sacerdoti, 1975a, p. 10). For now, the interesting aspect of the procedural net is the fact that its basic objects describe actions and ordering

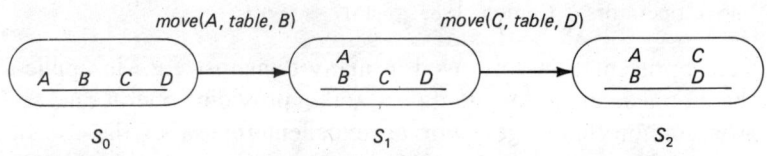

FIGURE 6-1
A state-space plan (committed order).

Start   move(A, table, B)   move(C, table, D)   Finish
☐—————————→☐—————————→☐—————————→☐

FIGURE 6-2
An action-ordering plan (total order).

relationships. A procedural net can be represented in a computer as a directed graph structure, where nodes describe actions and arcs describe ordering relations. An arc from one node α to another node β means that the action denoted by α must occur *before* the action denoted by β. When the order over planned actions is not fully specified, it is possible for some actions to be unordered with respect to each other. Because of this, these plans are often called *partially ordered plans,* and the systems which construct them are called *partially ordered planners.* See Fig. 6-3 for an example of a partially ordered plan.

NOAH and its descendants, such as NonLin (Tate, 1977), Deviser I (Vere, 1981), SIPE (Wilkins, 1983), O-Plan (Currie and Tate, 1985), TWEAK (Chapman, 1985), Forbin (Miller, Firby, and Dean, 1985), and Deviser II (Vere, 1985), all use partially ordered plans. The following comments are expressed principally in terms of the NOAH system, but apply equally to these later planners.

An action-ordering representation contains no explicit notion of *state*. Any given action-ordering plan can be considered to be in any of some number of possible states, but there is no state *explicitly represented* in the net, as is the case with a state-space representation. This lack of global state can make it difficult to reason about partially ordered plans, as we will see.

Despite this lack of state, action-ordering representations are often useful. A plan need not give the conditions that each action affects. A plan can simply order two actions, and it need not specify what the intermediate state looks like. State-based approaches typically require complete specification of intermediate states. As noted by Lansky (1985, p. 6),

> ...strictly state-based approaches to domain description can be awkward in describing behavioral properties; i.e., those that entail complicated causal and tem-

move(A, table, B)

Start ☐   ☐ Finish

move(C, table, D)

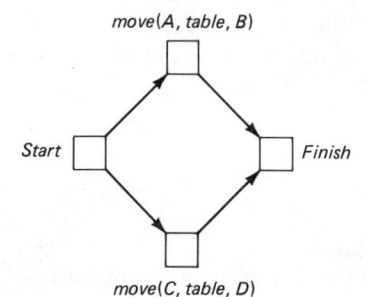

FIGURE 6-3
An action-ordering plan (partial order).

poral relationships *between* actions. Priority requirements, for example, fall into this category; they restrict future relationships between actions (for example, the order in which a service is performed) based on past relationships (the order in which requests for service were registered).

Many people now feel that the main strength of a partially ordered plan is its ability to describe parallel activity. However, the procedural net was originally introduced by Sacerdoti "with the intention of executing the actions serially" (Sacerdoti, 1980, p. 4). The procedural net was actually introduced to allow the use of *least-commitment* plans construction techniques. The term least commitment, in this context, means the practice of leaving actions unordered as long as possible.

Notice that a partially ordered plan does not *insist* that actions occur in parallel; instead, it simply *allows* parallelism. To see the distinction, consider the problem of moving a heavy trunk. Two people are needed to move the trunk, one at each end. The two people must lift in tandem, or else the trunk cannot be moved. It is useless for one of the two people to raise and lower an end before the other. Both ends *must* be raised simultaneously. A partially ordered plan cannot represent this in any obvious way. Parallelism typically occurs in a partially ordered plan only when its actions are free from interference; in the trunk example, the actions are clearly not free from interference. It is exactly their intimate interaction that permits moving the trunk at all.

3.3 Least Commitment in State-Space Plans

A planner can also carry out least-commitment planning by using state-space plans. Defined correctly, state-space plans allow postponement on action execution ordering. As in Section 3.1, a graph is used to represent the plan, but now *sets* of operators label the arcs. The interpretation is that in order to traverse an arc, we must select and execute each of the operators in its labeling set in any order. For example, say that some set contains the operators α, β, and γ and that an arc labeled by the set joins two states S_0 and S_1. The interpretation is that if in state S_0, we can arrive at state S_1 by executing α, then β, then γ, or by first executing α, then γ, then β, and so on. The order does not matter: as long as all the indicated operators are executed, the resulting state will be S_1. See Fig. 6-4 for the blocks world plan represented in this way.

To use a least-commitment strategy, a planner needs a representation that can leave action execution unordered. Some action-ordering plans can

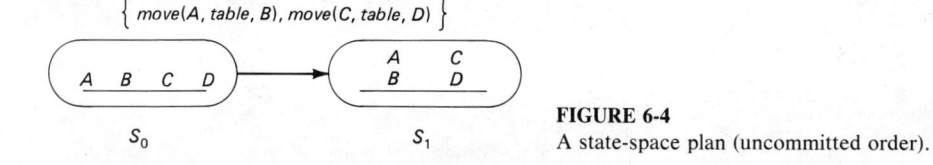

FIGURE 6-4
A state-space plan (uncommitted order).

do this, as can some state-space plans. For a formal treatment and definition of a state-space representation able to support least commitment, see Reisig (1985, p. 28).

3.4 Summary

Two main types of plan representation have seen frequent service in the planning community. First, we have state-space representations, as borrowed from the classical representations of physics and other domains in applied mathematics. A state-space *plan* contains descriptions of states of the world in which the plan is to be executed, and move operators which correspond to the actions to be carried out to execute the plan. A state-space solution is trivial to recognize: it is quite simply a *path* from some initial state to a state satisfying the goal specification.

Second, there are action-ordering representations, of which the "list of actions" plan is a simple case and the procedural net the most often cited example. Action-ordering plans allow us to describe the relationships among actions directly, rather than through states and predicates contained within states. Action-ordering plans suffer since it is not always obvious what "states" of the world are possible, given an initial state and an arbitrary plan. Unlike a state-space representation, a possible behavior in an action-ordering plan does not correspond to a simple path in a graph: traversing a path through an action-ordering plan only indicates the "before" relations that exist along that path. Behaviors are generated by plan executions that respect all such orderings. This can make plan generation quite hard.

We use the term *parallel plan* to refer to any representation that can leave actions unordered. Partially ordered action-ordering plans are one example, and state-space plans that allow arcs to be labeled with sets of operators are another. A *serial plan* is one which is not able to leave actions unordered. Totally ordered action-ordering plans are an example of such plans, as are state-space plans where the arcs can be labeled only with single operators.

In this section, we have addressed issues of plan representation. In the next, we turn to plan construction and consider the nature of the planner's search space.

4 CONSTRUCTING PLANS THROUGH SEARCH

Planners *search* to find plans that are solutions to problems. The search process is driven by the need to achieve a problem's goal. There are various ways to define the search space explored by a planner and various ways to define a planner's goal achievement mechanism. The topics of search and goal achievement are inextricably bound together. This section considers each individually and how they relate.

4.1 The Search Space

Problem solving is often viewed as search. Solving a problem by using the search metaphor can be understood as follows.

> We postulate some kind of space in which treasures are hidden. We build symbol structures (nodes) that model this space, and "move" operators that alter these symbol structures, taking us from one node to another. In this metaphor, solving a problem consists in searching the model of the space (selectively), moving from one node to another along links that connect them until a treasure is encountered. (Simon, 1983, p. 7)

What do the nodes in a search space describe? As noted by Charniak and McDermott (1985), the nodes are often interpreted as describing either *states of the world* (task domain) or *partially completed plans*. When describing a partially ordered planner, Charniak and McDermott (1985, p. 514) say that:

> The space searched by our planner is the space of "partial plans" or task networks. Operators reduce tasks to subtasks, and impose orderings on hitherto unordered tasks. A goal state is a plan that is guaranteed to work without any protection violations.
>   A rather different use of search, and historically prior to NonLin, is to search through world situations rather than through the space of alternative task networks. Under our stringent assumptions about time, we can think of planning as a search through possible states of the world. We can take the initial world model as the initial state; we can take primitive acts (described by addlists and deletelists) as operators; and the state of affairs to be made true (*achieved*) in the final world model as the goal description.

There are thus two types of search spaces that a planning system can explore. In one, the nodes examined contain descriptions of possible states of the world, and in the other, the nodes contain possible partial plans. We refer to the former as search through *world states,* or simply *state-space search,* and to the latter as search through *partial plans,* or *partial-plan search.*

How does this distinction relate to the comments of the last section? The issue is really one of *reasoning*. A planner which searches through world states can also have plan structures in each node of its search space. However, such a planner can only use each node's world-state information to guide its next move in the search space. If the planner does carry along a plan in each search space node, then it is natural to think of the plan as a path through the search space describing how the planner arrived at the node containing the plan. Thus the state information in the node describes the world state that will be obtained after execution of the plan. Retaining this path information might be a good idea, since if the plan is left implicit in the planner's state-space search graph, it may take extra work to extract the plan upon successful termination of the search. If each search-space node contains not only world-state information but also plan (or path) information, then when a node satisfying the goal is found, the path contained in the node will describe the plan to achieve the goal.

When a planner searches through possible world states, it is using its state-space navigation procedure to do its plan construction reasoning. This allows the planner to avoid any problems with goal achievement procedures, since all questions are asked with respect to the "current" search-space node. But by doing this, the planner confuses two conceptually distinct things: plan construction reasoning and search-space navigation. The projection of a plan and the search for it become identical.

Why would a planner perform search through a space of world states? Because the search procedure is almost all the *reasoning* that is required of the planning system. The reasoning required to build a plan is packaged into the planner's next-stage generator. This means that the planner will only generate *possible* plans and need perform no checking after extending a plan.

Why, then, would a planner ever search a space of partial plans? It would do this to *separate* its search-space exploration procedure from its plan construction reasoning: the search for a plan and the plan construction process are made distinct. Instead of having the plan construction reasoning encoded in the planner's search-space navigation procedure, we have plan construction reasoning operate over a plan contained in each search-space node. The results of the reasoning will guide the search through the space of possible plans.

If a planner searches through partial plans, it can add and remove operators at various points in the plan. When the planner searches through world states, plan modification is possible only at the tail end of the plan. Operators are added to the plan by trying out another operator application, and operators are removed from the plan when backtracking occurs. This is a result of confusing the plan construction reasoning with the search-space navigation control. However, if the planner searches through a space of partial plans, it can add and remove operators in the plan wherever it sees fit.

4.2 Making a Plan Achieve a Goal

The difficulty of analyzing a partially ordered plan has been known for some time. The problem is well put by Charniak and McDermott (1985, p. 504):

> [The conditions which are true after the execution of a task]...depend in general on the state of the world when the task is performed, and this state depends in turn on the ordering of tasks, which is only partially known. For example, consider the simple action (*move x y*), "Move block *x* to the top of block *y*." One effect of this action is that the block where *x* originally was now has nothing on it. But which block this is will depend on how the *move* action is ordered with respect to other planned actions.

Uncertainty with respect to what is true or false can cause problems in plan construction. In particular, there can be problems with finding goals, goal violations, and variable bindings. In general, these problems all relate to the fundamental business of a planner: ensuring that operator preconditions are true just before the operator must be executed.

If a planner searches a space of partial plans, then a planning problem can be given as a partial plan with operator preconditions that have not yet been achieved. Such initial plans typically contain two pseudo-operators: *Start* and *Finish*. The *Start* operator is used to assert the problem's initial state. This is accomplished by giving the facts that are true in the initial state as the postconditions of the *Start* operator. The *Finish* operator is used to declare the problem's goal: the goal is given as the precondition of the *Finish* operator. In some sense, *Start* represents the beginning of time, and *Finish* represents the end of time. The planner is expected to flesh out the bit in the middle.

The preconditions of the *Finish* operator are not true in the initial plan. The business of the planner is one of achieving operator preconditions. To do this, the planner must understand how the existence of operators, orderings, and bindings affects the truth of preconditions in a plan. Given a partially ordered plan and an operator precondition, the planner needs a procedure for determining whether the precondition holds just before the operator must be executed. If the precondition holds, then nothing need be done; if the precondition does not hold, then the planner must modify the plan in such a way that the precondition *does* hold just before the operator must be executed.

NonLin was the first planner to demonstrate this ability. It used a procedure called *question answering* (QA) to determine the truth of operator preconditions. If the precondition in question is true, the QA returns true. If the precondition is not true, then QA returns a description of the various ways in which the precondition can be made true. The alternatives include the addition of operators, addition of orderings, and binding of variables within the developing plan. NonLin had efficient algorithms for finding possible contributors and violators for a precondition, and it used these algorithms to suggest ("minimally") all possible ways of making a precondition true where required.

Chapman (1985) has formalized some of what NonLin does. He presents what he calls the *modal truth criterion* (MTC) as a statement of the necessary and sufficient conditions for a condition to be true at a point in a partially ordered plan. Essentially, the MTC says that a condition p is necessarily true at a point in a plan if and only if, first, there is a point necessarily before the required point where p is necessarily asserted and, second, for every operator that could possibly come between the point of assertion and point of requirement, if the intervening operator possibly deletes a condition which might turn out to be p, then there must be another (appropriately placed) operator which restores the truth of p whenever the intervening operator deletes it.

Although it sounds a bit complicated, the MTC is simply a statement of the conditions under which an operator's preconditions will be true. We can take a more "procedural reading" of the MTC to derive a goal achievement algorithm such as NonLin's QA. To see how such an algorithm might go about achieving an outstanding precondition, see Fig. 6-5. Operators are drawn as boxes, and conditions are drawn as circles. Operator C "contributes" condition p; that is, C has p on its add list. Operator R "requires" condition p to be

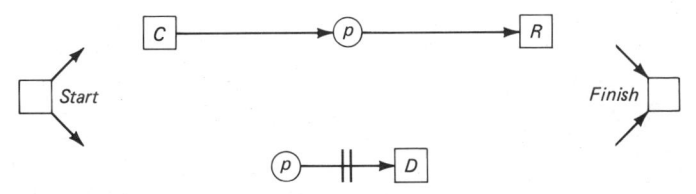

FIGURE 6-5
Goal achievement graphically.

true; that is, *R* has *p* as one of its preconditions. Operator *D* has *p* as a precondition *and* deletes it. This deletion is indicated in the diagram by the double line drawn through the precondition relation from *p* to *D*. In the diagram it is assumed that objects to the left are ordered before objects to the right. Objects above each other are not ordered in time. This means that *Start* comes before everything else and that *Finish* comes after everything else; *C* comes before *R*; *D* possibly comes after *C* and before *R*; *D* possibly comes after *C* and before *R*. Of course, *D might* come *before C* or *after R*. The orderings as shown in Fig. 6-5 do not preclude this. The diagram shows only part of a presumably larger plan. So some operators might be ordered after *Start* but before *C* and *D*; similarly, there might be some operators after *R* and *D* but before *Finish*. The diagram simply focuses on a small part of the overall plan.

We will focus on the achievement of precondition *p* for operator *R*. First, we must have a contributor for the condition that is necessarily before *R* in the plan. If there is no such contributor, we can create one by instantiating an appropriate operator schema and installing it in the net. If there is already a contributor for *p* but it is not yet ordered before *R*, then we can simply add the required ordering. After this we will have a guaranteed contributor for precondition *p* for *R*. Second, we must check the operators that possibly delete condition *p*. Condition *p* is possibly deleted if there is an operator in the plan that could come after *C*, the contributor, and before *R*, the point of requirement. Not all operators are a cause for worry: only those that explicitly mention *p* on their delete list can actually delete *p*. If no such deleters exist, then the truth of *p* at *R* is guaranteed. If there is such a deleter, then it must be ordered outside the range over which *p* is expected to hold. Assume that *D* is a deleter of *p*, as shown in the diagram. This deleter may be rendered harmless by ordering it *before C*, the point of contribution, or *after R*, the point of requirement. The story is only slightly more subtle than this: We have to guarantee that this work is being done for the last possible contributor of *p* for *R*. Defining this notion of last possible contributor is what the final part of the MTC is about. NonLin accomplished this by using specialized algorithms; see Tate (1977) for details.

The problem is that such a goal achievement procedure will not necessarily work for more expressive operator languages, because a full interleaving analysis is sometimes required to determine the truth of a condition at a point in the partial order. The set of all possible interleavings is used to completely determine the condition's truth. Whenever a plan can describe *synergy, dis-*

junction, or *deduction*, the MTC might fail. Let us look more closely at the specific difficulties with synergy and disjunction.

For the problem of synergy, consider the following example borrowed from Chapman (1986). Imagine an initial situation where three blocks of equal size are ready for placement, as shown in Fig. 6-6. The plan we are analyzing says that each of the three blocks should be picked up and put down on block *D*. This plan is shown in Fig. 6-7. The three planned actions are unordered, and the planner must find and remove all interactions between the actions which could prevent successful plan execution. A difficulty is that there is only space for *two* of the three blocks on the designated target. A precondition for each of the movement actions is that there be space for the block on *D*. This could be formalized by a precondition $space(D) \geq 1$, where 1 indicates that 1 unit of space is required to stack the block on *D*. Block *D* has only 2 units of space available.

Unfortunately, the MTC does not detect the fact that not all three stacking actions can occur. That is, *we know* that only *two* of the planned *three* actions are valid, but the MTC says that all can occur. The problem hinges on the way that operator preconditions are checked. Consider the case of the *space* precondition for the *place(A)* action. First, there must be a contributor which makes $space(D) \geq 1$ true. This condition is achieved by the operator *Start*. Second, there must not be any operator which might make the precondition false. Consider each possible operator in turn. Does *place(B)* make the precondition false? No, since after it occurs, $space(D) \geq 1$ is still true. Does *place(C)* make the precondition false? No, since after it occurs, $space(D) \geq 1$ is still true. So according to this analysis, the precondition $space(D) \geq 1$ is achieved by *Start* and maintained until required by *place(A)*. Of course, this is incorrect. We are not considering the possibility of *synergistic* interactions in the plan. And this is exactly what happens: no *one* of the actions acts by itself to disable the others; it is only two actions in concert which make the third impossible.

What about disjunction in a partially ordered plan? The MTC does not work here, since it is impossible to even *represent* disjunction in a partially ordered plan. Such a plan only specifies a set of actions and a set of *before* relationships on those actions. The understanding is that each of the specified actions *must* be executed. In this framework, it is impossible to express the fact that some number of actions are disjunctive. This means, e.g., that is it impossible to have a partially ordered plan which captures the following: "Go into Frank's office. If he's there, tell him the party starts at 18:00; otherwise, leave a message on his desk."

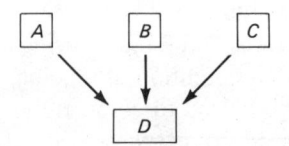

FIGURE 6-6
A tricky goal achievement problem.

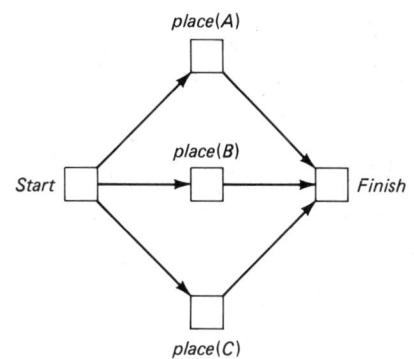

FIGURE 6-7
Plan for the tricky goal achievement problem.

4.3 The Linearity Assumption

One of the major contributions to problem solving by the general problem solver (GPS) project (Ernst and Newell, 1969) was the technique of means-ends analysis, often abbreviated MEA. Now MEA is a search heuristic which calls for applying operators that can be shown to be relevant to reducing the differences between the initial and goal states, as specified in the problem description. MEA reduces differences, one by one until no differences remain. Unfortunately this technique will not always arrive at a solution. The problem is well described by Simon (1983, p. 15):

> The search procedure of GPS is built on the implicit premise that if the present situation differs from the goal situation by features $A, B, C, \ldots$, then the goal situation can be attained by removing the differences, $A, B, C, \ldots$, in some order. Of course this premise is false unless the matrix of connections between differences and operators can be triangularized. This matrix can be triangularized just under those conditions when an appropriate composition axiom would be valid in the modal logic; that is, just when there is independence among the actions.

Sussman (1973, p. 58) has expressed the same idea earlier and had called it the *linear assumption:* "Subgoals are independent and thus can be sequentially achieved in an arbitrary order." Here, we refer to this as the *linearity assumption.* Sussman's planner, Hacker, could not solve the blocks-world *Sussman anomaly* problem (see fig. 6-8). This was because Hacker's search space was reduced by the linearity assumption so as to not include a solution. Other planners, which built totally ordered plans, *could* solve the Sussman

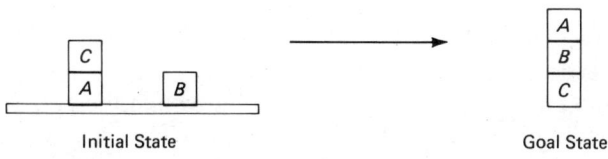

FIGURE 6-8
Sussman anomaly.

anomaly. For instance, Interplan (Tate, 1975) and Warplan (Warren, 1974) could both solve it. NOAH (Sacerdoti, 1977), the first partially ordered planner, solved the Sussman anomaly through its judicious selection of "goal to work on next," *not* through its use of partially ordered plans.

So why is the Sussman anomaly hard? Consider the problem's goal: $on(A, B) \wedge on(B, C)$. This means that A must be on B *and* that B must be on C. This goal conjunction can be split into its two components: $on(A, B)$ and $on(B, C)$. Each of these represents a subgoal that Hacker assumes may be independently achieved of the other. So consider achieving $on(B, C)$ from the initial state shown in Fig. 6-8. This can be done by simply stacking B directly on top of C. But then block A is stuck beneath the other two: we have made a mistake. So consider achieving $on(A, B)$ from the initial state. To do this, block A must be clear. To make A clear, we must take C off and put it somewhere. Let us assume that we choose to place C on the table. Now A is clear and may be placed on B. So we have A on B, and block C is clear and on the table. Once again, we made a mistake. Without backing up, we cannot make simple progress toward the conjunctive goal.

The Sussman anomaly highlights the goal-ordering problem. NOAH's success with the anomaly is a direct result of its approach to goal ordering. The successful goal-ordering strategy was simply built into NOAH's control structure. Inferring from NOAH's success, it is often supposed that partially ordered plans "solve" the linearity assumption. The literature is full of statements which perpetuate this confusion. For two representative views, see Barr and Feigenbaum (1982, p. 520) and Stefik (1981a, p. 134). Partially ordered plans do *not* usefully address the linearity assumption. The linearity assumption has to do with the way that a planner constructs its plans, not with the plan representation itself. The word *linear,* as used in the phrase *linear assumption,* refers to the way that a planner approaches plan construction: a linear planner will not interleave recursive subgoals with top-level goals. Such a planner proceeds *linearly* in the sense that it only considers different orderings of *goals* from the top-level goal specification.

5 A PLANNING SYSTEM REVIEW

Bearing in mind the comments of the last couple of sections, let us take a look at some planning *systems*.[1] A graphical taxonomy of most of these systems (plus a few others) is given at the end of the chapter.

5.1 Green's System

Green (1969) formulated planning problems in the predicate calculus and used a resolution theorem prover to generate plans. In Green's system, each condition included a state argument, which denoted the state in which the condition held. (Perhaps some still hold to this day. Who can say?) The idea of in-

---
[1] If we have misrepresented anyone here, please let us know!

cluding such a state argument in conditions comes directly from McCarthy (1958) and McCarthy and Hayes (1969). This technique is often referred to as the *situation calculus*. In Green's implementation of this idea, a goal is a formula with an existentially quantified state variable. The system attempts to prove that there exists a state in which the required goals are true.

The execution of actions and the states thus produced are considered equivalent. In Green's system, the function *do* maps from states to states. Action performance and states are represented by structures of the form: $do(action(...), S)$, where the ellipsis denotes the arguments of the action and S is another such structure or the initial state. For instance, $on(a, c, do(move(a, b, c), s_0))$ says that block a will be on block c in the state which results from moving block a from b to c, in state s_0.

Operators are given as implications. A frame axiom is required for each predicate that each operator does not affect. Thus, the number of frame axioms required is the product of the number of predicates and operators (i.e., very many).

SUMMARY

System	QA3 theorem prover
Reference	Green (1969)
Plan structure	Action ordering
Parallel/serial	Serial
Search space	World states
Search control technique	As used in the QA3 system for selecting resolvents

5.2 Kowalski's System

Kowalski (1979) offered a different formulation for planning that uses the predicate calculus. Kowalski uses a predicate *holds* to indicate that a given condition holds in a given state. For instance, $holds(on(a, c), s_0)$ is used for Green's expression $on(a, c, s_0)$. This means that the number of frame axioms required is equal only to the number of operators, since it is possible to quantify over all those conditions unaffected by an action by using the *holds* predicate. See Fig. 6-9 for a Prolog implementation of Kowalski's formalism. The "planner" has been given an operator and domain predicates describing a simple blocks-world domain. In the example given, there is only one operator and one frame axiom. The initial state given is, of course, just one of the many that are possible.

A goal state is specified as a conjunction of conditions containing an existentially quantified state variable. Actions and states are of the same form as in Green's system, but instead of maintaining the explicit state-space information regarding the holding of each predicate, Kowalski's system looks up the information only when required. In the Prolog example, this must be done since the predicate *poss* returns a possible plan each time it is backtracked over, and this plan must be examined to determine what does and does not

```
poss(s0).                  /* Initial State is the first possible one */

holds(on(c,a), s0).        /* Initial State specification: the Sussman Anomaly */
holds(on(a,t1),s0).
holds(on(b,t2),s0).
holds(clr(c),  s0).
holds(clr(b),  s0).
holds(clr(t3), s0).

holds(clr(Y),   do(move(X,Y,Z), S)). /* Action postconditions */
holds(on(X,Z),  do(move(X,Y,Z), S)).

pact(move(X,Y,Z), S) :-              /* Action preconditions */
      holds(clr(X),  S),
      holds(clr(Z),  S),
      holds(on(X,Y), S),
      X \== Z, Z \== Y, X \== Y.

poss(do(U,S)) :-                     /* Defines possible action sequences */
      poss(S),
      pact(U,S).

holds(V, do(move(X,Y,Z), S)) :-      /* Single frame axiom */
      holds(V,S),
      V \== clr(Z),
      V \== on(X,Y).
```

FIGURE 6-9
A simple Prolog implementation of Kowalski's formalism.

hold at its end. Instead of having a collection of conditions of the form *pred(args, state-spec)* for each *pred* true of *state-spec*, we have *state-spec* as a structure we can examine to find out whether the required conditions hold. In Fig. 6-9, this lookup function is performed by the predicate *holds*. This predicate defines a version of the MTC for a simple totally ordered plan.

SUMMARY

System	Untitled
Reference	Kowalski (1979)
Plan structure	Action ordering
Parallel/serial	Serial
Search space	Partial plans
Search control technique	Breadth first on the length of the plan

5.3 STRIPS

STRIPS (Fikes, Hart, and Nilsson, 1972a, 1972b) is probably the single most significant planning project to date. No short description can ever do justice to the contributions and impact of the STRIPS system on the AI planning world. However, we feel free to try.

STRIPS was an elegant adaption of the GPS (Ernst and Newell, 1969) to planning problems. It included an implementation of means-ends analysis based on the QA3 theorem prover. This use of a theorem prover is quite surprising, since STRIPS also introduced the idea of using nonlogical state-transforming operators. The nature of these operators, including the idea of using precondition lists, add lists, and delete lists, is well discussed in Charniak and McDermott (1985).

In terms of plan structure, STRIPS was tremendously innovative. It used *triangle tables* to encode all the essential features of a plan's teleological structure. The word *teleology* is generally used to mean "interpretation in terms of purpose or reason," and we use it here to refer to the *reason* for the introduction of specific actions in a plan. STRIPS supplied this information to the PLANEX system, to enable accurate runtime plan execution monitoring. There was a minor resurgence of interest in triangle tables as a programming language for robot actions (Nilsson, 1985).

STRIPS is one of the classic systems to apply the linearity assumption, as embodied in means-ends analysis. STRIPS searched through a space of world states to find a solution and employed MEA to guide its way. As a direct result, there are many problems for which STRIPS can provide no "optimal" solution. The Sussman anomaly is typical of the sort of problem STRIPS was unable to optimally solve.

SUMMARY

System	STRIPS
Reference	Fikes, Hart, and Nilsson (1972a, 1972b)
Plan structure	Action ordering
Parallel/serial	Serial
Search space	World states
Search control technique	MEA and best-first search

5.4 Hacker

Hacker (Sussman, 1973) was advertised primarily as a learning program, but did actually contain a significant planning element. The learning was of a procedural sort; i.e., the program learned new "skills." Hacker generated a plan by first composing elements of its plan library; second, it submitted this plan to its *gallery of critics*. Critics were designed to catch and correct known plan "bugs." After criticism, Hacker assumed that the plan would work. The plan was then simulated, and new bugs detected. The bugs were analyzed, and the result of this analysis was passed on to the learning element to enable Hacker to do better on its next attempt.

To help the bug detection phase operate, each plan had attached to it an account of its teleological structure. This account indicated where in the plan goals were achieved and where they were required to be true. The bug detector used this information to index into a library of bug types. Each bug type was in essence a "teleological template" and could be matched to any given plan to diagnose that plan's bug. As will become evident in our discussion of various planning systems, teleological information is vital in constructing plans. Hacker was the first program to make use of such information in plan construction.

Hacker searched through a space of partial plans and used its critics to suggest structural plan modifications. The critics were nonconstructive in the sense that they found reasons for existing faults and were unable to suggest fixes in advance of fault appearance. The interesting feature of Hacker as a *planner* is the way it turned attention to the issues of having a plan as a structure and using procedures to reason over the structure, modifying it as necessary. Thus, it was one of the first planners to actually search through a space of partial plans. It was also in this project that the linearity assumption was first acknowledged. In making this assumption, Hacker greatly reduced the size of its search space. Of course, a reduction in the search space often entails a loss of ability to solve many interesting problems.

The plan representation used by Hacker was essentially action-ordering. The plans were encoded as Lisp programs, but no interesting programming constructs could be handled. Only serial program execution was ever considered. For this reason, it seems reasonable to say that Hacker did actually employ an action-ordering plan representation.

SUMMARY

System	Hacker
Reference	Sussman (1973)
Plan structure	Action ordering with teleology add-on
Parallel/serial	Serial
Search space	Partial plans
Search control technique	Achieve a goal then debug; made linearity assumption

5.5 Interplan

Interplan continued where Hacker left off. Tate (1975) used a perspicuous representation for the teleological structure of each plan, called a *tick list*. The tick list allowed Interplan to pursue goal reorderings not considered by Hacker. Interplan could solve the Sussman anomaly (Fig. 6-8); Hacker could not. Recall that this problem cannot be solved by a planner that makes the linearity assumption.

Once again, Interplan searched through a space of partial plans; however, it used a procedure to guide its search which opened the space to include a solution to the Sussman anomaly. This procedure performed *subgoal promotion*. This can be viewed as a means of relaxing the linearity assumption not only by attempting permutations of the highest-level goals of the problem but also by considering *relevant* goal interleavings which include interacting subgoals introduced during plan construction.

Interplan was interesting because of the way it highlighted the fact that it is possible to reason about a very simple teleological account of a plan, with significantly improved problem-solving ability. It was the first planning system to use teleology to actually correct for interactions between parts of a plan initially assumed to be independent.

SUMMARY

System	Interplan
Reference	Tate (1975)
Plan structure	Action ordering, with teleology add-on
Parallel/serial	Serial
Search space	Partial plans
Search control technique	Teleologically relevant subgoal promotion

5.6 Waldinger's System

One problem with Interplan was the way it threw out much of the work achieved in the construction of an unsuccessful plan. Once a given goal order was found not to work, the plan was thrown away, and work began anew on a plan to achieve goals in the new order (in Interplan, this order was called an *approach*). Waldinger (1977) suggested a technique called *goal regression* which overcame this problem. Goal regression calls for building a plan to achieve one goal and then modifying that plan so that it also achieves another goal. A plan is modified with respect to a goal by finding a place in the plan where the goal can be achieved or is found to be compatible.

The basic idea amounts to "dragging" a goal back along the list of actions that is the plan until some interaction gap is found in which the goal can be inserted. Goal regression modifies plans more constructively and less wastefully than Interplan, since existing plans to achieve goals are not thrown away when a goal achievement attempt fails.

Once again, this technique can be applied only when a planner searches through a space of partial plans. Goal regression will call for trying a goal at various points along the plan, until a location is found which works. The search space considered by Waldinger's system excluded movements in which an action was deleted from a plan. Once an action was added, it was in for better or worse and would never be removed. Given a plan of length n, the technique would call for consideration of plans only of length $n + 1$. Although plans of length $n - 1$ could have been candidates for consideration, they were not.

SUMMARY

System	Untitled
Reference	Waldinger (1977)
Plan structure	Action ordering
Parallel/serial	Serial
Search space	Partial plans
Search control technique	Goal regression

5.7 Warplan

Warplan (Warren, 1974) is based on a technique much like Waldinger's goal regression. Warplan uses *action regression* to construct plans. This technique focuses on the actions required to achieve goals, rather than the goals themselves.

Given a partial plan and a goal, Warplan finds an action to achieve the goal and then regresses the action along the plan to a point at which the action can be peacefully inserted. Peaceful insertion means that the actions' preconditions can be simply achieved at the selected spot. Warplan suffered from the problems itemized above for Waldinger's goal regression and, in addition, would often fail to recognize when an action being regressed ceased to be necessary.

Warplan seems most noted for its size and efficiency. It was one of the first planners written in Prolog. Since Warren included the Prolog source code in the research report describing the system, the planner has been well received and reimplemented many times.

SUMMARY

System	Warplan
Reference	Warren (1974)
Plan structure	Action ordering
Parallel/serial	Serial
Search space	Partial plans
Search control technique	Action regression

5.8 NOAH

NOAH (Sacerdoti, 1977) is possibly one of the most misunderstood (and thus misrepresented) planning systems ever. It is often claimed that the procedural nets introduced by NOAH "solve" the linearity assumption. We argued above that this is not the case.

NOAH as a planner did not, in fact, make the linearity assumption. But as discussed above, this had nothing to do with its plan representation; rather, not making the linearity assumption was a direct result of NOAH's plan construction algorithm. NOAH solved the Sussman anomaly through its approach to goal ordering, *not* through its use of partially ordered plans. The Sussman anomaly highlights the goal-ordering problem. The successful goal-ordering strategy was simply built into NOAH's control structure. The problem of goal ordering is not addressed in NOAH, since it has no mechanism for backtracking out of a bad choice. If NOAH were to make an incorrect ordering choice, it would be fatally stuck since the choice could not be undone.

NOAH is most important for introducing the procedural net, the first partially ordered plan representation. These plans were important at the time for two reasons. First, they contained an explicit notion of action hierarchies. Second, they were the first plan structures which allowed "least-commitment" plan construction policies to be adopted.

Procedural nets allow least-commitment plan construction only because the order on actions in a net is partial. This means that a net can contain two (or more) unordered actions. As discussed above, this is not a feature unique to the procedural net, since state-space plan representations can represent unordered actions, if defined appropriately. As a matter of historical fact, however, the procedural net was the first plan structure in the AI planning literature which could represent unordered actions.

NOAH performed very little (if any) search and had no backtracking facilities for undoing bad plan construction decisions. NOAH primarily studied plan representation; only basic facilities were included for plan construction. NOAH never made a mistake in constructing the simple plans it produced. A small class of potential plan faults could be dealt with by *critics*. A critic was provided for a specific class of plan construction problems. As in Hacker and Interplan, each critic in NOAH made reference to a plan's underlying teleological structure when suggesting plan modifications. NOAH used a table of multiple effects (TOME) to record the effects of each action added to a plan so that conflicting action additions could be handled. Sacerdoti (1975b) explicitly acknowledges the influence of Interplan's tick list structures when explaining the function of NOAH's TOME.

SUMMARY

System	NOAH
Reference	Sacerdoti (1975)
Plan structure	Action ordering, with limited teleology (TOME)
Parallel/serial	Parallel
Search space	Partial plans, but extremely limited search space
Search control technique	None

5.9 NonLin

NonLin (Tate, 1977) was designed to correct certain problems found in the NOAH prototype planner. NonLin was able to search the space of partial plans, while NOAH performed only limited search, if any. NonLin was thus the first planning system to use partially ordered plans that was also able to really *search* the space of alternative plans for a solution.

The search heuristic used by NonLin for plan selection is called *one-then-best*. This heuristic calls for focusing on the choice currently being made and trying to select one of the local alternatives which seems most promising. If a failure occurs, the entire set of alternatives which has been generated is reconsidered to select the refocusing point. See Section 6.1 for more on how this works.

Another contribution made by NonLin came in the form of a more declarative task formalism (TF) for describing action-ordering operators to a planner. The operators in NOAH were represented as SOUP (semantics of user's problem) code and were extremely difficult to separate from the internals of the planner itself. In contrast, NonLin's TF allowed a person writing domain operators to specify actions, preconditions, and ordering relations in a simple, convenient formalism. TF also introduced the notion of *typed preconditions*, now found in many planning systems. This precondition typing mechanism allowed a user to specify an operator's teleological structure very simply and graphically. Tate (1984a) calls this information *goal structure,* and recent planners such as O-Plan (Currie and Tate, 1985) and SIPE (Wilkins, 1984) use it extensively, for both plan generation and plan execution.

NonLin exploited information about action duration and action start and finish times. This information was attached to each action within a plan and used during planning to ensure that the plan being generated satisfied certain overall metric temporal constraints. The system was one of the first to be applied to large, realistic domains. NonLin was used to generate turbine overhaul plans for the UK Central Electricity Generating Board, and it also produced house building plans of realistic complexity.

SUMMARY

System	NonLin
Reference	Tate (1977)
Plan structure	Action ordering, with explicit teleology (goal structure)
Parallel/serial	Parallel
Search space	Partial plans
Search control technique	One-then-best search

5.10 Deviser

Deviser (Vere, 1981, 1985) built on NonLin and extended it in many ways. Most notable is Deviser's ability to specify *time windows* as well as *durations* for actions and goals. An action start-time window is a triple of the form (E, I, L), where E is the action's earliest start time, I is its ideal start time, and L is its latest possible start time. A duration for an action is simply an account of how long the action will take to perform; a duration for a goal says how long the goal condition must be maintained. Deviser includes the time-window management routines required to manipulate a network consisting of actions and goals containing these temporal attributes.

Deviser breaks up "activities" into three sorts: *events* which are triggered by environmental circumstances; *inferences,* performable by the planner; and *actions,* which the actor in the plan can carry out. An inference in Deviser is a special sort of activity where the validity of the conditions it produces depends on the continuing truth of its preconditions. Deviser provides the required machinery to ensure that the preconditions of all inferences remain true.

Events are considered by Deviser to be spontaneous world changes. For instance, if the actor in a plan opens a valve on a tank containing water, an event will occur: the water in the tank, acted on by gravity, will drain from the tank. This sort of event is dealt with in Deviser by linking the specification of the event to an action which causes the holding of its preconditions. This link is achieved by the use of special *CONSECUTIVE preconditions on the event, made true by the preceding action. It is also possible in Deviser to describe scheduled external events, such as the delivery of some parts necessary to an assembly operation.

Vere uses information on *function literals* to improve efficiency. For instance, Deviser can exploit information about the predicate *color,* stating that no two literals $color(obj_1, c_1)$ and $color(obj_2, c_2)$ may coexist in the database if $obj_1 = obj_2$, unless $c_1 = c_2$. This basically expresses the constraint that each object must have one and only one color.

More recent work by Vere (1985) concentrates on techniques for attempting goals in different orders. His work makes it clear that partially ordered plans do not address the linearity assumption and that a partially ordered planner must still deal with the order in which it pursues its goals.

The Deviser system is also noteworthy by its application domain: it has been used to generate spacecraft command sequences for the NASA Voyager mission.

SUMMARY

System	Deviser
Reference	Vere (1981)
Plan structure	Action ordering
Parallel/serial	Parallel, with goal/activity windows and durations
Search space	Partial plans
Search control technique	Typed preconditions, goal splicing (see Vere, 1985)

5.11 MolGen

MolGen (Stefik, 1981a, 1981b) is notable not for its plan representation, but rather for its ability to perform object selection using least-commitment techniques. The major phases of MolGen's approach call for constraint formulation, constraint propagation, and constraint satisfaction. MolGen is interesting as the first planner to call attention to this method of object selection. The presentation of these ideas in MolGen preceded their application in other planners, such as SIPE and O-Plan.

The representation of time used by MolGen is a basic state-space model. In terms of describing time and change, MolGen is not very interesting. One of the areas for further research suggested by Stefik himself is the integration of least-commitment object selection with least-commitment action ordering.

SUMMARY

System	MolGen
Reference	Stefik (1981a, 1981b)
Plan structure	State space
Parallel/serial	Serial
Search space	Partial plans
Search control technique	Least-commitment object selection

5.12 SIPE

As in the case of Deviser, SIPE (Wilkins, 1984) is on the line of development rooted at NOAH and running through NonLin. SIPE extends the capabilities of NonLin in various ways. SIPE's major contributions are in resource management reasoning, provision of inferential operators, and user interaction. This latter feature is facilitated by the development of a declarative and descriptive task net formalism, similar to NonLin's TF. Like TF, Wilkins' net description formalism allows the user to easily give an account of an operator's teleology and condition types. Wilkins calls this teleological information *plan rationale*. The main goals of SIPE are easily understood from the derivation of its title: system for interactive planning and execution (SIPE) monitoring.

As with MolGen, SIPE can postpone commitment on the selection of objects by doing constraint satisfaction reasoning. Such reasoning calls for narrowing the set of objects which satisfy given requirements until all the requirements for the desired object are known. This can be contrasted with the standard technique of selecting an object arbitrarily and remaking the choice when the selected object is found to be unsatisfactory.

SIPE also includes some simple heuristics for reasoning about resources. The heuristics are used primarily to order actions in a developing plan. The ordering is performed on the basis of resource analysis: If one action requests a resource and does not return it, and another (parallel) action requests the same resource, and *does* return it, SIPE will order the nonconsuming action before the consuming one. Not all interaction problems fit this paradigm, but those that do are handled efficiently by SIPE. Unfortunately, SIPE can overconstrain the search space (by not considering interleavings within certain action clusters) and thus may fail to find a valid plan.

Inferential operators are also provided. Similar to Deviser, SIPE allows the user to write inferential operators which are used to derive action outcomes not explicitly stated in action descriptions. An example given by Wilkins (1984) uses an inferential operator to deduce some postconditions of a blocks-world move operator. But as noted by Wilkins when presenting the example, SIPE's deductive operators really provide only syntactic shorthand, and the result at the end of the planning day is essentially the same as would be achieved by STRIPS-like operators.

SUMMARY

System	SIPE
Reference	Wilkins (1984)
Plan structure	Action ordering
Parallel/serial	Parallel
Search space	Partial plans
Search control technique	Resource reasoning, also human-aided

5.13 PLANX10-D

The PLANX10-D system contains a plan generation component but is part of a larger cognitive psychology project concerned with *plan recognition*. PLANX10-D was applied to the problem of "commonsense planning," i.e., the problem of generating plans to deal with day-to-day activities and situations (Schmidt, 1985). As a planner it concentrates more on expressive plan representation than on search control.

PLANX10-D uses an action-ordering plan representation which is organized around the notion of goal reduction, i.e., the decomposition of goals based on operator preconditions. These plans are more expressive than standard partial orders. They allow constraints more general than the simple α before β; in particular, it is possible to constrain action ordering such that two actions cannot occur simultaneously. With a conventional partially ordered plan, this is impossible.

A state-space projection is also maintained to enable temporal reasoning. PLANX10-D maintains alternative "perspectives" on the plan, called *DSpaces*. There are four such DSpaces, describing object features, object roles and relations, action features, and action roles and relations. DSpaces help in critiquing solutions, verifying constraints, and revising plans.

SUMMARY

System	PLANX10-D
Reference	Sridharan and Bresina (1985)
Plan structure	Action ordering
Parallel/serial	Parallel
Search space	Partial plans
Search control technique	Best-first mixed initiative with user

5.14 PRS

The debate of what to call the procedural reasoning system (PRS) rages on (Georgeff, 1984; Georgeff, Lansky, and Bessiere, 1985; Georgeff and Lansky, 1986). Some claim it is a planner or at least a planning shell; others claim it is simply a graphical programming language. Whatever the final answer, PRS is an interesting system because of its commitment to real-time response and its expressive representation for plans. PRS uses a state-space (basically, a finite-state automata) formalism that allows description of iteration with termination and conditional execution. However, issues of representing underlying teleological structure have not been addressed.

SUMMARY

System	PRS
Reference	Georgeff and Lansky (1986)
Plan structure	State space
Parallel/serial	Parallel
Search space	Nil (does not generate plans)
Search control technique	Nil

5.15 O-Plan

O-Plan (Currie and Tate, 1985) is a planning system and overall architecture designed through experience gained with previous major planners. O-Plan stands for *open planning architecture,* and it is designed to exploit ideas about blackboard control architectures (Hayes-Roth and Hayes-Roth, 1979), least-commitment action ordering (Sacerdoti, 1975a), least-commitment object selection (Stefik, 1981a, 1981b), goal structure (Tate, 1977), and ideas on attaching temporal window information to actions in a plan (Vere, 1981). With regard to this last point, it is interesting that much of the metric temporal reasoning installed in the O-Plan architecture is borrowed from standard operations research theory (Bell and Tate, 1985). Because of this, O-Plan promises to be useful as a real planning system and more than a vehicle for experimentation. The architecture is designed to be sufficiently flexible as to act as a framework for the exploration of a variety of planning ideas, from AI *and* from other disciplines.

The current version of the O-Plan system focuses on dealing with realistic time and resource modeling, extending the basic window manipulation algorithms of Deviser, reasoning about consumable resources, and providing a limited reasoning ability for reusable resources. It is also being applied to the problem of scheduling the activities of communications and scientific satellites.

O-Plan is also extending NonLin's task formalism (TF) language. An example of O-Plan's TF is given in Fig. 6-10. This TF is used to describe a house-

```
schema goal_build_house;
  only_use_for_effects {goal} = true at 3;

  nodes 1 start   {null},
        2 finish  {null},
        3 action  {build house};

  orderings 1 ---> 3, 3 ---> 2;
endschema;

schema build;
  expands {build house};

  nodes     1 action {excavate, pour footers      },
            2 action {pour concrete foundations   },
            3 action {erect frame and roof        },
            4 action {lay brickwork               },
            5 action {finish roofing and flashing },
            6 action {fasten gutters and downspouts},
            7 action {finish grading              },
            8 action {pour walks, landscape       },
            9 action {install services            },
           10 action {decorate      };

  orderings 1 ---> 2,  2 ---> 3,  3 ---> 4,  4 ---> 5,
            5 ---> 6,  6 ---> 7,  7 ---> 8;

  conditions  supervised {footers poured       } at 2 from [1],
              supervised {foundations laid     } at 3 from [2],
              supervised {frame and roof erected} at 4 from [3],
              supervised {brickwork done       } at 5 from [4],
              supervised {roofing finished     } at 6 from [5],
              supervised {gutters etc fastened } at 7 from [6],
            unsupervised {storm drains laid    } at 7,
              supervised {grading done         } at 8 from [7];

  time_window start between 1~11/30/00 and 1~14/30/00 at 2,
              start between 1~12/00/00 and 1~14/00/00 at 3;
endschema;
```
FIGURE 6-10
An example of O-Plan's task formalism.

building domain. Briefly, the schema expands into three more primitive components. The first two are placeholders only and bracket the real action to be planned. This action is indicated by *build house* in the schema. The schema requires that *build house* occur between *start* and *finish*. The second schema, *build*, is one possibility for expanding this action. It contains 10 more primitive actions, constrained to execute in the order indicated by *orderings*. The *con-*

ditions section of the schema is used to provide an account of the expansion's teleology. This example is only presented to give a feeling for the language that a planner can use to specify problem domains. For more on the specific details of O-Plan's TF, see Currie and Tate (1985).

SUMMARY

System	O-Plan
Reference	Currie and Tate (1985)
Plan structure	Action ordering, with explicit teleology
Parallel/serial	Parallel
Search space	Partial plans
Search control technique	Least-commitment and opportunistic

6 RECURRENT AI PLANNING THEMES

A few themes are recurrent in AI planning. Unlike most other areas of AI, research has been based around a common set of "benchmark" problems and a growing armory of techniques. This has helped produce a (mildly) competitive spirit in the field and has been useful in fostering the production of realistic prototype planners. This section looks at some of these recurrent themes.

6.1 Search-Space Control

MEANS-ENDS ANALYSIS. To reduce the number of intermediate states considered, only those operators or activities that can satisfy some outstanding goal may be considered. These in turn can introduce (hopefully simpler) subgoals. This idea was introduced in GPS and used in many later system.

SEARCH REDUCTION THROUGH LEAST COMMITMENT. Some systems, especially ones which introduced important techniques, did not search through the possible alternatives at all (e.g., NOAH). Selections were made on the basis of the information available locally, and then a commitment to that solution path was made. Of course, this often means that some problems could not be solved. However, since these systems were typically demonstrated on particular applications for which one technique was most appropriate, the approach was often successful at arriving at a solution directly. Very often techniques demonstrated in isolation in this way were incorporated in later planners that could search alternatives and use the method alongside others, depending on its relevance.

DEPTH-FIRST BACKTRACKING. A simple method of considering alternative solution paths (especially when there are only a few to choose from due to the use of means-ends analysis, etc.) is to save the state of the solution at each point at which there are alternative ways to proceed and to keep a record of the alternative choices. The first is chosen, and the search continues. If there is any failure, the saved state at the last choice point is restored and the next alternative taken (if there are none, "backtracking" continues over previous decisions). Simple stack-based implementation techniques can be used for this process (e.g., as in Prolog).

BEAM SEARCH. This is a search method which considers all possible solutions within a preconstrained area so that they can be compared. Such a method is normally used when tight search-space constraints are known, and hence the solution space considered by the beam search is expected to be small. It is also normally employed along with other heuristic search methods so that a choice can be made of the "best" solutions found among the solutions to some subproblem which have been proposed by a beam search [e.g., in ISIS-II (Fox, Allen, and Strohm, 1981)].

ONE-THEN-BEST BACKTRACKING. Since often good local information is available to indicate the preferred solution path, it is appropriate to try the best choice indicated by local heuristic information before considering the many alternatives that may be available should the local choice prove faulty. Taken to the extreme, depth-first search gives something of the flavor of such a search strategy. However, gradual wandering from a valid solution path could entail backtracking through many levels when a failure is detected. An alternative is to focus on the choice currently being made and to try to select one of the local choices which seems most promising. This continues while all is going well (perhaps with some cutoff points to take a long, hard look at how well things are going). However, if a failure occurs, the *entire* set of alternatives which have been generated (and ranked by a heuristic evaluator) is reconsidered to select the refocusing point for the search [e.g., in NonLin (Tate, 1977)].

DEPENDENCY-DIRECTED SEARCH. It is well known that any backtracking system based on saved states and resumption points (whether depth-first or heuristically controlled) can waste much valuable search effort. There may be several unrelated parts to a solution. If backtracking on one part has to go back beyond points at which work was done on an unrelated part, all effort on the unrelated part will be lost. Examples of work that has explored this important area includes that by Stallman and Sussman (1977), NonLin + Decision Graph (Daniel, 1983) and MolGen (Stefik, 1981a, 1981b) to some extent. Some systems do not keep saved states of the solution at choice points. Instead they record the dependencies between decisions, the assumptions on which they are based, and the alternatives from which a selection can be made. Then these systems use methods of undoing a failure by propagating the undoing of all dependent parts of the solution. This leaves unrelated parts intact irrespective of whether they were worked on after some undone part of the solution. (See Hayes, 1975.)

OPPORTUNISTIC SEARCH. Some systems do not take a fixed (goal-driven or data-driven) directional approach to solving a problem. Instead, a current focus for the search is identified on the basis of the most constrained operation that can be performed. This may be suggested by comparison of the current goals with the initial world model state, by consideration of the number of likely outcomes of making a selection, by the degree to which goals are instantiated, etc. Any problem-solving component may summarize its requirements for the solution as constraints on possible solutions or restrictions of the values of variables representing objects being manipulated. The systems can then suspend their operation until further information becomes available on which a more definite choice can be made (e.g., MolGen). Many such systems operate with a "blackboard" through which the various components can communicate via constraint information [e.g., HEARSAY-II (Erman et al., 1980), and OPM (Hayes-Roth and Hayes-Roth, 1979)]. The scheduling of the various tasks associated with arriving at a solution may also be dealt with through (an area of) the blackboard.

METALEVEL PLANNING. A number of planning systems have an operatorlike representation of the different types of operations that a planner can perform. A separate search is made to decide which is best applied at any point; this happens before decisions are made about the details of the particular application plan being produced [e.g., MolGen (Stefik, 1981a, 1981b), and Wilensky (1981a)]. This technique is often used in opportunistic planners.

DISTRIBUTED PLANNING. Some systems have gone further in distributing the sources of problem-solving expertise or knowledge. They allow fully distributed planning with the subproblems being passed between specialized planning experts. The use of the experts or the type of things they can do may be controlled through a centralized blackboard and executive (with a system rather like priority scheduling of parallel processes) or may be controlled in a more distributed fashion via pairwise negotiation. Examples of relevant work include Smith's (1977) contract net, Corkill (1979), Kornfeld (1979), Konolige and Nilsson (1980), Georgeff (1982), and Corkill and Lesser (1983). Work is under way to interleave the time spent planning, sensing the state of the world to gather information, and execution [e.g., Doran and Trayner (1985); Drummond (1985a, 1985b)].

GENERAL PRUNING OF THE SEARCH SPACE. Besides the basic methods of reducing the search space by selection of relevant operators in means-end analysis, many other methods of reducing the search space size have been employed in planners:

- By considering "higher-priority" goals first in hierarchical planners [e.g., ABSTRIPS (Sacerdoti, 1973) and LAWALY (Siklossy and Dreussi, 1975)].
- By detecting and correcting for interactions in an intelligent fashion, e.g., Waldinger's (1975) goal regression and in Interplan (Tate, 1975), NOAH (Sacerdoti, 1975a), and NonLin (Tate, 1977).

- By rejection of states or plans that are known to be impossible or in violation of some rule about the state of plan [e.g., Warplan's *imposs* statements (Warren, 1974), Allen and Koomen's (1983) *domain constraints*, and Drummond and Currie's (1989) technique of *temporal coherence*].
- By checks on resource usage levels, time constraints on actions, etc. [e.g., NonLin, Deviser, and SIPE (Wilkins, 1983)].
- By "typing" the preconditions of operators. NonLin introduced the notion of an operator applicability precondition that the system should not attempt to achieve. Such a precondition served to constrain the search, since paths in the space corresponding to the introduction of operators to achieve the typed condition would never be considered. Other condition types were also used in NonLin to reduce the search space.

6.2 Executing Plans of Action

There have been some limited successes with attempts at closing the loop between planning and execution [Fikes, Hart, and Nilsson (1972b); Wilkins (1985); Smithers and Malcolm (1989); Drabble (1988)], but often the plans generated were rather limited and not very flexible. In general, the complexities of the individual tasks of plan representation, generation, execution monitoring, and repair have led to research into each of these areas separately. In particular, there is now a mismatch between the scale and capabilities of plan representations proposed for real-time execution systems [Nilsson (1988), Georgeff and Lansky (1986), and Kaelbling (1988)] and those that can be generated by AI planners.

However, the demand is for a system that can take a command request, generate a plan, execute it, and react to simple failures of that plan, either by repairing it or by replanning. Explicit knowledge about the structure of the plan, the contribution of the actions involved, and the reasons for performing plan modifications at various stages of the plan construction process provides us with much of the information required for dealing with plan failures.

The use of dependencies within planning promises great benefits for the overall performance of a planning system, particularly for plan representation, generation, execution, and repair.

Early work on decision graphs by Hayes (1975) at Edinburgh has shown how the explicit recording of the decisions involved in the planning process could be used for suggesting where and how much replanning should take place when unforeseen situations invalidate the success of the current plan. Some work to link these ideas with a parallel AI planner was undertaken during the mid-1970s by Daniel (1977).

As mentioned above, plan teleology, in the form of goal structure (Tate, 1977) or plan rationale (Wilkins, 1984), expresses the dependencies between the preconditions and postconditions of operators involved in the plan. Although such dependencies have been shown to be useful for describing the internal structure of the plan and for monitoring the execution of the plan [Fikes,

Hart, and Nilsson (1972a); Tate (1984a)], there has been no comprehensive discussion of their use in all aspects of plan generation, execution monitoring, and plan repair.

6.3 Exploiting Hierarchy and Abstraction Levels

HIERARCHY: STRICT SEARCH BY LEVELS. The early hierarchical systems [e.g., ABSTRIPS (Sacerdoti, 1973), LAWALY (Siklossy and Dreussi, 1975), NOAH (Sacerdoti, 1977)] formed a solution at the most abstract level and then made a commitment to this solution. The lower levels were then planned by using the preset skeleton plan formed at the upper levels. No backtracking to the higher levels was possible.

HIERARCHY: NONSTRICT BY LEVELS. Later systems [e.g., NonLin (Tate, 1977)] treat the abstraction levels as a guide to a skeleton solution, but are able to replan or consider alternatives at any level if a solution cannot be found or if problems with part of a solution indicate that a higher-level choice was faulty.

HIERARCHY: OPPORTUNISTIC BY LEVELS. Other hierarchical systems use the abstraction levels as one guide to the ordering of goals, but have other mechanisms that can be considered as well. Some systems are able to determine when a particular choice (at whatever level) is sufficiently constrained to be a preferable goal to work on at any time [e.g., MolGen (Stefik, 1981a, 1981b)].

6.4 Conditionals and Iterators

Most of the search-space control techniques, goal ordering, and interaction correction mechanisms developed in AI planners to date have been oriented toward the generation of plans which are fully or partially ordered sets of primitive actions. However, there has been some limited effort on the generation of plans which contain conditionals ("if...then...else...") and iterators ("repeat...until...").

BLACK-BOX APPROACH TO CONDITIONALS AND ITERATORS. A conditional or an iterator which can be modeled as a single entity and for which the differences between the multiple paths or the individual loops are not important can be handled by many of the planning systems and techniques already described. NOAH had a simple way of dealing with such packaged conditionals and iterators. SIPE is able to deal with iteration which repeated a planned action sequence on each member of a set of parameters (which could be objects or points on an aircraft flight trajectory, for instance). This latter sort of iteration happens at "plan time" and can be considered to be loop unrolling, in the sense that the possibilities must be enumerable while the plan is being generated.

CONDITIONALS: BRANCH AND CASE ANALYSIS. Conditionals were handled in Warplan-C (Warren, 1976) by branching the plan at the conditional and performing separate planning on the two branches, using the assumption that the statement in the conditional was true in one branch and false in the other. This led to a case analysis of the separate branches to produce a plan that was tree-structured. Notice that this sort of plan cannot be represented as a partially ordered set of actions.

CONDITIONALS: BRANCH AND REJOIN. There has been relatively little work in the mainstream AI planning literature which has attempted a full treatment of conditionals which can branch and rejoin later in the plan and for which the black-box approach was not sufficient. However, work on automatic programming and theorem proving has considered this area. See, e.g., Luckham and Buchanan (1974) and Dershowitz (1985). The latter reference has an extensive bibliography of papers on automatic programming and shows how AI planning research overlaps with this work.

6.5 Time and Resource Handling

It is becoming increasingly important in planning systems to perform on realistic applications where resources of various types are limited. Also planners are being used in domains where time considerations and matters beyond the control of the planning system itself must be accounted for. Several systems have explored this area.

COMPUTE TIME AND RESOURCE USAGE. In a project planning domain, NonLin maintained information about the durations of the various activities and used this to compute earliest and latest start times for each action. A critical path of actions could be found. A "cost" measure could also be kept with each activity. This information was used in an extension to the NonLin planner that could selectively remove costly activity or lengthy activities from a plan and replace them with others that reduced either the cost or the duration depending on the limitation that was exceeded (see Daniel, 1983). More recent work on NonLin (Tate and Whiter, 1984) has added the capability of representing multiple limited resources and making selection from appropriate activities on the basis of reducing some overall computed "preference" between them.

OBJECTS AS RESOURCES. SIPE is able to reason about competing demands for the use of scarce objects (such as a ladder or a key). An analysis of the objects used and returned by actions or objects "consumed" by them can be used to linearize the plan to prevent usage conflicts.

EVENT AND TIME SPECIFICATIONS FOR ACTIONS. Deviser allows a time window to be specified for any goal or action. External events can be described as having some effect at a particular time. The planner propagates the temporal links between these time windows, progressively narrowing them as they become constrained by other actions. It can detect when some plan step prevents a goal from being achieved or an activity from being executed at the required time. In such a case, backtracking will consider alternative solutions. Airplan (Masui, McDermott, and Sobel, 1983) was able to reason about time intervals and how actions suggested in them could interfere.

O-Plan (Bell and Tate, 1985) uniformly represents time constraints (and resource usage) by a numeric (*min, max*) pair which bound the actual values for activity duration, activity start and finish times, and delays between activities. The actual values may be uncertain for various reasons, such as the plan's being at a high abstraction level, not having chosen values for objects referred to in the plan, or uncertainty in modeling the domain. Constraints can be stated on the time and resource values which may lead to the planner's finding that some plans are invalid.

FLEXIBLE TIME HANDLING. Much more flexible handling of the propagation of temporal constraints between the steps of a plan is being considered in the widespread research effort on temporal logic [e.g., McDermott, 1982; Allen and Koomen, 1983)].

SOFT CONSTRAINTS. ISIS-II (Fox, Allen, and Strohm, 1981) allows a wide variety of constraints on the problem to be specified. Some can be in the form of preferences, or "soft" constraints, which are used to guide the search for acceptable solutions.

6.6 Support Languages and Data Management Systems

Some early planning systems were implemented in specially tailored languages designed to allow search and pattern-directed inference [e.g., PLANNER (Hewitt, 1972), POPLER (Davies, 1973), Conniver (Sussman and McDermott, 1972)]. The depth-first search strategy of the logic programming language Prolog (Clocksin and Mellish, 1981) was used in Warplan (Warren, 1974), probably the first planner to be implemented in that language. In other cases a base AI programming language was augmented by suitable language extensions [e.g., QA4 (Rulifson, Derkson, and Waldinger, 1972) and QLISP (Sacerdoti et al., 1976)].

However, most systems have been implemented in a combination of an AI language with a knowledge representation language, such as UNITS (Stefik, 1979) in LISP. Separate research and development on knowledge representation languages—such as KRL (Bobrow and Winograd, 1976), PEARL (Deering et al, 1981), LOOPS (Bobrow and Stefik, 1983), KEE (Intellicorp,

1985), SRL (Fox, 1983), Knowledge Craft (Carnegie Group, 1987), and ART (Inference Corporation, 1985), and large content-addressable structured memories, such as NETL (Fahlman, 1979) and FACT (McGregor and Malone, 1981), are important sources of support for new planning work.

7 CONCLUSION

Although the 1970s were productive years for AI planning, little applications work in the 1980s has capitalized on early results (see Fig. 6-11). Of course, some projects have taken the technical successes of the early AI planners and applied them to large, complex domains. The problem is that in many cases the search spaces are simply too large to be controlled automatically; when this happens, the search problem can be managed by a human user. In this context the planner becomes a decision support system, tracking the effects of decisions made by the user. There is still much work to be done in the application of AI planning research to realistic domains.

Work in AI planning must be integrated with other fields, such as operations research, control theory, and decision theory. The results could be impressive. AI has expressive representations for time and change. Many of these representations come complete with a theory of plan synthesis. But, contrary to some opinion, AI does not have all the answers. Operations research can provide tools to analyze plans. Control theory can provide a view of plan execution as closed-loop feedback control. Decision theory can better inform the decision of when to plan and when to act. Together, these areas of research can usefully address the synthesis and management of complex action in realistic domains.

ACKNOWLEDGMENTS

The authors would like to thank Mike Uschold of the Artificial Intelligence Applications Institute at Edinburgh University, and Mark Greenwood, of International Computers Ltd., for their careful reading and helpful comments on an early draft. Mike has always been around to talk about planning, and his view on things helped shape this document. Ken Currie has been a constant source of information and discussion; much of this work would not have existed without his help. In addition, we would like to thank John Bresina, Dave Thompson, Steve Minton, Susan Rose, and Kevin Thompson for comments on this, the most recent draft of an evolving document. Remaining opinions, errors, and omissions are, of course, the authors' responsibility.

8 BIBLIOGRAPHY

A two-part bibliography follows. The first part is concerned with pointers into core planning work, both historical and contemporary. The second part deals with secondary issues such as programming languages and database facilities.

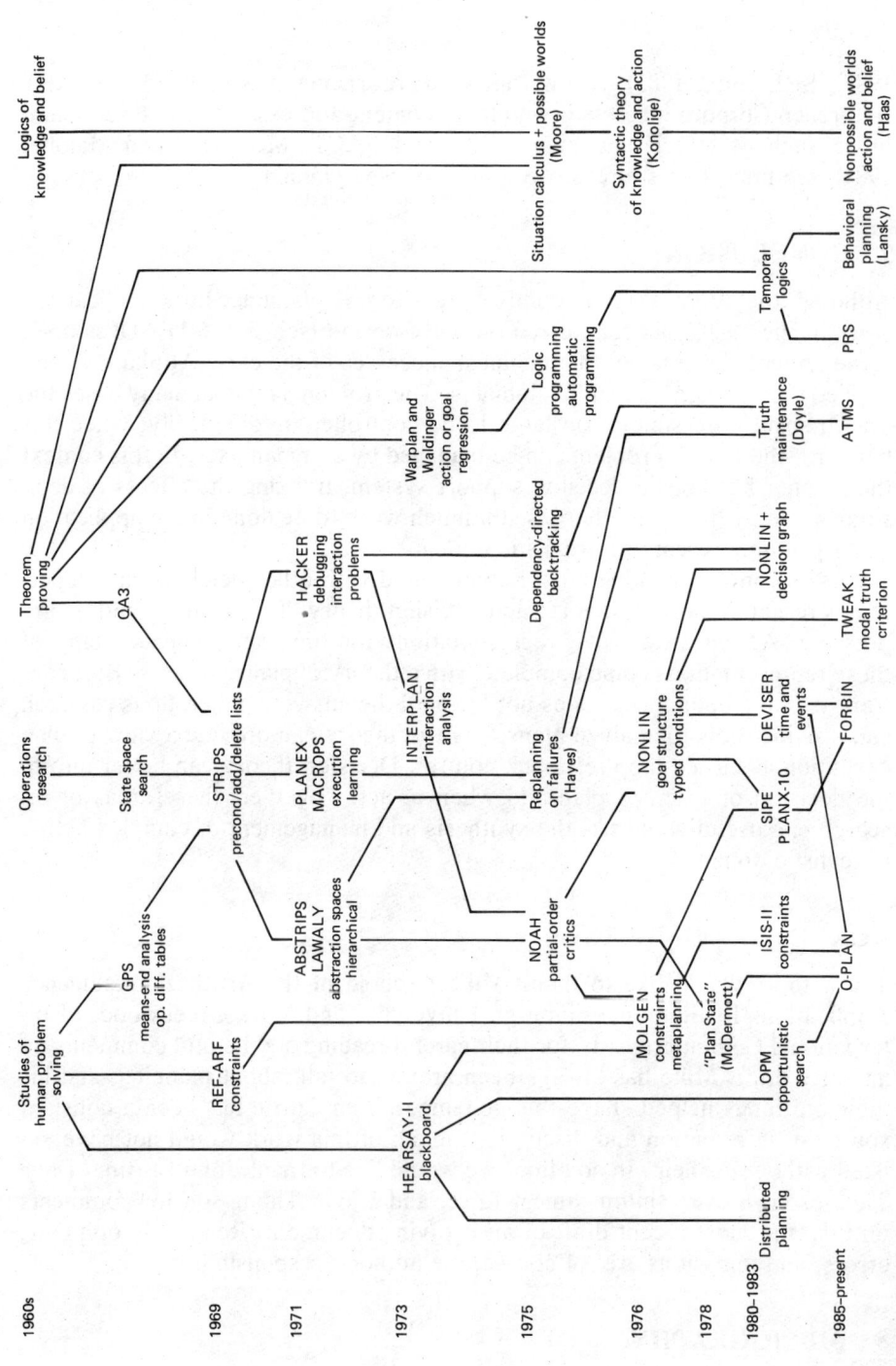

FIGURE 6-11
A taxonomy of AI planning systems and related techniques.

REFERENCES

Basic planning research

Allen, J. 1981 An interval-based representation of temporal knowledge. *Proc. of the International Joint Conference on Artificial Intelligence (IJCAI).* Vancouver, B.C., Canada.
Allen, J., and J. Koomen. 1983. Planning using a temporary world model. *Proc. of the 8th IJCAI.* Karlsruhe, West Germany, pp. 741–747. [TIMELOGIC]
Appelt, D. 1985. Planning English referring expressions. *Artificial Intelligence,* vol. 26, pp. 1–33. [KAMP]
Barr, A., and E. Feigenbaum. 1981. *The Handbook of Artificial Intelligence,* vol. 1, William Kaufmann Inc., Los Altos, Calif.
Barr, A., and E. Feigenbaum. 1982. *The Handbook of Artificial Intelligence,* vol. 3, William Kaufmann Inc., Los Altos, Calif.
Bell, C. E., and A. Tate. 1985. Using temporary constraints to restrict search in a planner. *Proc. of the Third Workshop of the Alvey IKBS Programme Planning Special Interest Group,* Sunningdale, Oxfordshire, UK. Available through the Institute of Electrical Engineers, London. [O-PLAN]
Brauer, W. (ed.). 1979. Springer-Verlag LNCS series. Net theory and applications. *Proc. of the Advanced Course on General Net Theory of Processes and Systems,* Hamburg, West Germany.
Bresina, J. L. 1981. An interactive planner that creates a structured annotated trace of its operation. Rutgers University, Computer Science Research Laboratory, Rep. CBM-TR-123. [PLANX10]
Chapman, D. 1985. Nonlinear planning: A rigorous reconstruction. *Proc. of IJCAI-9,* pp. 1022–1024. [TWEAK]
Chapman, D. 1986. Planning for conjunctive goals. M.I.T. AI Lab. Tech. Rep. 802. [TWEAK]
Charniak, E., and D. McDermott. 1985. *Introduction to Artificial Intelligence.* Addison-Wesley, Reading, Mass.
Cheeseman, P. 1983. A representation of time for planning. Tech. Note 278, SRI International, Menlo Park, Calif.
Corkill, D. D. 1979. Hierarchical planning in a distributed environment. *Proc. IJCAI-79,* pp. 168–175, Tokyo.
Corkill, D. D., and V. R. Lesser. 1983. The use of meta-level control for coordination in a distributed problem solving network. *Proc. IJCAI-83,* pp. 748–756, Karlsruhe, West Germany.
Currie, K., and A. Tate. 1985. O-Plan: Control in the open planning architecture. *Proc. of the BCS Expert Systems '85 Conference.* Warwick (December), Cambridge University Press.
Czaja, L. 1983. Making nets abstract and structured. Oxford University Computing Laboratory, Programming Research Group, 8-11 Keble Road, Oxford, U.K.
Daniel, L. 1977. Planning: modifying non-linear plans. Working Paper 24, Department of AI, Edinburgh University. [NONLIN]
Daniel, L. 1983. Planning and operations research. *Artificial Intelligence: Tools, Techniques and Applications.* Harper & Row, New York. [NONLIN]
Davis, P. R. and R. T. Chien. 1977. Using and re-using partial plans. *Proc. IJCAI-77.* Cambridge, Mass.
Davis, R., and R. Smith. 1983. Negotiation as a metaphor for distributed problem solving. *Artificial Intelligence* vol. 20, pp. 63–109.
Dean, T. 1985. Temporal reasoning involving counterfactuals and disjunctions. *Proc. IJCAI-85.* Los Angeles. [TMM]
Dershowitz, N. 1985. Synthetic programming. *Artificial Intelligence* vol. 25, pp. 323–373.
Doran, J. E., and D. Michie. 1966. Experiments with the graph traverser program. *Proc. of the Royal Society,* vol. A, pp. 235–259. [GRAPH TRAVERSER]
Doran, J. E., and C. Trayner. 1985. Distributed planning and execution—teamwork 1. *Computer Science Technical Report.* University of Essex, U.K.
Doyle, R. J., D. J. Atkinson, and R. S. Doshi. 1986. Generating perception requests and expectations to verify the execution of plans. *Proc. of AAAI-86.* Philadelphia.

Drabble, B. 1988. Intelligence execution monitoring and error analysis in planning involving processes. Ph.d. thesis, University of Aston in Birmingham.

Drummond, M. 1985a. Non-linear planning in terms of cause and effect. In: *Proc. of SCS multiconference: AI, Graphics and Simulation.* San Diego (also Department AI Research Paper 251, Edinburgh University).

Drummond, M. 1985b. Refining and extending the procedural net. *Proc. of IJCAI-85*, pp. 1010–1012.

Drummond, M. 1986. A representation of action and belief for automatic planning systems. *Proc. of CSLI/AAAI Workshop on Planning and Action*, Oregon (also Artificial Intelligence Applications Institute, Edinburgh University, Tech. Rep. AIAI-TR-16).

Drummond, M., and K. W. Currie. 1988. Exploiting temporal coherence in non-linear plan construction. *Computational Intelligence Journal*, vol. 4 no. 4, pp. 341–348.

Drummond, M., K. W. Currie, and A. Tate. 1987. Contingent plan structures for spacecraft. *Proc. of the JPL/NASA Workshop on Telerobotics*, Pasadena, Calif.

Duffay, P. and J-C. Latombe. 1983. *An Approach to Automatic Robot Programming Based on Inductive Learning.* IMAG, Grenoble, France. [TROPIC]

Erman, L. D., F. Hayes-Roth, V. R. Lesser, and D. R. Reddy. 1980. The HEARSAY-II speech-understanding system: Integrating knowledge to resolve uncertainty. *ACM Computing surveys*, vol. 12, no. 2.

Ernst, G., and A. Newell. 1969. *GPS: A Case Study in Generality and Problem Solving.* Academic Press, New York.

Fahlman, S. E. 1974. A Planning system for robot construction tasks. *Artificial Intelligence*, vol. 5, pp. 1–49.

Faletti, J. 1982. PANDORA—A program for doing commonsense reasoning planning in complex situations. *Proc. of AAAI-82.* Pittsburgh.

Fikes, R. E. 1970. REF-ARF: A system for solving problems stated as procedures. *Artificial Intelligence*, vol. 1, pp. 27–120.

Fikes, R. E. 1982. A commitment-based framework for describing informal cooperative work. *Cognitive Science*, vol. 6, pp. 331–347.

Fikes, R. E., P. E. Hart, and N. J. Nilsson. 1972a. Learning and executing generalized robot plans. *Artificial Intelligence*. vol. 3. [STRIPS/PLANEX]

Fikes, R. E., P. E. Hart, and N. J. Nilsson. 1972b. Some new directions in robot problem solving. *Machine Intelligence*, vol. 7 (B. Meltzer, and D. Michie, eds.), Edinburgh University Press. [STRIPS]

Fikes, R. E., and N. J. Nilsson. 1971. STRIPS: A new approach to the application of theorem proving to problem solving. *Artificial Intelligence*, vol. 2, pp. 189–208.

Fox, M. S., B. Allen, and G. Strohm. 1981. Job shop scheduling: An investigation in constraint-based reasoning. *Proc. IJCAI-81.* Vancouver, B. C., Canada, August. [ISIS-II]

Georgeff, M. 1983. Communication and interaction in multi-agent planning systems. *Proc. of AAAI-83*, pp. 125–129.

Georgeff, M. 1984. A theory of action for multiagent planning. *Proc. of AAAI-84*, Austin, pp. 121–125.

Georgeff, M., and A. Lansky. 1986. Procedural knowledge. *Proc. of IEEE: Special Issue on Knowledge Representation.* vol. 74, pp. 1383–1398. [PRS]

Georgeff, M., A. Lansky, and P. Bessiere. 1985. A procedural logic. *Proc. of IJCAI-85*, pp. 516–523, Los Angeles.

Green, C. C. 1969. Theorem-proving by resolution as a basis for question-answering systems. In: *Machine Intelligence*, vol. 4 (B. Meltzer and D. Michie, eds.), American Elsevier Publishing Co., New York. [QA3]

Haas, A. R. 1986. A syntactic theory of belief and action. *Artificial Intelligence*, vol. 28, pp. 245–292.

Hayes, P. J., 1975. A representation for robot plans. In: *The Advance Papers of IJCAI-75*, Tbilisi, U.S.S.R.

Hayes, P. 1981. The frame problem and related problems in artificial intelligence. In: *Readings in Artificial Intelligence* (B. L. Webber and N. J. Nilsson, eds.), Tioga Publishing Co., Palo

Alto, Calif. pp. 223–230. Originally appeared in *Artificial and Human Thinking*. (A. Elithorn and D. Jones, eds.), Jossey-Bass Inc. and Elsevier Scientific Publishing Company, 1973, pp. 45–59.

Hayes-Roth, B. 1983a. The blackboard architecture: A general framework for problem solving? Heuristic programming project, Rep. HPP-83-30. Stanford University, Stanford, Calif.

Hayes-Roth, B. 1983b. A Blackboard model of control. Heuristic Programming Project, Rep. HPP-83-38. Stanford University, June. [OPM]

Hayes-Roth, B., and F. Hayes-Roth. 1979. A cognitive model of planning. *Cognitive Science*, pp. 275–310. [OPM]

Hendrix, G. 1973. Modeling simultaneous actions and continuous processes. *Artificial Intelligence*, vol. 4, pp. 145–180.

Hewitt, C. 1972. Description and theoretical analysis (using schemata) of PLANNER: A language for proving theorems and manipulating models in a robot. Ph.d. thesis, Cambridge, Mass.

Kaelbling, L. P. 1988. Goals as parallel program specifications. *Proc. of AAAI-88*, St. Paul, Minn., pp. 60–65.

Kahn, K., and G. A. Gorry, 1977. Mechanizing temporal knowledge. *Artificial Intelligence*, vol. 9, pp. 87–108.

Konolige, K. 1980. A first order formalization of knowledge and action for a multiagent planning system. Artificial Intelligence Center, Tech. Note 232, SRI International, Menlo Park, Calif. (also in *Machine Intelligence*, vol. 10, 1981).

Konolige, K. 1983. A deductive model of belief. *Proc. of IJCAI-83*, pp. 377–381, Karlsruhe, West Germany.

Konolige, K., and N. J. Nilsson. 1980. Multi-agent planning systems. *Proc. of AAAI-80*, pp. 138–142, Stanford, Calif.

Kornfeld, W. A. 1979. ETHER: A parallel problem solving system. *Proc. of IJCAI-79*, pp. 490–492, Tokyo.

Kowalski, R. 1979. *Logic for Problem Solving*. North-Holland.

Kowalski, R., and M. Sergot. 1984. A logic-based calculus of events. Alvey IKBS planning special interest group, 3rd Planning workshop, Sunningdale, U.K. (available from IEE, London).

Lansky, A. 1985. Behavioral specification and planning for multiagent domains. Tech. Note 360, SRI International, Menlo Park, Calif.

Latombe, J-C. 1976. Artificial intelligence in computer-aided design—the TROPIC system. Stanford Research Institute AI Center Tech. Note 125, Menlo Park, Calif.

Lenat, D. B. 1975. BEINGS: Knowledge as interacting experts. *Proc. of IJCAI-75*, pp. 126–133, Tbilisi, U.S.S.R. [PUP]

London, P. 1977. A dependency-based modeling mechanism for problem solving. Dept. of Computer Science, University of Maryland, Memo. TR-589.

Luckham, D. C., and J. R. Buchanan. 1974. Automatic generation of programs containing conditional statements. AISB Summer Conference, University of Sussex, UK, pp. 102–126.

McCarthy, J. 1958. Programs with common sense. In: *Semantic Information Processing* (M. Minsky, ed.), pp. 403–418. M.I.T. Press, Cambridge, Mass.

McCarthy, J., and P. J. Hayes. 1969. Some philosophical problems from the standpoint of artificial intelligence. *Machine Intelligence*, vol. 4 (B. Meltzer and D. Michie, eds.) pp. 463–502. Edinburgh University Press.

McDermott, D. V. 1978. Planning and acting. *Cognitive Science*, vol. 2.

McDermott, D. V. 1982. A temporal logic for reasoning about processes and plans. *Cognitive Science*, vol. 6, pp. 101–155.

McDermott, D. 1983. Generalizing problem reduction: A logical analysis. *Proc. of IJCAI-83*, pp. 302–308.

McDermott, D. V., and J. Doyle. 1979. An introduction to non-monotonic logic. *Proc. of IJCAI-79*, pp. 562–567, Tokyo.

Masui, S., J. McDermott and A. Sobel, 1983. Decision-making in time critical situations. *IJCAI-83*, pp. 233–235, Karlsruhe, West Germany. [AIRPLAN]

Miller, D., J. Firby, and T. Dean. 1985. Deadlines, travel time, and robot problem solving. *Proc. of IJCAI-85*, pp. 1052–1054. [FORBIN]

Miller, G. A., E. Galanter, and K. H. Pribram. 1960. *Plans and the Structure of Behavior*. Henry Holt and Company, New York.
Moore, R. C. 1977. Reasoning about knowledge and action. *Proc. of IJCAI-77*, pp. 223–227.
Moore, R. C. 1980. Reasoning about knowledge and action. Artificial Intelligence Center, Tech. Note 191, SRI International, Menlo Park, Calif.
Moore R. C. 1985. A formal theory of knowledge and action. In: J. R. Hobbs and R. C. Moore (eds.), *Formal Theories of the commonsense World*. Ablex series in artificial intelligence. Ablex Publishing Company, Norwood, New Jersey.
Mostow, D. J. 1983. A problem solver for making advice operational. *Proc. of AAAI-83*, pp. 179–283.
Newell, A., and H. A. Simon. 1963. GPS: A program that simulates human thought. In: E. A. Feigenbaum, and J. Feldman. (eds.), *Computers and Thought*. McGraw-Hill, New York. [GPS]
Nielsen, M., and P. S. Thiagarajan. 1984. Degrees of non-determinism and concurrency: A Petri net view. *Proc. of the 4th Conference on Foundations of Software Technology and Theoretical Computer Science,* December.
Nilsson, N. 1980. *Principles of Artificial Intelligence*. Tioga Publishing Co., Palo Alto, Calif.
Nilsson, N. 1985. Triangle tables: A proposal for a robot programming language. Artificial intelligence Center, Tech. Note 347, SRI International, Menlo Park, Calif.
Nilsson, N. J. 1988. Action networks. *Proc. of the Rochester Planning Workshop*.
Pednault, E. P. D. 1985. Preliminary report on a theory of plan synthesis. Artificial Intelligence Center, Tech. Note 358, SRI International, Menlo Park, Calif.
Peterson, J. L. 1977. Petri nets. *Computing Surveys*, vol. 9, no. 3, September.
Peterson, J. L. 1980. *Petri Net Theory and the Modeling of Systems*. Prentice-Hall, Englewood Cliffs, N.J.
Pippenger, N. 1980. Pebbling. RC 8258 (#35937), IBM TJ Watson Research Center, Yorktown Heights, N.Y.
Reiger, C., and P. London. 1977. Subgoal protection and unravelling during plan synthesis. *Proc. of IJCAI-77*. Cambridge, Mass.
Reisig, W. 1985. *Petri nets: An Introduction,* vol. 4. Springer-Verlag, EATCS monographs on theoretical computer science.
Rich, C. 1981. A formal representation for plans in the programmer's apprentice. *Proc. of IJCAI-81*, pp. 1044–1052.
Rich, C., H. E. Shrobe, and R. C. Waters. 1979. Overview of the programmer's apprentice. *Proc. of IJCAI-79*, pp. 827–828, Tokyo.
Rosenschein, S. J. 1980. Synchronization of multi-agent plans. *Proc. of AAAI-80*.
Rosenschein, S. J. 1981. Plan synthesis: A logical perspective. *Proc. of IJCAI-81*. Vancouver, B. C., Canada.
Sacerdoti, E. D. 1973. Planning in a hierarchy of abstraction spaces. In: *Advance Papers of IJCAI-73*. Palo Alto, Calif. [ABSTRIPS]
Sacerdoti, E. D. 1975a. The non-linear nature of plans. *Proc. of IJCAI-75*. [NOAH]
Sacerdoti, E. D. 1975b. A structure for plans and behavior. Stanford research Institute, Tech. Note 109. [NOAH]
Sacerdoti, E. D. 1977. *A Structure for Plans and Behavior*. American Elsevier. Also SRI AI Tech. Note 109, August 1975. [NOAH]
Sacerdoti, E. 1979. Problem solving tactics. *Proc. of IJCAI-79*, pp. 1077–1085.
Sacerdoti, E. D. 1980. Plan generation and execution for robots. Stanford Research Institute, Tech. Note 209.
Sathi, A., M. S. Fox, and M. Greenberg. 1985. Representation of activity knowledge for project management. *IEEE Special Issue of Transactions on Pattern Analysis and Machine Intelligence,* July. [CALLISTO]
Schank, R. C., and R. P. Abelson. 1977. *Scripts, Plans, Goals and Understanding*. Lawrence Erlbaum Press, Hillsdale, N.J.
Schmidt, C. G. 1985. Partial provisional planning: Some aspects of commonsense planning. In: J.

R. Hobbs and R. C. Moore, (eds.), *Formal Theories of the Commonsense World*, chap. 6, pp. 227–250. Albex, Norwood, N.J.

Siklossy, L., and J. Dreussi. 1975. An efficient robot planner that generates its own procedures. *Proc. IJCAI-73*, Palo Alto, Calif. [LAWALY]

Siklossy, L, and J. Roach. 1973. Proving the impossible is impossible is possible: Disproofs based on hereditary partitions. *Proc. of IJCAI-73*, Palo Alto, Calif. [DISPROVER/LAWALY]

Simon, H. A. 1983. Search and reasoning in problem solving. *Artificial Intelligence*, no. 21, pp. 7–29.

Smith, R. G. 1977. The contract net: A formalism for the control of distributed problem solving. *Proc. of IJCAI-77*, Cambridge, Mass.

Smith, R. G. 1979. A framework for distributed problem solving. *Proc. of IJCAI-79*. Tokyo.

Smithers, T., and C. Malcolm. 1989. Programming assembly robots in terms of task achieving behavioural modules. *Journal of Structural Learning*, vol. 2, no. 10.

Sridharan, A., and A. Bresina. 1982. Plan formation in large, realistic domains. In: *Proc. of 4th Biennial Conf. of Canadian Society for Computational Studies of Intelligence* (CSCSI), Saskatoon, Sask.

Sridharan, A., and J. L. Bresina. 1985. Knowledge structures for planning in realistic domains. *Computers and Mathematics with Applications (Special Issue on Knowledge Representation)*, vol. 11, no. 5, pp. 457–480.

Stallman, R. M., and G. J. Sussman. 1977. Forward reasoning and dependency directed backtracking. *Artificial Intelligence*, vol. 9, pp. 135–196.

Steele, G. L., and G. J. Sussman. 1978. Constraints. M.I.T. AI Lab. Memo. 502, Cambridge, Mass.

Stefik, M. J. 1981a. Planning with constraints. *Artificial Intelligence*, vol. 16, pp. 111–140. [MOLGEN]

Stefik, M. J. 1981b. Planning and meta-planning. *Artificial Intelligence*, vol. 16, pp. 141–169. [MOLGEN]

Sussman, G. A. 1973. A computational model of skill acquisition. M.I.T. AI Lab. Memo. AI-TR-297, Cambridge, Mass. [HACKER]

Tate, A. 1974. Interplan: A plan generation system which can deal with interaction between goals. MIP-R-109, Machine intelligence research unit, Edinburgh University.

Tate, A. 1975. Interacting goals and their use. In: *Proce. of IJCAI-75*, pp. 215–218, Tbilisi, U.S.S.R. [INTERPLAN]

Tate, A. 1976. Project planning using a hierarchical non-linear planner. Dept. of Artificial Intelligence Rep. 25, Edinburgh University. [NONLIN]

Tate, A. 1977. Generating project networks. *Proc. of IJCAI-77*, Boston. [NONLIN]

Tate, A. 1984a. Goal structure: Capturing the intent of plans. *Proc. of ECAI-84*, Pisa, Italy, September. [NONLIN]

Tate, A. 1984b. Planning and condition monitoring in a FMS. *Proc. of International Conference on Flexible Automation Systems*, Institute of Electrical Engineers, London, July. [NONLIN]

Tate, A. 1985. A review of knowledge-based planning techniques. *Proc. of Expert systems 85* (5th technical Conference of the British Computer Society, Specialist group on expert systems). University of Warwick, pp. 89–111. Cambridge University Press.

Tate, A., and A. M. Whiter. 1984. Planning with multiple resource constraints and an application to a naval planning problem. *Proc. of First Conference on the Applications of Artificial Intelligence*, Denver, December. [NONLIN]

Tenenberg, J. 1986. Planning with abstraction. *Proc. of AAAI-86*, Philadelphia, pp. 76–80.

Vere, S. A. 1981. Planning in time: Windows and durations for activities and goals. *IEEE Transactions on Pattern Analysis and Machine Intelligence*, vol. PAMI-5, no. 3, pp. 246–267. [DEVISER]

Vere, S. A. 1985. Splicing plans to achieve misordered goals. *Proc. of IJCAI-85*, pp. 1016–1021. [DEVISER]

Vilain, M. B. 1980. A system for reasoning about time. *Proc. of AAAI-80*.

Waldinger, R. 1977. Achieving several goals simultaneously. In: E. W. Elcock and D. Michie (eds.) *Machine Intelligence,* vol. 8, Halstead/Wiley, New York.
Warren, D. H. D. 1974. Warplan: A System for generating plans. Memo 76, Computational Logic Dept., School of Artificial Intelligence, University of Edinburgh. [WARPLAN]
Warren, D. H. D. 1976. Generating conditional plans and programs. *Proc. of AISB Summer Conference,* pp. 344–354, University of Edinburgh, July [WARPLAN-C]
Wegner, P. 1970. Programming language semantics. In: R. Rustin (ed.), *Formal Semantics of Programming Languages.* Prentice-Hall, Englewood Cliffs, N.J., pp. 149–248.
Wilensky, R. 1978. Understanding goal-based stories. Dept. of Computer Science, Yale University, Research Rep. 140.
Wilensky, R. 1981a. Meta-planning: Representing and using knowledge about planing in problem solving and natural language understanding. *Cognitive Science,* vol. 5, pp. 197–233.
Wilensky, R. 1981b. A model for planning in complex situations. Electronics Research Lab. Memo. UCB/ERL M81/49, University of California, Berkeley.
Wilensky, R. 1983. *Planning and Understanding.* Addison-Wesley, Reading, Mass.
Wilkins, D. E. 1983. Representation in a domain independent planner. *Proc. of IJCAI-8,* pp. 733–740. [SIPE]
Wilkins, D. E. 1984. Domain independent planning: Representation and plan generation. *Artificial Intelligence,* no. 22. [SIPE]
Wilkins, D. E. 1985. Recovering from execution errors in SIPE. *Computational Intelligence,* vol. 1, pp. 33–45.
Wilkins, D. E., and A. E. Robinson. 1981. An intertactive planning system. SRI Tech. Note 245. [SIPE]

Support languages and databases

Barrow, H. G. 1975. HBASE: A fast clean efficient data base system. D.A.I. POP-2 library documentation. Edinburgh University. [HBASE]
Bobrow, D. G., and M. J. Stefik. 1983. *The LOOPS Reference Manual.* Xerox, Palo Alto Research Center, Calif.
Bobrow, D. G., and T. Winograd. 1976. An overview of KRL—A knowledge representation language. Xerox PARC Rep. CSL-76-4, Xerox Palo Alto Research Center, Calif.
Carnegie Group Inc. 1987. Knowledge Craft. Commerce Court at Station Square, Pittsburgh.
Clocksin, W., and C. Mellish. 1981. *Programming in Prolog.* Springer-Verlag, New York.
Davies, D. J. M. 1973. POPLER 1.5 Reference Manual. D.A.I. Theoretical Psychology Unit Rep. 1, Edinburgh University. [POPLER]
Deering, M., J. Faletti, and R. Wilensky. 1981. PEARL: An efficient language for artificial intelligence programming. *Proc. of IJCAI-81,* Vancouver, B. C., Canada, August. [PEARL]
de Kleer, J. 1984. Choices without backtracking. *Proc. of AAAI-84,* pp. 79–85. [ATMS]
de Kleer, J. 1986a. An assumption-based truth maintenance system. *Artificial Intelligence,* vol. 28, pp. 127–162. [ATMS]
de Kleer, J. 1986b. Extending the ATMS. *Artificial Intelligence,* vol. 28, pp. 163–196.
de Kleer, J. 1986c. Problem solving with the ATMS. *Artificial Intelligence,* vol. 28, pp. 197–224.
de Kleer, J., and B. Williams. 1986. Back to backtracking: Controlling the ATMS. *Proc. of AAAI-86,* pp. 910–917, Philadelphia.
Doyle, J. 1979. A truth maintenance system. *Artificial Intelligence,* vol. 12, pp. 231–272. [TMS]
Doyle, J. 1982. Some theories of reasoned assumptions: An essay in rational psychology. Department of Computer Science, Carnegie-Mellon University (CMU-CS-83-125).
Doyle, J. 1983. The ins and outs of reason maintenance. *Proc. of IJCAI-8,* pp. 349–351.
Doyle, J. 1985. Reasoned assumptions and pareto optimality. *Proc. of IJCAI-9,* pp. 87–90.
Elcock, E. W., J. M. Foster, P. M. D. Gray, J. J. McGregor, and A. E. Murray. 1971. ABSET, a programming language based on sets: Motivation and examples. In B. Meltzer and D. Michie (eds.), *Machine Intelligence,* vol. 6, Edinburgh University Press.

Fahlman, S. E. 1979. *NETL: A System for Representing Real World Knowledge.* M.I.T. Press, Cambridge, Mass.
Fox, M. 1983. SRL User's Manual Tech. Rep. Robotics Institute, Carnegie-Mellon University, Pittsburgh.
Hendrix, G. 1975. Expanding the utility of semantic networks through partitioning. *Proc. of IJCAI-75,* Tbilisi, U.S.S.R.
Hewitt, C. 1972. Description and theoretical analysis (using schemata) of PLANNER. M.I.T. AI Lab. Memo. MAC-TR-256, Cambridge, Mass.
Hillis, W. D. 1985. *The Connection Machine.* M.I.T. Press, Cambridge, Mass.
Inference Corporation. 1985. *ART Manual.* 5300 West Century Blvd., Los Angeles.
Intellicorp. 1985. *KEE System Manual.* 707 Laurel Street, Menlo Park, Calif.
Jiang, Y. J., and S. H. Lavington. 1985. The qualified binary relationship model of information. University of Manchester, Department of Computer Science, Internal Rep. IFS/2/85.
McDermott, D.V., and G. J. Sussman. 1972. The CONNIVER reference manual. M.I.T. AI Lab. Memo. 259, Cambridge, Mass.
McGregor, D. R., and J. R. Malone. 1981. The FACT database: A system using generic associative networks. Research Rep. 2/80, Department of Computer Science, University of Strathclyde, UK.
Nii, H. P., and N. Aiello. 1979. AGE (attempt to generalize): A knowledge-based program for building knowledge-based programs. *Proc. of IJCAI-79,* Tokyo.
Reiter, R. 1978. On closed world data bases. In: H. Gallaire and J. Minker (eds.), *Logic and Data Bases,* Plenum Press, New York.
Rulifson, J. F., J. A. Derkson, and R. J. Waldinger. 1972. QA4: A procedural calculus for intuitive reasoning. Tech. Note 73, SRI International Menlo Park, Calif.
Sacerdoti, E. D., R. E. Fikes, R. Reboh, D. Sagalowicz, and R. J. Waldinger. 1976. QLISP: A language for the interactive development of complex systems. SRI Tech. Note, SRI International, AI Center, Stanford, Calif.
Sloman, A. 1983. POPLOG—A multi-purpose, multi-language program development environment. Cognitive Studies Programme, University of Sussex, UK.
Stefik, M. J. 1979. An examination of a frame-structured representation system. *Proc. of IJCAI-79,* pp. 845–852, Tokyo.
Sussman, G. A., and D. V. McDermott. 1972. Why conniving is better than planning. M.I.T. AI Lab. Memo. 255A, Cambridge, Mass.
Sussman, G. J., T. Winograd, and E. Charniak. 1971. Micro-planner reference manual. M.I.T. Tech. Rep. 203a, Cambridge, Mass.

CHAPTER 7

KNOWLEDGE IN THE FORM OF PATTERNS AND NEURAL NETWORK COMPUTING

Y. H. PAO

1 INTRODUCTION

Knowledge-based systems tend to be rule-based systems, but sometimes it is not appropriate to express knowledge primarily in the form of rules. Instead, sometimes it is more appropriate to represent knowledge in the form of associations between patterns.

The concern is that of being able to use available knowledge to arrive at correct decisions. In rule-based systems, inference proceeds a step at a time through sequential logic, in a search mode, until a conclusion is inferred or a hypothesis is confirmed. In part, this is possible because the propositions and hypotheses are all discrete-valued. In terms of trees or graphs, we can think of a finite number of alternate paths available at every node. Although errors can be made in the choice of a path, the task reduces to the navigation of a large but usually finite network. Recovery from error might be tedious but is, nevertheless, feasible.

Circumstances are different when a situation is described in terms of a pattern of compensating continuous-valued features. If knowledge is represented in terms of mappings between such pattern spaces, then rules may become seriously inadequate for describing such knowledge, and it may become essentially impossible to make decisions on the basis of sequential logical inferences.

To clarify the conceptual issues involved, we consider the situations illustrated in Fig. 7-1a and b. The state of a system is described in terms of two features x_1 and x_2, and from the training set of patterns we know that a system is of class A if it falls within the shaded area and of class $\neg A$ if it is out of the shaded area. This is true of both situations.

More precisely, there is enough information to make us think that the situations are as depicted in Fig. 7-1a and b. Now the task is to find a representation of that presumed knowledge in a form, so that we can verify it with additional test cases and can both generalize and specialize that representation, as necessary.

For this particular example, the two features are continuous-valued in both cases, but the difference is that the two features are not of a compensating nature in Fig. 7-1a and are of such compensating nature in Fig. 7-1b. Thus, in the former case, it is possible to express the knowledge in the form of a rule, namely,

If x_1 is greater than a and less than b, and if x_2 is greater
than c and less than d, then the system X is of class A. (7.1)

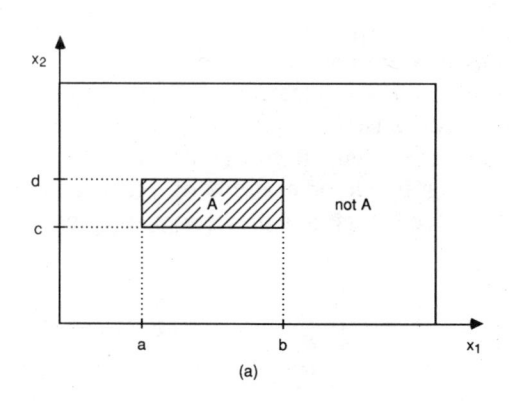

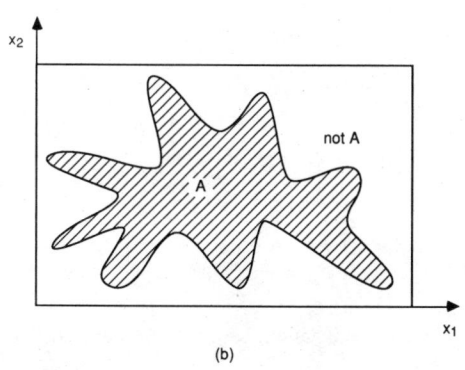

FIGURE 7-1
Representation of knowledge in terms of rules is suitable in (a) and unsuitable in (b).

It is not feasible to construct a similar representation for the knowledge represented in Fig. 7-1*b*.

We might try, for example, to synthesize such rules with the aid of quantized intervals, as indicated in Fig. 7-2. In essence, we divide pattern space into a number of boxes and assign class membership for patterns in a box on the basis of the class membership of the majority of the population in that box. A rule is obtained for each box on the boundary, and a number of additional rules are required to account for the totally internal and external spaces. However, a very large number of erroneous decisions will be made if the quantization is coarse, as shown in Fig. 7-2. And the number of rules will be intolerably large if quantization is fine, as suggested in Fig. 7-3.

In view of these realities, various research communities have concentrated on trying to understand how knowledge in the form of patterns might be represented and manipulated. We discuss some of these activities in this chapter.

At least three of the topics are inspired by biology. The thought is that since humans have truly amazingly powerful pattern recognition capabilities, computational approaches based on utilization of massive networks of neuronlike elemental processors might be endowed with comparable capabilities. This approach seems to be valid, and we will describe some of the progress made. The fourth topic represents an opportunistic attempt made to capitalize on some unique properties of the network associative memories. The inspiration there is more that of need, rather than biology. In other words, there is no evidence that biological networks are good at optimizing.

The four topics are supervised and unsupervised learning with neural networks, associative memories for recall in response to partial or distorted cues, and use of network associative memories for optimization. In each case, there has been progress in understanding as well as demonstrations of exciting results. These topics are not disjointed. Instead, together they promise to pro-

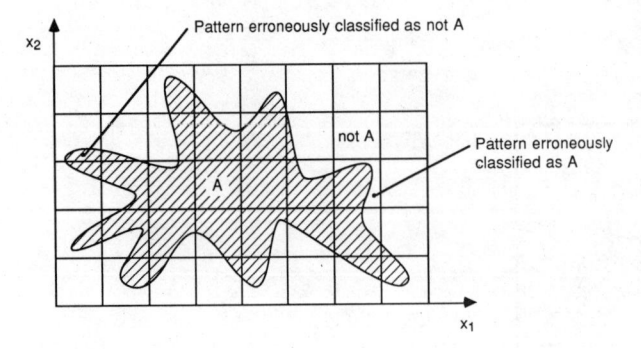

FIGURE 7-2
Dealing with complex distribution through use of coarse quantization, with high incidence of errors.

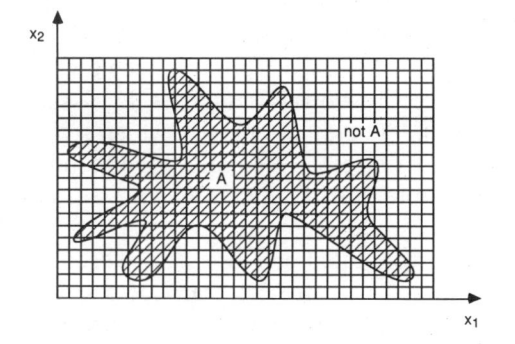

FIGURE 7-3
Dealing with complex distribution through use of fine quantization, with excessively large number of rules.

vide practitioners in knowledge-based systems with a collection of methods for representing and using knowledge in the form of patterns.

Before going on to discuss these topics, we digress to address a related point which is often mispresented or only acknowledged in a confused manner. It is helpful if we articulate that point clearly.

Sometimes it is mistakenly assumed that pattern-formatted information processing is synonymous with parallel processing. This is not so. Parallel processing is not the essence of pattern information processing, nor is it even necessary. Of course, parallel processing might be used, but it is neither a sufficient nor a necessary aspect of pattern information processing. Instead in pattern-formatted information processing, knowledge is represented in the form of relationships between *patterns* and *patterns,* or between *patterns* and *class indices,* or attribute values. Although patterns are described in terms of smaller entities, namely the features, *knowledge* refers to the pattern as a *whole.* All the aspects of the pattern have to be considered before any decision can be made, and it is this gestalt nature of knowledge processing which is the very heart of pattern information processing.

Returning to the four topics covered in this chapter, we address each briefly to place them in perspective relative to each other.

The first topic involves learning associations between input/output pairs of patterns and then generalizing on an association so that correct outputs can be inferred for new input patterns, even though the latter had never been encountered previously. Outstanding progress has been made in recent years through improvement of the original Perceptron idea.

The generalized Perceptron with the *generalized delta rule* (GDR) and back propagation of error can autonomously synthesize a mapping from input pattern space to output pattern space. This mapping is a network representation of the knowledge contained in the training set of input/output pattern pairs and is the equivalent of an infinite number of rules. The generalized Perceptron is truly a clever and powerful concept which has come into its own after a latency period of nearly 25 years. It is a powerful tool for representing and utilizing knowledge in the form of patterns. It constitutes a prime example of authentic, meaningful, and powerful neural net computing.

However, unfortunately, the conventional GDR net does not scale well, and the rate of learning can often be intolerably slow. To address that issue, we describe briefly an innovation of ours, the functional-link net, which scales well and has much higher rates of learning. It also has a simpler architecture. It represents an extension and generalization of some high-order net results reported previously by others.

The second topic concerns neural networks that can be used to discover similarities among patterns and to form these into clusters, each of which is represented by a prototype. In a sense, there is nothing new or surprising about unsupervised learning, because the ISODATA and K-means algorithms of pattern recognition have been known and used for many decades. However, unsupervised learning, or self-organization of data, is readily implemented with neural nets and can be of considerable use in the representation and manipulation of pattern-formatted knowledge. Clustering can be used for data reduction, for filtering out faulty data, for classification, or for associative recall. It can also be used recursively to build up the hierarchical structure of classifiers. Used skillfully, in combination with supervised learning, unsupervised learning can play a unique role in pattern knowledge processing, a role which is not adequately assumed by any other methodology.

The third topic is that of associative memories. Humans seem to operate on the basis of some mysteriously powerful and flexible associative memory. A stimulus produces a response, but even a partial cue or distorted cue can sometimes evoke a lengthy, complete response. There is medical evidence to suggest that these memories are distributed spatially. As a result of all these factors, computational interest has tended to focus on distributed associated memories which can be implemented in fault-tolerant forms.

Distributed associative memories may be of matrix, holographic, or network form and play a role significantly different from the GDR networks or clustering networks.

In the matter of capabilities, we note that the Perceptron can learn associations between input/output pattern pairs and then generalize on these. In other words, a Perceptron-like structure can generate an estimated response for a new input, neither of which it had encountered before. In contrast to this, associative memories of matrix or holographic form (even when implemented in network form) respond by retrieving exactly one of the previously stored patterns, even though the stimulus or cue might be partly distorted or missing in part. The nature of the generalization is different in the two cases.

As a fourth topic, we note that the network form of the distributed associative memory has also been used to deal with scheduling tasks, such as the traveling salesperson problem. That effort represents a creative exploration of an idea that pattern recognition might allow us to find a different and better way for dealing with the "truly difficult" problems, so intractable by naive algorithmic means. Unfortunately, the results, to date, do not suggest that globally optimum results are ensured. However, the topic remains an active research issue, and our attitude is that even possible solutions are of interest as long as they satisfy all specified constraints. Indeed, after all, humans can plan, sometimes "wisely," but not necessarily always optimally.

The four topics are discussed separately in the following sections. The discussions are brief and schematic. The reader interested in further details might want to refer to the book by Yoh-Han Pao (1989a).

2 GENERALIZING ON LEARNED ASSOCIATIONS

In the practice of traditional pattern recognition, the central task was that of classification. Given a pattern, the task of principal interest was to be able to decide what class it belonged to.

Normally, we would be provided with a training set of patterns, i.e., with a set of patterns with known class memberships. As illustrated in Fig. 7-4, a classifier would learn a procedure which it could use to correctly classify all the patterns in the training set. The subsequent issue would be to test whether the classifier was able to correctly classify other new patterns of that same population. These two phases of traditional pattern recognition activity, namely, the training and test phases, are shown in Fig. 7-4.

The estimation task is a generalization of the classification task. The viewpoint in this case is that we learn to estimate the values of an *attribute* for the various patterns in a training set of patterns and then generalize on that learned ability so as to be able to estimate the value of that same attribute for patterns which had not been encountered previously, but which belong to the same population.

Training set patterns {X} and associated class membership label → Procedure for learning a classifier → Classifier

(a) Training Phase

New pattern (class membership unknown) → Classifier (previously synthesized) → Class membership index

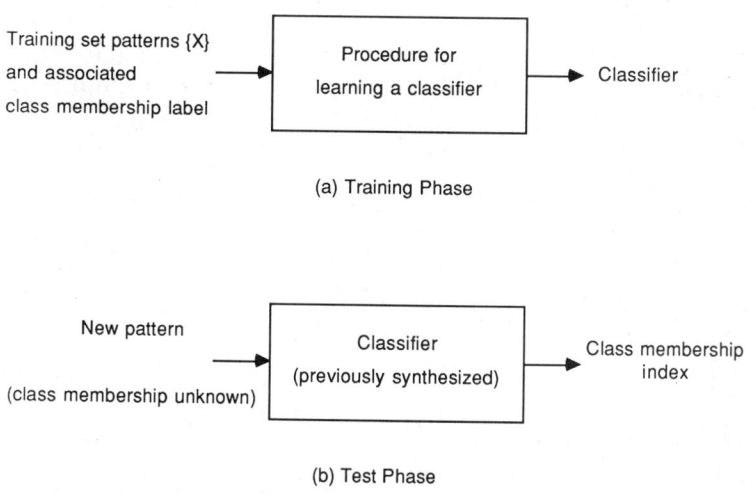

(b) Test Phase

FIGURE 7-4
The two phases of traditional pattern recognition methodology.

To generalize even further, it is useful to think of the training procedure as that of learning a *mapping*. For patterns which are in the nature of a list of N numbers, the pattern itself can be thought of as being a point in an N-dimensional space. There may be M attributes, and so that pattern of attribute values may be regarded as a point in an M-dimensional space. The mapping can be viewed as a procedure which maps a point in a *pattern space* onto a point in an *image space,* as illustrated in Fig. 7-5.

If the image space is unidimensional and there are only a few discrete target points, then we have the classification model. If the image space is unidimensional but ranges of values are allowed as target values, then we have the single-attribute estimation model. However, in general, the mapping is from one space to another space, and the learning to be carried is for a finite set of associated vector pairs. The pattern recognizer *learns* a procedure which maps each pattern, i.e., each point in a pattern space, onto its associated point in image space and then *generalizes* on that so that it is able to cope with any and all other points in the pattern domain of interest, even if these had not been encountered before. In principle, the pattern recognizer should also be adaptive enough that once an error is made, it *specializes* to modify its mapping procedure so as to maintain correct mapping. In practice, in traditional pattern recognition the learning procedures were often so formidable that dynamic on-line adaptation was well-nigh impossible. We will see that implementation of pattern recognition in the form of neural networks accommodates generalization specialization needs more readily.

Reverting to the classification model, we note that real-world situations are often incompletely described or are confused by extraneous effects beyond our control. Statistical pattern recognition concepts and techniques have been developed for coping with such circumstances.

As an example, consider the task of classifying patterns in the two-dimensional domain, shown in Fig. 7-6, as that of deciding whether a pattern is a member of class A or $\neg A$, and let the training patterns be those points shown in Fig. 7-6. The circles are the class A patterns, and the crosses are the $\neg A$ patterns.

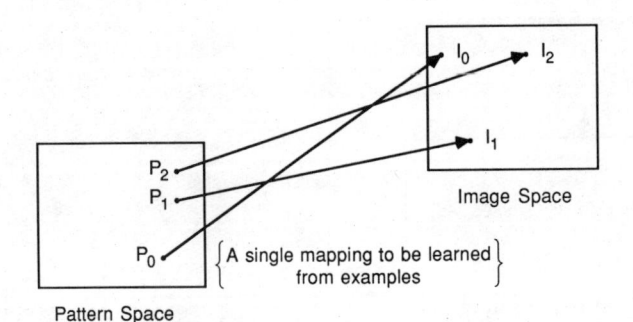

FIGURE 7-5
Pattern information processing viewed in terms of mapping operations.

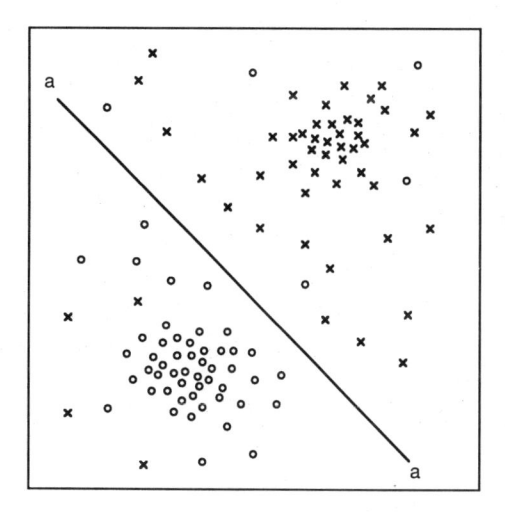

FIGURE 7-6
Schematic illustration of use of discriminant for optimal classification for stochastic distributions.

The distributions are stochastic in the sense that class A patterns can be found everywhere and anywhere and so can patterns of $\neg A$ class membership. For such a situation we cannot decide with certainty that any specific pattern is either A or $\neg A$. For example, we cannot expect to find a line such as a-a in Fig. 7-6 and use that as a discriminant in the sense that all patterns to the lower left of it are of class A and all patterns to the upper right are of class $\neg A$. The problem is not deterministic, and we cannot hope to avoid all errors. In fact, we can be reasonably sure that we will make errors in the classification task. The statistical pattern recognition task is, then, to learn a recognition or mapping procedure which is *optimum* in the sense that the least amount of error is incurred.

Somewhat ironically, once we are resigned to striving for *optimum* performance, rather than *perfection,* we recover once more a semblance of simplicity which hitherto could be had only in the deterministic case.

It is possible, e.g., in the illustration of Fig. 7-6, to find a discriminant such as line a-a which we can use seemingly in a deterministic way. As far as classification decisions are concerned, we decide quite definitely that it divides the class A patterns from the class $\neg A$ patterns. Errors will be made, but those are unavoidable, and the objective is to minimize the incidence and consequences of these.

In statistical pattern recognition, we arrive at such discriminant surfaces in many different ways, including those of learning decision functions or learning prototypes, but the net result is actually always the same. Eventually we have to make definite decisions, and we arrive at criteria on which we base our decisions. These criteria can be viewed to be hypersurfaces in N-dimensional space.

The situation can be quite complex even in low-dimensional tasks. For example, in the two-class classification task for patterns with only two features, the class A patterns may be distributed in bimodal form so that class A patterns are more or less grouped in two widely spread clusters of shape, as shown in Fig. 7-7 (ignore the straight lines for the time being).

A decision function approach would concentrate on the synthesis of three or more decision functions, say, $g_{A1}(x)$, $g_{A2}(x)$, and $g_{\neg A}(x)$ such that

$$X \text{ belongs to } \begin{cases} \text{class } A & \text{if } g_{A1}(X) > g_{\neg A}(X) \text{ or } g_{A2}(X) > g_{\neg A}(X) \\ \text{class } \neg A & \text{otherwise} \end{cases} \quad (7.2)$$

A discriminant synthesis approach, however, might yield the result shown in Fig. 7-7, where we have this rule:

If X is on the positive side of line a-a
and on the positive side of line b-b
and on the negative side of line c-c
and on the negative side of line d-d (7.3)
or X is on the positive side of line e-e
and is on the positive side of line f-f
and is on the negative side of line g-g

Then X belongs to class A

This approach, though feasible, was essentially not practicable until recently. There was simply no way for learning these discriminants.

The Perceptron scheme originally proposed by Rosenblatt was supposed to handle this type of learning task, but it was shown to be inadequate for the task envisioned for it. However, in recent years, it has been shown that

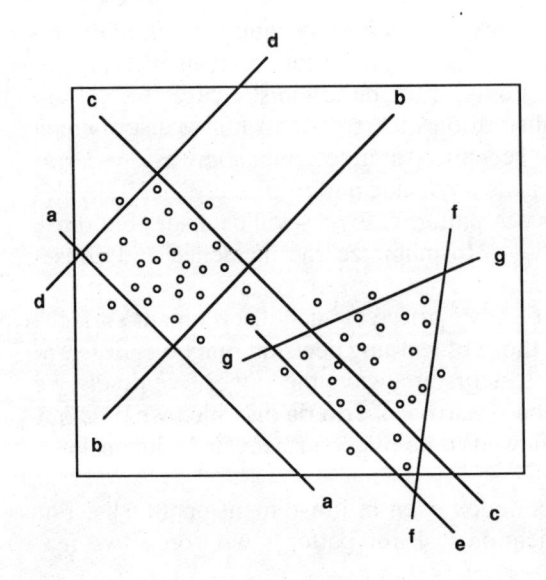

FIGURE 7-7
Use of discriminant in classification of multimodal distributions.

Perceptrons with additional intermediate (or "hidden") layers can learn such mappings autonomously. Qualitatively the action may be viewed in terms of learning all the requisite linear discriminants, suggested in Fig. 7-7, but in actual practice the nature of the result corresponds more closely to learning a decision function.

The Generalized Perceptron

The GDR can be used with a Perceptron-like neural net for learning a mapping between input and output pattern populations. The training is achieved by having the net learn a finite discrete set of associated pairs of input/output patterns. The result is that the net will have learned a mapping which generalizes on the examples provided by the training set of associated pairs and is able to generate an output for any new input pattern even if the input had never been encountered before.

The connectivity of such a net is shown schematically in Fig. 7-8. To fix ideas, we show an input layer, an internal layer (a "hidden" layer), and an output layer. In practice, if need be, there may be more than one hidden layer, but both theoretical considerations and practical experience indicate that rarely would one want to use more than two hidden layers.

The multilayered network functions in a repeated two-phase manner. The feed-forward phase propagates the input pattern through the network to produce an output pattern which initially will be different from the correct pattern to be learned. The values of the weights on the links and the thresholds in the nodes are then adjusted in the back-propagation-of-error phase so as to make the output more nearly correct on the next try. It is not difficult to train a net to generate the correct output for a specific input, but the actual learning task needs to go beyond that, to ensure that the net performs well for the *entire* set of associated input/output pattern pairs provided by the training set of patterns. Therefore, in practice, the adjustments in weights and thresholds need to be averaged over all patterns as well as repeated iteratively until the same

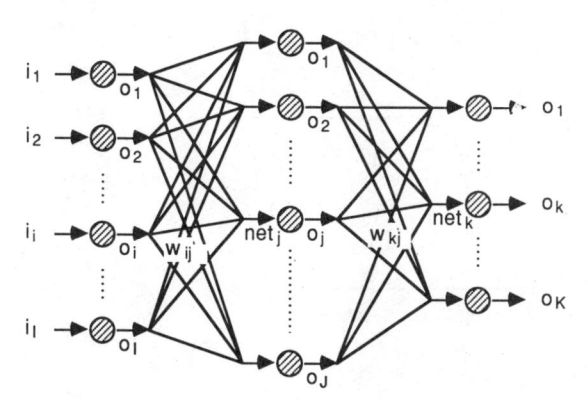

FIGURE 7-8
Connectivity of a Perceptron network with a hidden layer.

criterion is reached. One practice is to iterate until the least mean square error evaluated over the entire ensemble of training set examples is less than a prescribed value.

The input layer nodes are primarily to be used as registers in which case the output of node i is equal to the input to node i. They may also be used to normalize the input feature values, to map them into the interval [0, 1].

In the feed-forward phase, the output of units in layer i is multiplied by various appropriate weights w_{ji}, and these are fed as input to the next layer, the hidden layer.

Thus if the o_i are the outputs of units in layer i, the total input to a unit in layer j is

$$\text{net}_j = \sum_i w_{ji} o_i \qquad (7.4)$$

and the output of a unit in layer j is

$$o_j = f(\text{net}_j) \qquad (7.5)$$

where f is the activation function or transfer function.

The sigmoidal relationship illustrated in Fig. 7-9 is a convenient function for modeling the nonlinear transformation carried out by a node, and for that relationship we have

$$o_j = f(\text{net}_j) = \frac{1}{1 + \exp[-(\text{net}_j + \theta_j)/\theta_0]} \qquad (7.6)$$

where θ_j is the value of the "threshold" parameter for node j. This activation function yields an output which varies continuously from 0 to 1. The quantity θ_j serves as a threshold and positions the transition region of the f function. The quantity θ_0 determines the abruptness of the transition. For the input node layer, there is generally no reason to have threshold values other than zero.

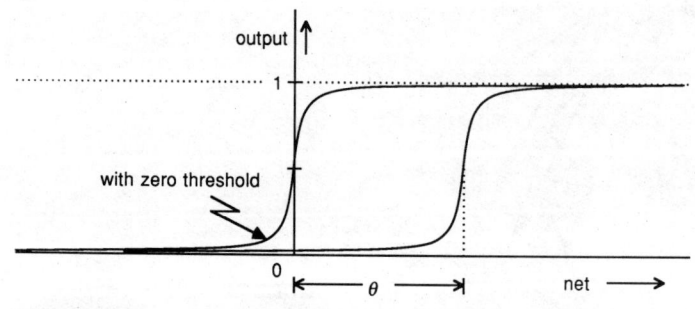

FIGURE 7-9
Sigmoidal node activation function.

As stated previously, in the learning process the network is presented with a pair of patterns, an input pattern and a corresponding output pattern. Using its (possibly incorrect) weights and thresholds, the network produces its own output pattern in accordance with procedures described in expressions 7.4 and 7.6. This output is compared with the desired output pattern.

The error at any output unit in layer k is

$$e_k = t_k - o_k \tag{7.7}$$

where t_k is the desired output for that unit in layer k and o_k is the actual output. A total error function might be written as

$$E = \tfrac{1}{2} \sum_k (t - o_k)^2 \tag{7.8}$$

Learning consists of changing the weights and thresholds so as to minimize this error function in a gradient descent manner. The analytic nature of the activation function allows errors to be traced backward so that we can have an estimate of an "error" even at an internal node. This is carried out as follows.

By using the so-called delta rule, and in our case the GDR, the convergence toward improved values for the weights and the thresholds may be stated in general as

$$\Delta w_{kj} = \eta \delta_k o_j \tag{7.9}$$

where the error signal δ_k at an output unit k is given by

$$\delta_k = (t_k - o_k) f'(net_k) = (t_k - o_k) o_k (1 - o_k) \quad \text{for } \theta_0 = 1 \tag{7.10}$$

It can also be shown that the values of the deltas (δ) at nodes in the hidden layer j can be inferred from delta values at an upper layer and that the estimated value is a linear combination of the higher-layer deltas. In this way we think of errors being propagated backward, to yield

$$\delta_j = o_j(1 - o_j) \sum_k \delta_j w_{kj} \tag{7.11}$$

and

$$\Delta w_{ji} = \eta \delta_j o_i \tag{7.12}$$

In expression 7.9, η is called the *learning rate parameter*. In the past, it has been suggested that one way to increase the learning rate without causing oscillations is to modify expression 7.9 to include a momentum term, i.e.,

$$\Delta w_{kj}(n + 1) = \eta \delta_k o_j + \alpha \, \Delta w_{kj}(n) \tag{7.13}$$

where the parameter n indexes the presentation number or the number of times for which a set of input patterns have been presented to the network. The parameter α is a constant which determines the effect of past weight changes on the current direction of movement in weight space. This ad hoc practice is not desirable. If need be, the rate of change of the weight values might be monitored and η varied accordingly in an adaptive manner. However, the α term is nevertheless used in practice.

The thresholds are learned by taking θ_j to be equivalent to another weight w_{jo} connecting the unit j to a lower-layer unit which is always turned on to give unity output.

In choosing values for the target or desired outputs, one should be aware that output values of 1 or 0 should be avoided; otherwise, convergence would be slow indeed. Values which approximate 1 or 0, that is, 0.95 and 0.05, would suffice; or equivalently acceptable residual error margins might be set to be rather large.

There is some question whether the change in the value of a weight or a threshold in any one iteration in the training phase should be carried out for each pattern individually or whether $\Delta w_{kj}^{(p)}$ should be evaluated for all the set of P patterns and the correction be taken as the average of all the individual changes. Both theoretical and practical considerations suggest that changes be made after all the patterns have been presented. However, in practice we do what can be accomplished realistically. For example, if a net is initially trained with a set of 250 patterns and specialization calls for correcting the weights to exclude a specific additional pattern, it is doubtful that anyone would start from some arbitrary random set of values and train the net with all 251 patterns in a *ab initio* manner.

The supervised learning paradigm of adaptive pattern recognition as implemented with neural nets is useful and powerful. Use of higher-order terms as advocated by Giles and Maxwell (1987) and as implemented in the functional-link net described by Pao and his collaborators provides support for even more powerful realizations of this paradigm.

In the context of knowledge-based systems, supervised learning has been shown to be useful for capturing *compiled experimental knowledge* of experts. Such knowledge is not easily combined out into the form of a body of rules, and perhaps it is impossible to do that. However, a GDR net or a functional-link net can be used in the supervised learning mode to capture such knowledge. That knowledge, represented in the form of a net, can be subsequently used to provide guidance regarding appropriate action on situations encountered subsequently.

In the case of the functional-link net, links as well as nodes may be nonlinear. In fact, links generate functions which are evaluated at instances of the input pattern. These newly generated features make additional facts explicitly known to the net, and many new aspects of information processing become available in this manner. As shown in Fig. 7-10, the functional link enhances the original pattern by augmenting it with functions of the pattern or

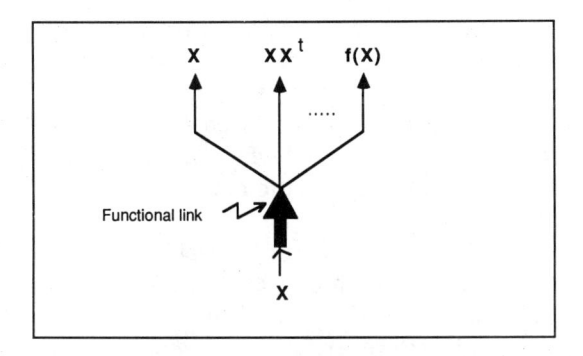

FIGURE 7-10
Schematic illustration of action of functional-link net.

with functions of the individual features. The first mode corresponds to the approach of Giles and his collaborators, and our work on the functional link constitutes an extension of that previous work as well as articulating some specializations.

In Fig. 7-11, we show the different architectures of a functional-link net and a GDR net devised to deal with the parity 2 classification task, for which

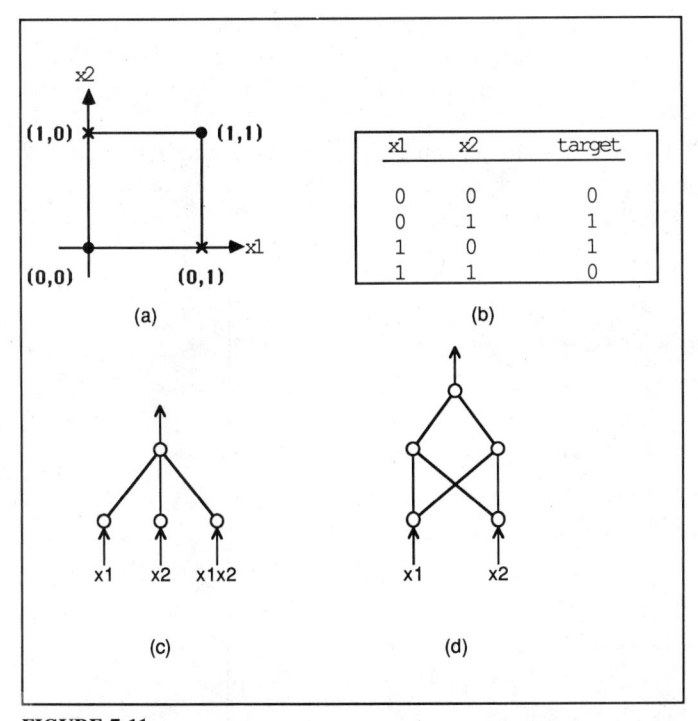

FIGURE 7-11
The parity-2 problem. (*a*) Patterns in two-dimensional space; (*b*) patterns with associated targets; (*c*) the functional-link net architecture; (*d*) the GDR net architecture.

all triplets with an even number of 1s are of one class and all others are of the other class. Both nets can perform well, but the functional-link net typically learns at a much higher rate, as shown in Fig. 7-12. For this particular case, the functional-link net is merely a restatement of results reported previously by Giles and collaborators in their studies of the effect of high-order terms. But, in general, the functional-link net is a generalization of the notion of including high-order terms.

The functional-link net often results in a flat net or a "slab" net architecture and can be processed with the delta rule rather than the GDR. This seemingly simple architecture is actually the abstraction of many underlying complex operations in the functional link. However, at the training level the procedure is much simplified. One of the advantages which result is that the weights on the links in such a net can provide explanations of why certain results are obtained at the outputs.

3 UNSUPERVISED LEARNING

Unsupervised learning or clustering is another important pattern information processing paradigm. In this instance we see revealed again the intrinsic difference between pattern information processing and sequential rule-based processing. In the clustering of patterns in feature space, the low-level processing step may indeed be carried out sequentially, but the essence of the matter is that no conclusion can be drawn or decision made until all features of a pattern have been considered as a whole.

The idea behind unsupervised learning is that much might be learned about the reason for the occurrence of certain phenomena, as described by

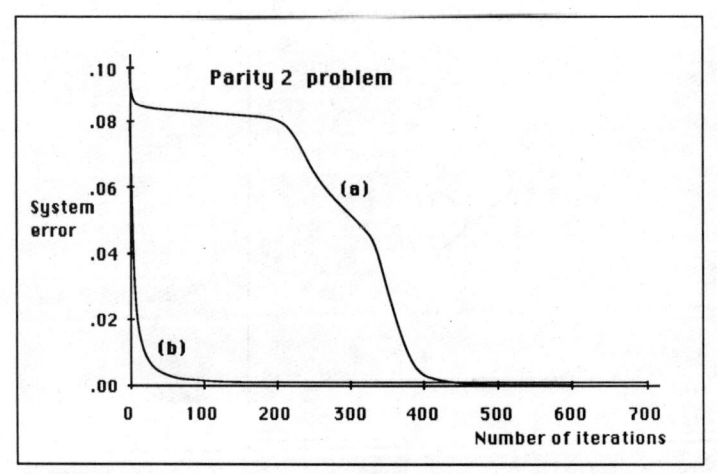

FIGURE 7-12
Comparison of rates of learning. (*a*) GDR net with one hidden layer; (*b*) functional-link net.

patterns, if we just observe where these patterns occur in pattern space. If we limit ourselves to a few observations, then there might be no higher-level spatial patterns in the manner in which these few simple patterns are distributed. However, in essentially all cases, as the number of observations is increased, it is observed that the incidence of occurrence is much higher in some locations in pattern space than in others. In fact, if each pattern is represented as a small billiard ball, then we see that these billiard balls are not distributed as a random gas of billiard balls, but in fact might be distributed in the form of several clusters with relatively dense centers and a decreasing density as we leave the centers. The distributions are not always necessarily radially symmetric.

The learning is considered to be "unsupervised" because the clusters were formed and learned, so to speak, without recourse to class labels or to information on attribute values. However, this clustering can yield many benefits.

If a very large number of patterns in one cluster can be represented by one prototype, the cluster center, then we have achieved *data compression*. If nearly all patterns within a cluster are of one class, then discovery of that cluster can provide the basis for *classification* and can provide an "explanation" for the occurrence represented by that pattern.

Unsupervised learning can be carried out successively in a hierarchical manner and can also be combined with supervised learning.

This learning program is very close to that of ISODATA algorithm of traditional pattern recognition. The idea is to "follow the leader" until one finds that the new pattern is much too dissimilar, in which case one forms a new leader, so to speak. That is, one forms a new cluster. This implies that we must have a metric for measuring similarity.

The algorithm ART1 is suitable for binary-valued patterns. In this algorithm the similarity between an incident pattern and the computer prototype is measured on the basis of the number of coincidences in feature values between prototype and pattern. A device known as the MAXNET detects that cluster center node for which the similarity measure is the greatest. That incident pattern is automatically indexed to that cluster without need for any further explicit indexing. However, the prototype as described by the values of the weights in the bottom-up links is updated. Since the assignment of cluster membership are made on a relative basis rather than an absolute basis, there is need in ART1 for an additional top-down verification step. For example, pattern X might have been assigned to cluster k on the basis of similarity, but if the vigilance ratio is too low, that assignment is invalidated and a new cluster is started.

The connectivity of the MAXNET layer is exhibited in Fig. 7-13. The net input to each of the MAXNET nodes generates a reinforcing signal to itself and inhibiting signals to all other nodes. This action is exercised repeatedly until the outputs from all but one of the nodes become nonpositive, and the node with maximum initial input is thus identified automatically.

An extension of ART1 to cover circumstances of continuous-valued features has also been described. That algorithm, ART2, strives to cope with a

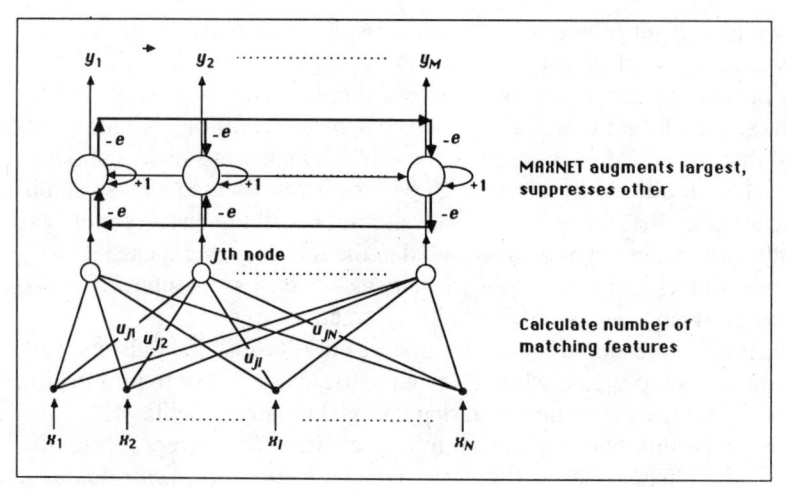

FIGURE 7-13
Connectivity of MAXNET.

more general measure of similarity and yet conform with network computation ideas.

We believe that other metrics, such as the Euclidean metric, can be used in artificial neural networks, and indeed the Euclidean metric procedure works well. A clustering procedure based on the Euclidean metric is described in Fig. 7-14.

4 ASSOCIATIVE STORAGE AND RECALL OF PATTERNS

In many respects, humans function cognitively in the manner of an associative memory. Cues stimulate responses. The mechanism of association seems to be remarkably fault-tolerant and robust, and even partial or distorted cues can still evoke full and accurate responses. These associative recall traits are exhibited by humans all the time, but especially dramatically and almost effortlessly in our ability to read handwritten messages, to comprehend speed, and to recognize images. Often a partial or distorted cue not only is identified, but also can trigger the recall of a much larger context of which it is a part. This fascinating trait of human behavior has long been of interest to those who would understand human behavior as well as to those who would like to reproduce such capabilities in machines.

The distributed associative memory is a computing concept which exhibits some of the same characteristics, even though only within certain severely limited circumstances, and we describe that concept in this section. In a sense, a device based on this concept produces order out of disorder. What is distorted or incomplete is recognized, reshaped, and returned as a standardized

> 1. Activate the first clusternode ($j = 1$) whose member is the first input pattern $\{y_i^1\}$ for $i = 1, \ldots, N$, where N is a number of input features.
> 2. Initialize the first cluster node weights $b_{1i} = y_i^1$ for $i = 1, \ldots, N$.
> 3. For each input pattern $\{y_i^p\}$:
>   3.1. Calculate euclidean distance ED_{jp} for all j: $ED_{jp} = \sqrt{\Sigma_i (b_{ji} - y_i^p)^2}$.
>   3.2. Determine that j for which $ED_{jp} < ED_{kp}$ for all $k = 1, \ldots, j$, $k \neq j$.
>   3.3. If $ED_{jp} \leq$ threshold, then:
>   3.3.1. Assign pattern $\{y_i^p\}$ to node j.
>   3.3.2. Update $b_{ji}(n + 1) = [n/(n + 1)]b_{ji}(n) + [1/(n + 1)]y_i$, where n is a number of members in cluster j.
>   3.4. If $ED_{jp} >$ threshold, then form a new cluster.
>   3.4.1. Increment number of clusters: $j = j + 1$.
>   3.4.2. Initialize weights for new cluster $b_{ji} = y_i^p$ for $i = 1, \ldots, N$.

FIGURE 7-14
A clustering algorithm based on the Euclidean distance metrics.

whole. The clustering action of self-organized unsupervised learning could also produce the same final results, but through significantly different means. In the latter case, the distorted cue would be recognized as belonging to a certain cluster, and the prototype would then be returned.

As the name suggests, the matrix associative memory can be implemented in the form of a matrix, formed from the outer products of the stimuli and associated responses. Thus, if X is a cue for which the response is to be Y, then the matrix memory for this pair is

$$M = YX^t \quad (7.14)$$

Clearly in recall, for an input X, matrix multiplication yields the response vector:

$$MX = YX^tX = \langle X^tX \rangle Y \quad (7.15)$$

The output is the associated response Y multiplied by a scalar $\langle X^tX \rangle$. In most cases it would not be necessary to normalize the quantity $\langle X^tX \rangle$ because the pattern Y would be of interest and only the relative magnitudes of the components would be of significance.

The cleverness of this scheme is better appreciated if we write the memory explicitly in terms of the components of the stimulus and response vectors. For X a three-element column vector and Y a four-element column vector, the matrix memory would be

$$M = \begin{pmatrix} y_1 x_1 & y_1 x_2 & y_1 x_3 \\ y_2 x_1 & y_2 x_2 & y_2 x_3 \\ y_3 x_1 & y_3 x_2 & y_3 x_3 \\ y_4 x_1 & y_4 x_2 & y_4 x_3 \end{pmatrix} \qquad (7.16)$$

It is clear that

$$MX = \begin{pmatrix} y_1 x_1 & y_1 x_2 & y_1 x_3 \\ y_2 x_1 & y_2 x_2 & y_2 x_3 \\ y_3 x_1 & y_3 x_2 & y_3 x_3 \\ y_4 x_1 & y_4 x_2 & y_4 x_3 \end{pmatrix} \begin{pmatrix} x_1 \\ x_2 \\ x_3 \end{pmatrix} = \begin{pmatrix} y_1 \\ y_2 \\ y_3 \\ y_4 \end{pmatrix} (x_1^2 + x_1^2 + x_3^2) \qquad (7.17)$$

a perfect recall multiplied by a scalar coefficient.

We note that a partial cue would still result in a correct response

$$MX' = \begin{pmatrix} y_1 x_1 & y_1 x_2 & y_1 x_3 \\ y_2 x_1 & y_2 x_2 & y_2 x_3 \\ y_3 x_1 & y_3 x_2 & y_3 x_3 \\ y_4 x_1 & y_4 x_2 & y_4 x_3 \end{pmatrix} \begin{pmatrix} x_1 \\ x_2 \\ 0 \end{pmatrix} = \begin{pmatrix} y_1 \\ y_2 \\ y_3 \\ y_4 \end{pmatrix} (x_1^2 + x_2^2) \qquad (7.18)$$

and loss of part of the memory would not destroy associative recall capability, even if the cue were also incomplete at the same time, as shown for the following drastic circumstance:

$$M'X' = \begin{pmatrix} 0 & y_1 x_2 & y_1 x_3 \\ 0 & y_2 x_2 & y_2 x_3 \\ 0 & y_3 x_2 & y_3 x_3 \\ 0 & y_4 x_2 & y_4 x_3 \end{pmatrix} \begin{pmatrix} x_1 \\ x_2 \\ 0 \end{pmatrix} = x_2^2 \begin{pmatrix} y_1 \\ y_2 \\ y_3 \\ y_4 \end{pmatrix} \qquad (7.19)$$

A perfect recall, nevertheless!

Such a memory is capable of remembering many such associations in parallel. For those circumstances we have

$$M = \sum_k Y_k X_k^t \qquad (7.20)$$

The response to a cue X_j could be

$$MX_j = Y_j X_j^t X_j + \sum_{k \neq j} Y_k X_k^t X_j \qquad (7.21)$$

It is clear that the response would still be accurate, if the second grouping of terms on the right half side of the equation were small relative to the first. This would be the case if all the X vectors were orthogonal to each other, that is, $X_k^t X_j = 0$ for all $k \neq j$, or if there were enough accidental cancellations among the various incorrect responses. In practice, such a memory has a rather low storage efficiency, and errors become troublesome as we attempt to store and recall multiple parts. Such memories are called *heteroassociative* when $Y \neq X$ and *autoassociative* when $Y = X$.

Performance is much improved when we use an intermediate set of orthonormal vectors (patterns) as internal references. The storage and recall are accomplished as shown in the following, at first for just one pair of stimulus/response patterns.

Let Z be the internal reference associated with the stimulus X. Then the first part of the associative memory is

$$M_1 = ZX^t \qquad (7.22)$$

And let Y be the corresponding response. The second part of the associative memory is then

$$M_2 = YZ^t \qquad (7.23)$$

For many such pairs,

$$M_1 = \sum Z_k X_k^t \qquad (7.24)$$

and

$$M_2 = \sum Y_m Z_m^t \qquad (7.25)$$

In recall, if the cue is X_j, we form at first

$$M_1 X_j = Z_j X_j^t X_j + \sum_{k \neq j} Z_k X_k^t X_j \qquad (7.26)$$

Then we form the terms $Z_w^t M_1 X_j$ and identify the value of w for which this term is maximum. In this way we retrieve the value of index j.

After that we form

$$M_2 = \sum Y_m Z_m^t Z_j = Y_j Z_j^t Z_j + \sum_{m \neq j} Y_m Z_m^t Z_j \qquad (7.27)$$

$$= Y_j \quad \text{a perfect recall!}$$

This type of memory is called the *holographic associative memory,* and it can be both robust and of high capacity.

The Walsh functions are binary-valued discrete-form analogs of the sine and cosine functions and can function well as internal references in associative memories. There are several ways of generating such functions, but a way which is easily visualized and remembered consists of recursively substituting the following H matrix in place of each of its components. Thus, the H matrix

$$H^{(2)} = \begin{pmatrix} 1 & 1 \\ 1 & -1 \end{pmatrix} \qquad (7.28)$$

may be understood to be a display of the two Walsh functions which span 2-space, i.e.,

$$W_0 = (1 \quad 1) \quad W_1 = (1 \quad -1) \tag{7.29}$$

To obtain the four Walsh functions which span 4-space, we substitute H back into itself, to obtain

$$H^{(4)} = \begin{pmatrix} H^{(2)} & H^{(2)} \\ H^{(2)} & -H^{(2)} \end{pmatrix} = \begin{pmatrix} 1 & 1 & 1 & 1 \\ 1 & -1 & 1 & -1 \\ 1 & 1 & -1 & -1 \\ 1 & -1 & -1 & 1 \end{pmatrix} \tag{7.30}$$

Then the corresponding Walsh functions are

$$W_0^{(4)} = (1 \quad 1 \quad 1 \quad 1) \quad W_1^{(4)} = (1 \quad 1 \quad -1 \quad -1)$$

$$W_2^{(4)} = (1 \quad -1 \quad -1 \quad 1) \quad W_3^{(4)} = (1 \quad -1 \quad -1 \quad -1) \tag{7.31}$$

The Walsh functions in 7.31 have been reordered so that the index corresponds to the number of times the function changes sign as one proceeds from the first element (on the left) to the last element (on the right).

The Walsh associative memory is of remarkably high capacity and is suitable for implementation in many interesting forms, including optical computing forms. For two-dimensional image processing purposes, internal reference matrices may be formed from outer products of the Walsh function. These are known as *Hadamard matrices* and have long been used in space exploration in data reduction schemes.

Incidentally, the operation of finding that value of w for which $\{Z_w^t M_1 X_j\}$ is maximum can be implemented by the fast Walsh transform, an extremely fast algorithm.

5 OPTIMIZATION, OR SOLVING "TRULY DIFFICULT" PROBLEMS

It has been known for some time that the matrix memory can be implemented in network form. More recently, Hopfield and Tank (1985) revived interest in matrix memories in network form by discussing the convergence and stability properties of such networks in greater detail and by suggesting that they might be used to solve "truly difficult problems," such as the traveling salesperson type of problem. We address that possibility as our fourth topic of discussion.

Following Hopfield and Tank (1985), we discuss the traveling salesperson problem in terms of a search for the global minimum in the value of an energy function characterizing both the objective function and the constraints of the problem.

We preface our discussion with a disclaimer: Although the solution of such a problem can indeed be stated in terms of a search for a global minimum of an objective function, and although it can be proved that the retrieval algorithm always converges to a minimum, it is nevertheless regrettable that it cannot be shown that the minimum attained is the global minimum. All statistic-guided random searches and even the simulated annealing procedures represent ad hoc attempts to increase our chances of falling within the neighborhood of the global minimum. There is no guarantee that we will ever chance upon it except by exhaustive search.

Nevertheless, the approach holds great interest, and some researchers, including Takefuji et al. (1988), have gotten very encouraging results for specific problems. Therefore, we include this important topic in neural net computing.

To fix ideas, let us consider a five-city, 5-day tour by the traveling salesperson. The operational constraints are that the salesperson should visit all the cities and should not visit any city more than once. In addition, there is the commonsense physical constraint that she or he cannot be in two places simultaneously.

The task can be described in terms of search of an optimum 5×5 pattern, such as that shown in Fig. 7-15. In that figure we show an itinerary for which the salesperson is in city D on the first day, in city B on the second day, and so on. The task is to find that pattern of 1s and 0s for which the value of the appropriate energy function is a minimum.

The energy function given by Hopfield and Tank (1985) is

$$E = \sum_X \sum_i \sum_{j \neq i} V_{Xi}V_{Xj} + \frac{B}{2} \sum_i \sum_X \sum_{X \neq Y} V_{Xi}V_{Yi}$$
$$+ \frac{C}{2} (\sum_X \sum_i V_{Xi} - n)^2 + \frac{D}{2} \sum_X \sum_{Y \neq X} \sum_i d_{XY} V_{Xi}(V_{Y,i+1} + V_{Y,i-1}) \quad (7.32)$$

where the first term discriminates against any solution for which any one city is visited more than once, the second term represents the constraint that the salesperson not be in more than one city on any one day, the third term represents the constraint that the total number of cities visited be η (5 in our case), and the fourth term is the actual mileage traveled.

When translated to a Hopfield net, the values of the internode links are

$$T_{Xi,Yj} = - A\delta_{XY}(1 - \delta_{ij}) - B\delta_{ij}(1 - \delta_{XY}) - C - D\, d_{XY}(\delta_{j,i+1} + \delta_{j,i-1}) \quad (7.33)$$

City\Day	1	2	3	4	5
A	0	0	1	0	0
B	0	1	0	0	0
C	0	0	0	1	0
D	1	0	0	0	0
E	0	0	0	0	1

FIGURE 7-15
Representation of a five-city traveling salesperson itinerary.

The external input currents are

$$I_{Xi} = C_n \quad \text{excitation bias} \tag{7.34}$$

The outputs of the nodes V_{Xi} are expressed in terms of the inputs u_{Xi} by

$$V_{Xi} = g(u_{Xi}) = \frac{1}{2}\left(1 + \tanh\frac{u_{Xi}}{u_0}\right) \tag{7.35}$$

and the equations of motion for the inputs are

$$\frac{du_{Xi}}{dt} = -\frac{u_{Xi}}{t} - A\sum_{j \neq i}V_{Xj} - B\sum_{Y \neq X}V_{Yi}$$
$$- C(\sum_X \sum_j V_{Xj} - n) - D\sum_Y d_{XY}(V_{Y,i+1} + V_{Y,i-1}) \tag{7.36}$$

The relative magnitudes of parameters A, B, C, and D represent the importance that we place on the respective constraints. The common scaling factor is determined by how rapidly we wish to carry out the convergence.

Hopfield and Tank (1985) report good results obtained with this approach, but the question of local minima remains a troublesome issue. Wilson and Pawley (1988) reported various difficulties involved with the approach, some of which can be taken care of. The real difficulty is the question of local minima.

At present there is a lack of data on comparison of the performance of such nets relative to other dynamic programming approaches.

6 COMMENTS ON THE USE OF PATTERN-FORMATTED KNOWLEDGE

Undoubtedly, Perceptron-like learning is one of the most useful knowledge acquisition procedures which has evolved from activities in neural nets. It is useful for representing knowledge by learning a mapping which represents many acts of cause and effect. Thus, in principle, it could represent the physics of a mechanistic process just as well as the subjective judgment of a group of experts.

This obvious application is being overworked. Important elementary mathematical matters are often blithely neglected. In this connection, we note that the GDR back propagation algorithm does not scale well and has low rates of learning. The work on higher-order nets by Giles and his collaborators and by Pao and his collaborators is very important in finding ways out of these difficulties.

There is no question that the Walsh holographic memory is a high-capacity heteroassociative memory and may be used in matrix or net forms as desired. In our opinion, the use of the network associative memory for suboptimization is still a worthwhile task.

7 BIBLIOGRAPHIC COMMENTS

The delta rule Perceptron was first proposed by Rosenblatt during the 1959–1961 period (Rosenblatt, 1962), and the description of the layered machine by Nilsson (1965) in his book on learning machines indicated clearly that the ideas of hidden layers and successive transformations were well understood by that time. The eclipse of that concept for supervised learning was due, in part, to misinterpretation and misapplication of the results of the analysis by Minsky and Papert (1969) of the limitations of the delta rule device. More to the point was the fact that no one knew of a good way to train such devices at that time, one hurdle being the use of hard-limiting on/off threshold logic units for nodes. Under those circumstances, if and when a node did not turn on, it was difficult to assess the reason. In other words, it could not be determined whether the input signal was slightly understrength or very much understrength. Accordingly, there was no rational procedure for improving the weights and thresholds.

In the intervening years, several researchers thought of using analytic functions for representing the activation function, but Rumelhart, Hinton, and Williams (1986) described specifically a procedure for utilizing the analyticity for back propagation of error.

It is unfortunate that the GDR in its present form does not scale well and often learns very slowly or wanders seemingly aimlessly in weight space. For this and for other reasons, researchers have probed to see if net performance might be improved through the inclusion of some various forms of nonlinear terms. These efforts include the proposed sigma-pi structures, the metageneralized delta rule net (Pomerleau, 1987) in which some link weights are also multiplied by feature values, introduction of higher-order effects (Giles and Maxwell, 1987; Chen et al., 1986; Lee et al., 1986) and the functional-link net. These developments are also discussed in *Adaptive Pattern Recognition and Neural Networks* by Pao (1989a). The functional-link net (Pao and Beer, 1988; Pao, 1989b) is a continuation and generalization of the seminal work by Giles and his collaborators, and there are indications that many interesting and powerful improvements will result from further investigation of that functional-link approach. We note in passing that artificial neural networks are patterned more closely after the schematic ideas of the Perceptron rather than the more complex channels and variable conductances of neurobiology. The functional-link net approach is probably the mode which will most easily accommodate the languages and concepts of the two views and areas of activity.

Combined use of supervised and unsupervised learning can provide powerful methodology for the processing of pattern-formatted information. Some discussions of this may be found in the book by Pao (1989a) and in the *AI Expert* article (Pao, 1989b). Accounts of the ART1 and ART2 algorithms may be found by Carpenter and Grossberg (1985, 1987). Discussions of the use of the Euclidean metric in unsupervised learning may be found in the book by Pao (1989a).

Good accounts of matrix, holographic, and network associative memories may be found in the books by Kohonen (1977, 1980). The Walsh transform associative memory was first advocated by Pao and Merat (1975) for a configuration which was particularly economical as far as memory requirements are concerned. That idea was further developed in subsequent publications (Pao and Hartoch, 1982). However, a truly remarkable high-capacity and robust associative memory can be implemented by using Walsh functions as internal references in a straightforward manner (Pao, 1989a).

Hopfield and Tank (1985) suggested that perhaps difficult optimization problems might be approached in a nonalgorithmic manner with use of the network associative memory. This aroused a great deal of interest initially, but since then some of that enthusiasm has been dampened by our realization that it is difficult to ensure true optimization. Local minima might, indeed, be attained, but global optimization is not ensured. In addition, it would seem that every problem needs to be formulated anew by itself, and difficulties might arise because of the manner in which the energy function is expressed (Wilson and Pawley, 1988). Nevertheless, we believe that useful results can be achieved through this approach for specific cases. The report by Takefuji, Lee, and Lin (1988) on their study of the four-color problem supports this view.

REFERENCES

Carpenter, G. A., and S. Grossberg. 1985. Category learning and adaptive pattern recognition, a neural network model. *Proceedings of the Third Army Conference on Applied Mathematics and Computing,* Army Research Office Rep. 86-1, pp. 37–50.

Carpenter, G. A., and S. Grossberg. 1987. ART2: Self-organization of stable category recognition codes for analog input patterns. *Applied Optics,* vol. 26, pp. 4919–4930.

Chen, H. H., Y. C. Lee, G. Z. Sun, H. Y. Lee, T. Maxwell, and C. L. Giles. 1986. High order correlation model for associative memory. *American Institute of Physics Conference Proceedings,* no. 151: *Neural Networks for Computing,* Snowbird, Utah, pp. 398–403.

Giles, C. L., and T. Maxwell. 1987. Learning, invariance and generalization in higher-order neural networks. *Applied Optics,* vol. 26, pp. 4972–4978.

Hopfield, J. J., and D. W. Tank. 1985. "Neural" computation of decisions in optimization problems. *Biological Cybernetics,* vol. 52, pp. 144–152.

Kohonen, T. 1977. *Associative Memory: A System Theoretical Approach.* Springer-Verlag, New York.

Kohonen, T. 1980. *Content-Addressable Memories.* Springer-Verlag, New York.

Lee, Y. C., G. Doolen, H. H. Chen, G. Z. Sun, T. Maxwell, H. Y. Lee, and C. L. Giles. 1986. Machine learning using a higher-order correlation network. *Physica,* vol. 22D, pp. 276–306, North-Holland, Amsterdam.

Minsky, M., and S. Papert. 1969. *Perceptron: An Introduction to Computational Geometry,* M.I.T. Press, Cambridge, Mass.

Nilsson, N. J. 1965. *Learning Machines: Foundations of Training Pattern Classifying Systems,* McGraw-Hill, New York.

Pao, Y. H. 1989a. *Adaptive Pattern Recognition and Neural Networks,* Addison-Wesley, Reading, Mass.

Pao, Y. H. 1989b. Eliminating the hidden layers. *AI Expert,* vol. 4, pp. 60–68. April.

Pao, Y. H., and R. D. Beer. 1988. The functional-link net: A unifying network architecture incorporating higher order effects. International Neural Network Society First Annual Meeting, Boston, The Institute of Electrical and Electronics Engineers, catalog no. 89CH2765-6, September 6–10.

Pao, Y. H., and G. P. Hartoch. 1982. Fast memory access by similarity measure. In: J. Hayes, D. Michie, and Yoh-Han Pao (eds.), *Machine Intelligence,* Wiley, New York.

Pao, Y. H., and F. L. Merat. 1975. Distributed associative memory for patterns. *IEEE Transaction Systems, Man and Cybernetics,* vol. 5, pp. 620–625.

Pomerleau, D. A. 1987. The meta-generalized delta rule: A new algorithm for learning inconnectionist networks. Carnegie-Mellon University, Computer Science Dept. Rep. CMU-CS-85-185, Pittsburgh.

Rosenblatt, F. 1962. *Principles of Neurodynamics: Perceptrons and the Theory of Brain Mechanisms,* Spartan, New York.

Rumelhart, D. E., G. E. Hinton, and R. J. Williams. 1986. Learning internal representations by error propagation. In: D. E. Rumelhart and J. L. McClelland (eds.), *Parallel Distributed Processing: Explorations in the Microstructures of Cognition,* vol. 1: *Foundations,* M.I.T. Press, Cambridge, Mass.

Takefuji, Y., K. C. Lee, and C. W. Lin. 1988. Neural networks solving four color problems. Case Western Reserve University, CAISR Tech. Rep. TR88-139.

Wilson, G. V., and G. S. Pawley. 1988. On the stability of the travelling salesman problem algorithm of Hopfield and Tank. *Biological Cybernetics,* vol. 58, pp. 63–70.

CHAPTER 8

MACHINE LEARNING

YVES KODRATOFF

1 INTRODUCTION: AI APPROACH TO MACHINE LEARNING

This chapter presents a state-of-the-art discussion of the AI approach to machine learning (ML). This approach is defined, and its main achievements are presented. Some lead to methodologies applicable to the automatic building of rules for expert systems. Among them, one can cite the inductive building of deduction trees, several conceptual clustering techniques, and explanation-based learning. All approaches to machine learning raise important research issues, but some are more research-oriented like the use of analogy, still awkward (but which, once achieved, is going to provide tremendous payoff), the improvement of inductive and abductive techniques, and the merging of numeric data analysis and symbolic techniques.

The AI approach shows that we now need symbols conveying more precise information than simple numbers, since the need for explanations is the kind of information required by modern applications of science.

This approach relies on several concepts (induction, abduction, generalization) that are often misunderstood. We therefore give a precise definition of them.

1.1 Induction and Deduction

Let us recall as briefly as possible what induction and deduction are; more details are given in Kodratoff (1988).

Deduction is the process by which one *infers while preserving the truth* of the data already stored. Applying the well-known modus ponens rule is truth-preserving. For instance, suppose that a theorem tells us that all green leaves are young leaves. While meeting green leaves, I can deduce that they are young.

Induction is the process by which one *infers while preserving the falsity* of the data already known. Michalski's generalization rules (Michalski, 1984) are falsity-preserving. For instance, while meeting green leaves I can induce that I am meeting green things (that is, forgetting that they are leaves), or I can induce that I am meeting young things, by using the above theorem. Suppose also that another theorem tells us that anything green is not red; then, conversely, I cannot induce that I am meeting red things since "green" implies "not red," i.e., red = false, and I must preserve this falsity during my induction.

1.2 Knowledge-Intensive Induction

Many authors link deductive learning with knowledge-intensive approaches (which they actually are, in existing implementations) and inductive learning with knowledge-poor approaches [there exist implementations of induction that are knowledge-intensive as well, e.g., Michalski (1983), Kodratoff et al. (1984), and Kodratoff and Ganascia (1986)]. It is clear that "pure induction" is knowledge-poor. The point is that pure induction would mean induction from raw data, and raw data do not exist in reality since the world representation we get in the machine, as raw as it can be, is bound to already bear much interpretation. A clear definition of abduction will help us to better define induction and to show the role of background knowledge in the global inductive process.

1.2.1 WHAT IS ABDUCTION? Technically, abduction is the process which "reverses" the implication symbol, i.e., the process by which, knowing that $A \Rightarrow B$, one infers A, knowing that B is true. Of course, this is not truth-preserving since $A \Rightarrow B$ tells that one can infer B from the knowledge of the truth of A, as classical modus ponens does. In practice, this amounts to the process by which we are able to complete a proof that just failed, by doing hypotheses that, if they were valid, would indeed lead the proof to a success.

Let us illustrate abduction by the following example, drawn from DeJong and Mooney (1986). In this example, our aim is to learn a definition of the concept of suicide KILL(x, x). Suppose, then, that we have been learning so far that suicide requires a weapon, as expressed by the rule

```
KILL(x, x) :- DEPRESSED(x), BUY(x, y), WEAPON(y)
```

Suppose now that we know of Mary who commits suicide. Suppose that we have the following knowledge about her

```
DEPRESSED(MARY).
BUY(MARY, OBJ1).
SLEEPING-PILLS(OBJ1).
PRICE(SLEEPING-PILLS, 20).
```

```
BUY(MARY, OBJ2).
BOOK(OBJ2).
...
```

where OBJ1 and OBJ2 are constants. We will be unable to prove KILL (MARY, MARY) because she has no weapon. Cox and Pietrzykowski (1986) present an algorithm that would automatically look for "weapons" in such a situation where WEAPON(y) fails; it catches as weapons the first object that has been satisfactory so far [here, this that has been satisfying BUY(x, y)]. It therefore abducts WEAPON(SLEEPING-PILLS) or WEAPON(BOOK) depending on which is met first. In the absence of any other kind of knowledge, it is clear that anything that has been bought might be taken as a weapon. We (Duval and Kodratoff, 1989) support the fact that another abduction is also possible, namely, by adding rules in which the objects that have been satisfying BUY(x, y) are added to the clause, without hypothesizing that they are "weapons." Here, we could abduct as well

```
KILL(x, x) :- DEPRESSED(x), BUY(x, y), SLEEPING-PILLS(y)
```

or

```
KILL(x, x) :- DEPRESSED(x), BUY(x, y), BOOK(y)
```

All this shows that even in very constrained environments, many abductions are possible, all of them solving the problem of completing the failed proof. In Addis (1987, 1988) more precise definitions are given; the above is enough to illustrate the difference between abduction and induction.

Abduction is the logical mechanism by which one can recover from failure. Since there are many such possible recoveries, choosing one and arguing why this is the best one (which can be called "an explanation of the recovery") make up the part which we would like to call here "choice and justification." Full induction requires handling them all—the abduction, the choice, and its justification. In many cases, the possible abductions are so many that it is quite hard to draw a sharp line between pure abductive and choice steps, since not all the abductions can be studied.

1.2.2 KNOWLEDGE-INTENSIVE ABDUCTION AND INDUCTION. Most people tend to call "pure induction" what we have been calling abduction. Abduction can be knowledge-poor, even though this is not always the case. For instance, as soon as you propose, as we did, to recover from failure by adding rules, any trivial rule may solve the problem, such as adding

```
KILL(x, x) :- PRICE(SLEEPING-PILLS, 20)
```

to the knowledge base, as a particularly stupid instance (from the knowledge point of view; pure logic without knowledge would find this abduction as good as any). The reason why we do not consider the above clause is that we have already been making an implicit choice by adding clauses that have some analogy with the one that failed to produce the proof. This failed proof, when completed, must have the same structure as the proofs that have been done before.

In our example, the proof that Mary committed suicide must have the same structure as the proofs proving suicides with the use of a weapon. More details can be found in Kodratoff (1988) and Duval and Kodratoff (1989). In that case, the knowledge we are using is implicit, which is often confused with "poor." We therefore claim that when abduction and choice are intertwined, their coupling relies on the background knowledge.

Remember that we defined induction as the triple of abduction, choice, and justification. The last is typically knowledge-intensive since the justification needs to "explain" the chosen abduction by already existing knowledge. In most induction systems, the justification step is intertwined with the others. It follows that in most cases all components of the triple depend on each other, which makes them all knowledge-intensive.

1.2.3 DISCUSSION. Classically, abduction is considered as knowledge-poor. A detailed analysis of the way it is performed shows that this opinion is due to a confusion between implicit and poor knowledge. In most cases, the coding of the knowledge contains much information, so that induction may appear independent of knowledge. Finding the good coding is by itself very much knowledge-intensive. As we shall see, rejecting implicit representations is typical of AI. It is therefore not surprising that knowledge-intensive induction comes to be acknowledged in an AI environment.

1.3 Generalization and Particularization; Bottom-Up and Top-Down Generalization

Tom Mitchell's version spaces (Mitchell, 1982; Genesereth and Nilsson, 1985; Kodratoff, 1988) maintain the generalization state of an operator. They give the exact generalization state in which a descriptor used by an operator must be kept in order to optimize the problem-solving efficiency. Given a set of positive and negative examples, their "version space" is the set of consistent formulas, i.e., the set of formulas that is both complete (they recognize all the positive examples) and coherent (they recognize none of the negative examples).

When the version space of an operator is built, positive examples are used *bottom-up*, i.e., by finding a formula which is more general than any of the examples and such that each example is an instantiation of this formula (also, none of the negative examples should be an instantiation of it, but this is not the point here). Conversely, the negative examples are mainly used *top-down*, since they are used to particularize a formula which covers none of the negative examples (also all the positive examples should be an instantiation of it, but this is not the point here). This should convince people that generalization (from positive examples) and particularization (from negative examples) are two facets of the same process, known under the name *generalization* because earlier works did not sufficiently recognize the importance of negative examples and the symmetry between positive and negative examples (Nicolas, 1988).

There are other approaches than those inspired by the version spaces. Michalski's (Michalski and Stepp, 1983; Michalski, 1984) work is typical of those who progressively generalize from examples (his algorithms perform bottom-up generalization). These algorithms cluster examples together and simultaneously find a function that recognizes these clustered examples together and simultaneously find a function that recognizes these clustered examples and rejects all counterexamples. The progressive growth of these clusters generates more and more general recognition functions. On the contrary, all the ID3 family inspired from Quinlan (1983) starts from a set that contains all the examples and progressively applies descriptors that cut it into subsets, according to the value of the descriptor. For instance, if we start with a set of containing persons with all types of eye colors, we form as many subsets as there are colors of the eye, each containing one color. Therefore, these algorithms progressively particularize; they perform top-down generalization.

Top-down or bottom-up, generalization is always an inductive process. It is therefore very wrong to confuse particularization and deduction. To make this precise, let us describe in a similar way generalization and particularization. When we are generalizing (bottom-up generalization), the problem is to induce the common features existing between several examples. These common features characterize the concept that they are all an instance of. When we are particularizing (top-down generalization), the problem is to induce the features that make the difference between several examples. Each example is characterized by the features not shown by the others.

1.4 What Is the AI Approach to ML?

We present here five features characterizing AI that make the AI approach to ML original. When they are used in a very strict way, i.e., when all are requested to be fulfilled, one defines something like the core of AI. A softer definition of AI, which will include all the existing AI systems, will be to request that some of the features given below be partly fulfilled.

First, contrary to the "intelligent machine" definition (Barr and Feigenbaum, 1981), AI refuses to build intelligent black boxes. For instance, Roger Schank points out (Schank, 1986) that the chess-playing machines presently available on the market are not a product of AI (actually, they have been built without using any AI techniques) even though they start off being good players. The reasons why they play so well are known to their designers only, not to their users, and therefore they cannot be considered as AI. Anyhow, it would be somewhat strange to attribute to AI the success of these chess machines since they result from operational research, not from AI. We do not claim here that temporary stoppage of communication is impossible in AI. Rather we point out that a definitive loss of any communication abilities is contrary to the spirit of AI. AI systems are open to their users who must understand them.

Second, AI tends to avoid using implicit knowledge. A clear illustration is the marked preference AI shows for declarative knowledge representation, in which knowledge and the way to use it are clearly differentiated. AI tends to avoid procedural representations where knowledge and its use are merged. For instance, AI favored the birth of rather strongly declarative languages like Lisp and Prolog. Another well-known example is an expert system (ES). An ES is nothing but a kind of decision tree, except that the tree has been flattened in a declarative form by writing out each rule that led to a decision, without telling in advance in which order they must be called. This generates the so-called conflict resolution problem, which is the price that must be paid for declarativity. In an AI system, knowledge should be stated explicitly, preferably in a declarative way. Procedural knowledge is expressed in a programming language as little as possible. Similarly, fancy ad hoc codings are very much in opposition to the spirit of AI.

Third, contrary to what the early workers believed, a system cannot pass the above tests without already carrying a huge amount of preliminary knowledge. At least, this knowledge must carry the user's understanding of the problem. This is illustrated by the importance given to knowledge bases in AI. Knowledge is a kind of datum: We do not speak of databases because our bases have to be declarative and understandable by their users. An AI system is supposed to make heavy use of background knowledge, provided by the user in a knowledge base.

Fourth, in accordance with the need for user understandability, an AI system has to speak its users' language, instead of speaking a "language of AI," which, by the way, does not exist. This feature distinguishes AI from other sciences whose first concerns are the creation of their own jargon, to be imposed on users as a part of their scientific formation. For instance, statistics describes interesting natural phenomena and does it in terms of mean square distances, etc., which is its own language, instead of expressing the knowledge in its users' language, as AI attempts (and still partly fails) to do.

Fifth, we believe (even if this is far from being accepted by all our colleagues) that AI is concerned with the definition, analysis, measurement, and comparison of explanations relative to an automatically executed reasoning session and provided in the user's language. Our main argument follows from the recognition that explicability fulfills the needs expressed above since the concern for understandability is perfectly met by a system able to provide itself, its users, with explanations. Moreover, a huge amount of background knowledge is necessary to be able to find relevant explanations.

2 EXPLANATION-BASED LEARNING IN STRONG THEORY DOMAINS

The most rigorous explanation-based learning (EBL) learns only generalizations; this is why it has been called *explanation-based generalization* (EBG).

2.1 EBG: An Intuitive Presentation

EBG is given a complete theory of the domain in which the learning takes place and a training instance of the concept to be learned. Since the system knows the complete theory, it knows a definition of the concept to be learned. The goal of the learning process is to learn a new, "better" definition of this concept. To achieve this goal, EBG works as follows. Using its knowledge of the domain theory, it proves that the training instance is actually an instance of the concept learned. The proof trace is then pruned by using an operationality criterion that tells which descriptors must be left out. This amounts to telling which part of the proof is considered as being of interest. Notice the importance of this criterion on which all the efficiency of the process relies. The pruned proof trace is then generalized into an explanation structure, which contains the parts of the original clauses that have been used. As noticed in DeJong and Mooney (1986), this amounts to keeping a part of the unifications needed to achieve the proof. A version of the training instance with variables is then regressed (Waldinger, 1977) through the explanation structure. The result of this regression is a new, more operational definition of the concept.

To illustrate EBG, we use a Prolog version of the SAFE-TO-STACK example, taken from Mitchell, Keller, and Kedar-Cabelli (1986). Several versions of EBG have been implemented, such as in Kedar-Cabelli and McCarty (1987), Puget (1987), and Siqueira and Puget (1988). The SAFE-TO-STACK example shows how to learn a more efficient rule to stack, given a definition, a theory of stacking, and an example of a particular box, BOX1, stacked on a particular table, ENDTABLE1. In this very particular example it happens that one needs to know the default value of the type TABLE.

The first kind of information is relative to the specific box BOX1 and the specific table ENDTABLE1.

```
C1 ON(BOX1, ENDTABLE1).
C2 COLOR(BOX1, RED).
C3 COLOR(ENDTABLE1, BLUE).
C4 VOLUME(BOX1, 10).
C5 DENSITY(BOX1, 1).
C6 FRAGILE(ENDTABLE1).
C7 OWNER(ENDTABLE1, CLYDE).
C8 OWNER(BOX1, BONNIE).
C9 ISA(ENDTABLE1, TABLE).
```

The second kind of information is relative to the background knowledge about stacking things, also called the *domain theory*.

```
C10 SAFE-TO-STACK(x, y)  :-  NOT FRAGILE(y)
C11 SAFE-TO-STACK(x, y)  :-  LIGHTER(x, y)
C12 WEIGHT(x, w)         :-  VOLUME(x, v), DENSITY(x, d), w is v * d
C13 WEIGHT(TABLE, 50)    :-
C14 LIGHTER(x, y)        :-  WEIGHT(x, w1),WEIGHT(y, w2), LESS(w1, w2)
C15 LESS(x, y)           :-  x < y
```

The third kind of information expresses that one can safely stack BOX1 on ENDTABLE1. We want to prove that by a Prolog refutation procedure this knowledge will be given as a question to the Prolog interpreter. It reads

```
C16                :- SAFE-TO-STACK(BOX1, ENDTABLE1)
```

The proof proceeds as shown by the trace in Fig. 8-1. This trace is provided by most Prolog interpreters, though in less readable form. In this trace the descriptors we have chosen to call nonoperational have been circled. By pruning the nonoperational descriptors and generalizing, we obtain the explanation structure shown in Fig. 8-2.

EBG then regresses a general goal through the explanation structure and propagates the obtained instantiations through the whole set of regressed subgoals. The final set of subgoals becomes the new conditions for the achievement of the general goal. Figure 8-3 on p. 235 shows the explanation structure for SAFE-TO-STACK.

Around each part of the explanation structure are the goals and their regressions indicated in boldface. At each level of regression, the substantiations are transmitted in all the subgoals. The last line of subgoals is the regression of SAFE-TO-STACK(OBJ1, OBJ2) through the explanation structure. The rule learned is therefore

```
SAFE-TO-STACK(OBJ1, OBJ2) :- VOLUME(OBJ1, v1), DENSITY (OBJ1, d1),
                    LESS-THAN(v1 * d1, 50), ISA(OBJ2, TABLE)
```

where v1 and d1 are variables and the 50 comes from C13.

2.2 EBL and the Refinement of Strong Theories

For the management of incomplete theories, the central idea is to complete the failure proofs by an abduction mechanism (Duval and Kodratoff, 1989). This

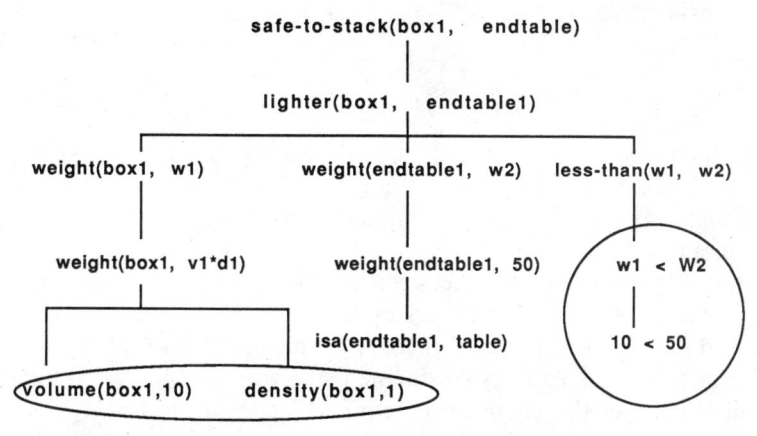

FIGURE 8-1
Tree of the proof that BOX1 can be stacked on ENDTABLE1.

234 KNOWLEDGE ENGINEERING: FUNDAMENTALS

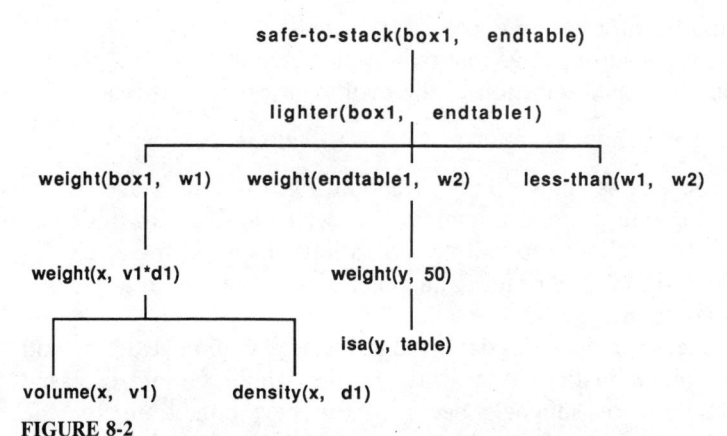

```
                    safe-to-stack(box1,  endtable)
                                |
                    lighter(box1,  endtable1)
        ┌───────────────────────┼───────────────────────┐
weight(box1, w1)     weight(endtable1, w2)      less-than(w1, w2)
        |                       |
weight(x, v1*d1)           weight(y, 50)
   ┌────┴────┐                   |
   |         |                isa(y, table)
volume(x, v1)  density(x, d1)
```

FIGURE 8-2
Pruned tree of the proof; nonoperational descriptors have been removed.

abduction process is guided by analogical reasoning about explanations and enables one to discover new rules that refine the domain theory.

In Section 1.2.1 some training instances led us to learn

```
KILL(x, x) :- DEPRESSED(x), BUY(x, y), WEAPON(y)
```

Recall also the story of Mary's suicide, which comes with this information:

```
DEPRESSED(MARY).
BUY(MARY, OBJ1).
SLEEPING-PILLS(OBJ1).
PRICE(SLEEPING-PILLS, 20).
BUY(MARY, OBJ2).
BOOK(OBJ2).
```

We then proposed to explain Mary's suicide by adding to the knowledge base

```
KILL(x, x) :- DEPRESSED(x), BUY(x, y), SLEEPING-PILLS(y)
```

or

```
KILL(x, x) :- DEPRESSED(x), BUY(x, y), BOOK(y)
```

These two rules have been built because, when they are applied to this story, they lead to a resolution tree analogous to the one obtained to explain the earlier stories. In that way, the theory is completed since we are now able to prove Mary's suicide as well.

We are still left with the choice between the two possible clauses. One way to choose involves considering theories of different granularities that support each other. By the *granularity* of a theory, we mean the level of detail it copes with. For instance, a sociology of suicide usually does not need to take into account the details of the chemical reactions that result in death. Even though social details can hold much subtlety, we say that, in this context, mo-

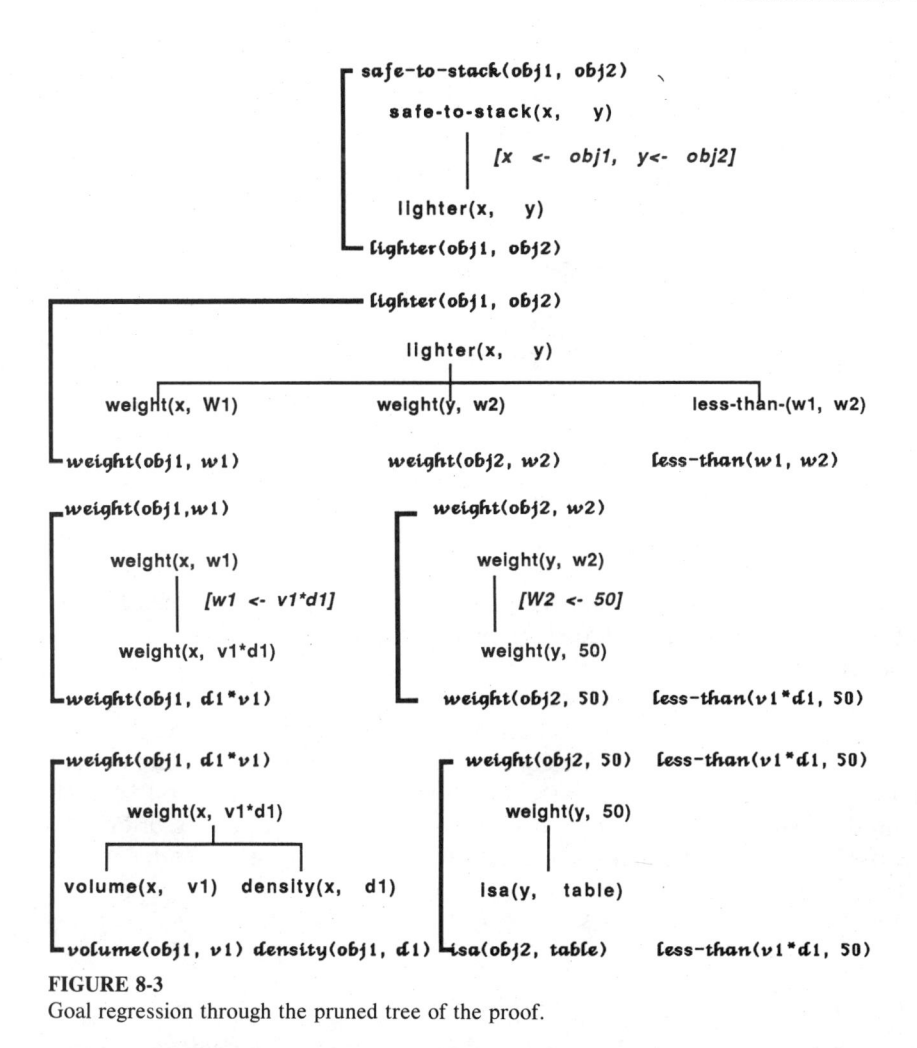

```
          ┌ safe-to-stack(obj1, obj2)
          │   safe-to-stack(x,  y)
          │       │ [x <- obj1, y <- obj2]
          │   lighter(x,  y)
          └ lighter(obj1, obj2)

┌──────────────── lighter(obj1, obj2)
│                    lighter(x,  y)
│   weight(x, W1)      weight(y, w2)        less-than-(w1, w2)
└ weight(obj1, w1)   weight(obj2, w2)       less-than(w1, w2)

┌ weight(obj1, w1)    ┌ weight(obj2, w2)
│   weight(x, w1)     │   weight(y, w2)
│       │ [w1 <- v1*d1]│     │ [W2 <- 50]
│   weight(x, v1*d1)  │   weight(y, 50)
└ weight(obj1, d1*v1) └ weight(obj2, 50)    less-than(v1*d1, 50)

┌ weight(obj1, d1*v1)      ┌ weight(obj2, 50)   less-than(v1*d1, 50)
│   weight(x, v1*d1)       │   weight(y, 50)
│        │                 │        │
│ volume(x, v1)  density(x, d1)  isa(y, table)
└ volume(obj1, v1) density(obj1, d1) └ isa(obj2, table)   less-than(v1*d1, 50)
```

FIGURE 8-3
Goal regression through the pruned tree of the proof.

lecular chemistry is of finer grain and will be discarded in a first step. Suppose that our theory (theory1) so far has been of granularity 1, and suppose that we have also a theory (theory2) of granularity 2 in which the "deeper" properties of the descriptors of theory1 are given. Theory2 will contain knowledge about the use of sleeping pills, books, and other objects that Mary may have been buying. Theory2 will have also more detailed knowledge about the way killing is performed, e.g., about the potential danger of all weapons. Thus, we should be able to prove that the potential danger of a book is of pure psychological nature, while the potential danger of an excess of sleeping pills is physical. We could then come back to theory1 with an explanation of why the sleeping pills, provided they are used in excess, are suitable tools for suicide. Using theory2, we could then revise theory1 by adding

```
KILL(x, x) :- DEPRESSED(x), BUY(x, y), SLEEPING-PILLS(y), EXCESS-
              INGEST(x, y)
```

If we are concerned only about choosing among the possible abductions, then we could also put aside all the stories unaccounted for by

```
KILL(x, x) :- DEPRESSED(x), BUY(x, y), WEAPON(y)
```

and, after many are found, cluster those that present common features. If the stories are varied enough, then SLEEPING-PILLS should be the only common feature among several examples.

2.3 Storing Chunks of Knowledge

After a problem-solving session, the trace of the solution obtained, or part of it, can be kept. Since such solutions are then stored in the knowledge base, the trace can be used to obtain immediate solution in further applications. At this stage, the problem consists of keeping only the most interesting chunks of information.

In practice, many other problems arise when the chunks are simply piled together and the system starts crumbling under its own weight. One then has to recognize when two chunks are equivalent or can be generalized into a more efficient one and to organize the chunks into "chunks of chunks," which amounts to using metaknowledge to sort them. The system Soar (Laird et al., 1986, 1987), which implements some of these ideas, is currently under test in the industrial environment.

3 EBL IN SOFT THEORY DOMAINS

Three different approaches have extracted explanations from soft deep theories (by *soft,* we mean incomplete and/or incoherent and/or intractable). One is illustrated by the system Protos (Bareiss, Porter, and Wier, 1989), the second one by the system DISCIPLE (Kodratoff and Tecuci, 1987a, 1987b), and the third one by Schank's theory of XPs (Schank, 1987).

3.1 Definition of an Explanation for Protos

In Protos, the production rules are associated to elementary explanations, which describe what kind of relation takes place between the antecedent and the consequent of each production rule. There are six such elementary explanations. Among them, one finds the classical ISA (generality relationship) and PART-OF (part-to-whole mappings), but also ENABLES which tells that a feature enables the function of an object (e.g., wings enable flight).

An explanation is a combination of such elementary explanations as given by a path in the relationship frame among objects. For example, one can say that the relation between engine and car is PART-OF. But a better explanation can be provided by considering a less direct path: the engine ENABLES the movement which is the FUNCTION of vehicles and car ISA vehicle. This knowledge is part of the training provided by the teacher to underline the explanations that are significant.

In the above example, suppose that the system is taught that the second path provides a good explanation, but not the first path. Then, in the future, when trying to detect if some unknown x is an engine, it will not try to prove that x is a part of an engine, but it will try to see if x can enable the movement of some vehicle.

In Protos a good explanation is a path in the frame of the relationships among objects, the path provided by the user.

3.2 Definition of an Explanation for DISCIPLE

DISCIPLE is an interactive learning apprentice system for weak theory fields. In DISCIPLE, the learning process starts with an example of a problem-solving step. Take, for instance, the following rule for the construction of loudspeakers.

EXAMPLE1

```
ATTACH sectors  ON chassis-membrane-assembly  ⊢
        APPLY mowicoll ON sectors
        PRESS sectors ON chassis-membrane-assembly
```

where $A \vdash B$ means "in order to achieve action A, perform actions B," where mowicoll is a kind of glue, and sectors and chassis-membrane-assembly are parts of a loudspeaker. This is interpreted as an example of a general rule indicating a way of performing the ATTACH action:

GENERAL RULE1

```
ATTACH x ON y  ⊢
        APPLY z ON x
        PRESS x ON y
```

Being completely instantiated, example1 implicitly contains the proof of its validity in the supposedly known properties of sectors, chassis-membrane-assembly, and mowicoll. On the contrary, since we have no knowledge of the possible properties of the variables found in General Rule1, it does not contain any implicit explanation.

Nevertheless, it is essential to obtain this knowledge, because it tells us when it is valid to perform an ATTACH operation as a sequence of APPLY and PRESS. To achieve this goal, the system overgeneralizes the examples (by transforming Example1 in General Rule1, for instance) and applies this overgeneralization to the data basis of the user to find instances of the overgeneralization. These instances are proposed to the user as tentative rules. When the user validates such a system-produced rule, we say that the system is provided with a positive example of its overgeneralization. When the user rejects such a rule, we speak of a negative example. This set of positive and negative examples is used by a classical generalizer such as AGAPE (Kodratoff et al., 1984) to produce a more refined generalization. Finally, from these interactions with the user, DISCIPLE may find what is the correct generalization of the proofs of the examples. This gives a correct set of conditions for the application of the generalized rule and therefore a good explanation.

For instance, the final rule DISCIPLE will learn, from Example1 and further interactions with the user, is the following:

GENERAL RULE2
```
IF
            z ISA adhesive
            z TYPE pure
            z GLUES x
            z GLUES y
THEN
        ATTACH x ON y ⊢
                APPLY z ON x
                PRESS x ON y
```

Since the structure of the rule is precisely the structure of the example, what is to be learned is just the explanation of the rule.

An explanation in DISCIPLE is a set of conditions that authorizes the application of a general rule. For instance, the conditions of General Rule2 are an explanation of General Rule1. General Rule2 is always obtained as the generalization of the particular proofs of the example rules provided to the system.

The definition of an explanation is therefore very similar to that of Section 2, the difference being in the generalization process used. DISCIPLE is just in the middle between strong theory EBL and Protos. In strong theory EBL, the explanations are provided by the system to the user, in Protos they are provided by the user to the system, and in DISCIPLE they are obtained as the final result of a consultation of the user with the system.

3.3 Prestored Explanations: Schank's XPs

Roger Schank has recently proposed his own vision of ML and AI, of which creativity is the central theme (Schank, 1987). Nevertheless, in this theory, the essential creative information is constituted by a set of explanation patterns (XPs). These XPs are prestored explanations that can be applied in a variety of situations. Scripts (Schank and Abelson, 1977) are classical in AI; they contain prestored information about standard situations. Let us define an XP as being a script for an explanation.

An XP is made of four parts:

- An index that allows us to get at the XP
- A set of states of the world under which the XP can expect to be activated
- A scenario which is a causal chain of states and events
- The resultant state following from the scenario application

For instance, in the case of the unexpected death of a promising racehorse named Swale, Schank (1986) proposes several XPs containing the following scenarios:

"Early death comes from being malnourished as a youth."
"High living brings early death."
"An inactive mind can cause the body to suffer."
"High-pressure jobs cause heart attacks."

The first scenario leads to ask whether Swale underwent bad treatments during its youth, etc. In the case of Protos and DISCIPLE, the problem could be viewed as one of obtaining new XPs. In this case, the explanations are prerecorded, and the problems are (1) the retrieval of the relevant explanations from a certainly huge number of them, (2) the application through some analogical instantiation process to the particular case under consideration, and (3) their modifications to fit into unexpected situations (tweaking them, as Schank puts it).

In practice, we can guess that some kind of XPs will have to be given beforehand to the system, but also that no system will be able to run a real-life problem without a mechanism's allowing it to generate its own explanations.

4 ANALOGICAL REASONING

First, let us briefly recall what analogical reasoning is. For more details, see Chouraqui (1985), Davies and Russell (1987), or Kodratoff (1988). Let us suppose that we dispose of a piece of information, called the source S (which can be viewed as a knowledge base) and that this information can be put into the form of a doublet (A, B) in which B depends causally on A. Let us suppose now that another piece of information, called the target T, can be put into the same form (A', B') with the requirement that there exist some resemblance between A and A'. See Fig. 8-4.

This scheme can be used in different ways to define analogical reasoning. In most cases, we consider that we know more or less precisely A, B, A', α, and β and that we try to invent B' or to justify its plausibility.

Let us illustrate a different use of analogy (Winston, 1982) in which the result of the analogy is finding the causality relation β, allowing an analogy to be drawn. Consider the following text.

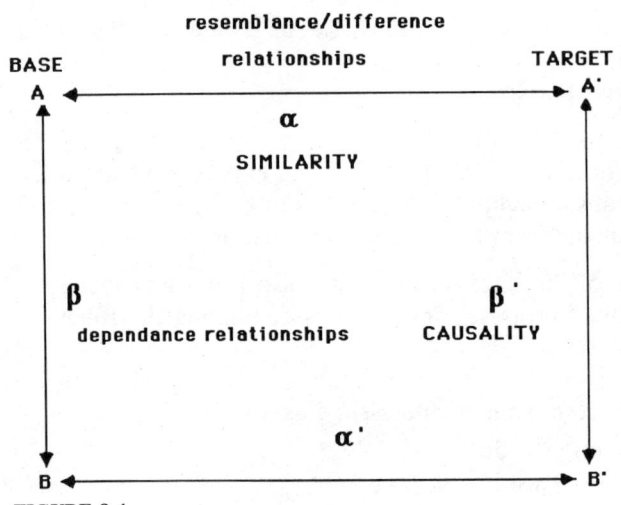

FIGURE 8-4
The general scheme of analogy.

> Let A = Earth has an atmosphere of 1013 mbar containing oxygen, B = Earth is inhabited by humans, A' = Mars has an atmosphere of 5 mbar containing oxygen, α = both Earth and Mars have an atmosphere containing oxygen, but the atmospheric pressure is 1013 mbar on Earth and 5 mbar on Mars.

The analogy problem is to complete the diagram in Fig. 8-5, in which β and β' are missing.

In that case, to be realistic we need a knowledge base from which A can be extracted. Let us therefore suppose that we have access to a huge knowledge base KB about earth life, in which the A above is contained together with a lot of information about life conditions on Earth. Suppose that we know for sure that A' = *the atmosphere of Mars contains oxygen, and its atmospheric pressure is 5 mbar.* Suppose also that we are asking the question: Is there any reason for us to believe that some algae live on Mars? Therefore, B' is assumed to be *algae live on Mars.* Now the process of analogy will consist of finding some information within KB that is relevant to A'. In this simple example, we can easily find that A = *Earth atmosphere contains oxygen, and its pressure is 1013 mbar,* which obviously is a very good match for A'. In a more realistic example, we can imagine that finding A within KB is a tedious process.

Now that A has been found, we try to determine B by finding in KB "something" that lives on Earth *and* has a causality relationship with oxygen and atmospheric pressure. Suppose that there is a causality relationship between oxygen, atmospheric pressure, and the presence of humans on Earth. Thus β is discovered to be *humans need oxygen and a 1013-mbar atmospheric pressure to live,* and B is discovered to be *humans live on Earth.* Now, since we are assuming that $\beta = \beta'$, we can also claim that algae live on Mars be-

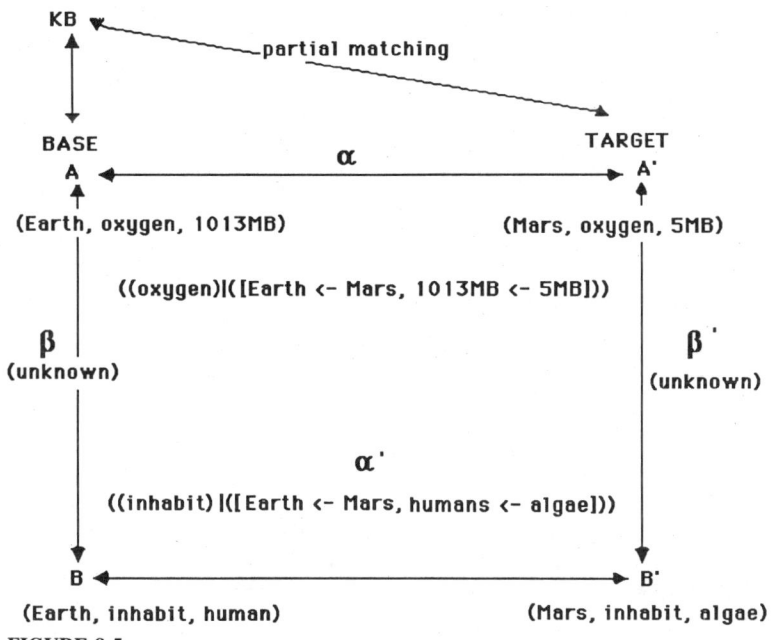

```
KB ←────── partial matching ──────┐
↑↓                                ↓
BASE           α              TARGET
 A  ←─────────────────────────→  A'
 ↑                                ↑
(Earth, oxygen, 1013MB)    (Mars, oxygen, 5MB)

    ((oxygen)|([Earth <- Mars, 1013MB <- 5MB]))

β  |                              β'
(unknown)                      (unknown)

                  α'
    ((inhabit)|([Earth <- Mars, humans <- algae]))

 ↓                                ↓
 B  ←─────────────────────────→  B'
(Earth, inhabit, human)    (Mars, inhabit, algae)
```

FIGURE 8-5
An application of the general scheme to a specific analogy.

cause they need oxygen (as humans do on Earth) and that they need a 5-mbar atmospheric pressure (unlike the humans that need a 1013-mbar pressure).

To summarize, the scheme for finding the causality underlying an analogy is as follows (see Fig. 8-6). From a knowledge base *KB* (as in Fig. 8-5) extract the base *A* matching the target *A'*. Find the causal relationships in *A* that can be translated to *A'*. Then the "underlying causality" is the one just found. For instance, consider the case where one would like to understand the causality hidden behind the argumentation of a lawyer about a case. Then *KB* is the general knowledge about law, *A'* is the present argumentation, *A* is the part of general knowledge of law that fits with the present argumentation, and β is the hidden causality one is trying to discover. The scheme in Fig. 8-6 summarizes this view of analogy, which stems from Winston (1982).

The process we just described requires a partial pattern matcher. It is quite easy to implement in Prolog.

Notice that there exist several other approaches to analogy that we did not study here. The interested reader should refer to Michalski, Carbonell, and Mitchell (1983, 1986), Priedetis (1988), and Kodratoff (1988).

5 INDUCTIVE LEARNING: ID3, INDUCE, AND STRUCTURAL MATCHING

We present here three induction methods ordered by the amount of background knowledge they rely on. ID3 is typically knowledge-poor. Neverthe-

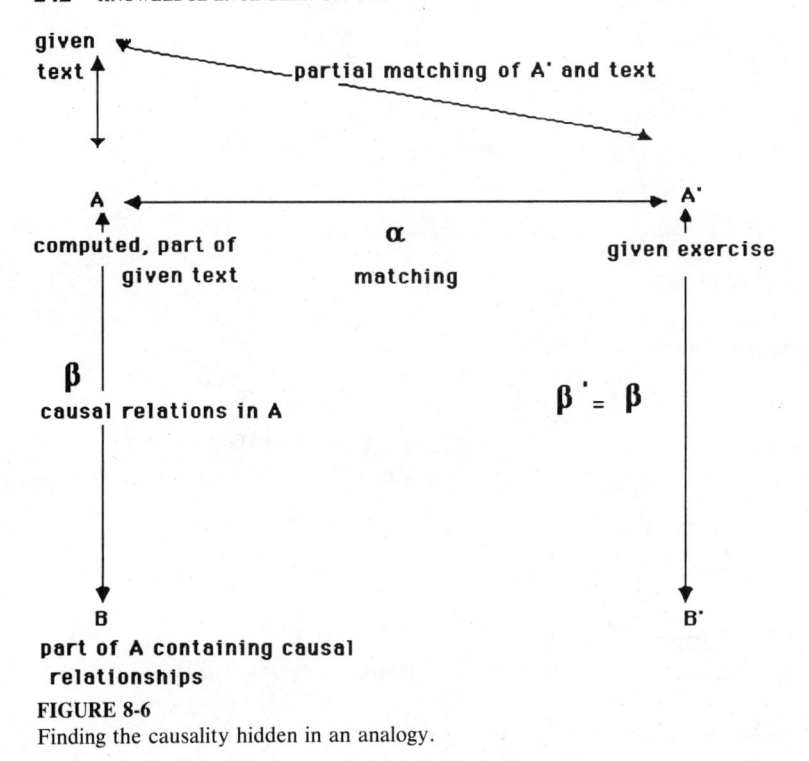

```
given text  ──partial matching of A' and text──▶
   ↕

A ◀────────── α ──────────▶ A'
↑          matching          ↑
computed, part of          given exercise
  given text

β |                         β' = β
causal relations in A

▼                           ▼
B                           B'
part of A containing causal
  relationships
```

FIGURE 8-6
Finding the causality hidden in an analogy.

less, all the work done on ID3 from the original version (Quinlan, 1983) to its recent improvements (see Bratko and Lavrac, 1987) is centered on understandability rather than efficiency, which gives to these works their deep originality as well as their belonging to AI. Surprisingly enough, most improvements in understandability have been followed by an increase in efficiency as well (Bratko, 1988).

5.1 ID3

5.1.1 STATING THE PROBLEM. The aim of this method is as follows. Given a set of descriptors, examples, and concepts to which the examples belong, find the most efficient way to "recognize" the examples. The method relies on information theory; it measures the amount of information associated with each descriptor and chooses the most informative one. By applying descriptors in succession, a decision tree is built that will recognize the examples. As opposed to similar numerical techniques, ID3 preserves the understandability of its results by avoiding the introduction of linear combinations of descriptors.

Example 1. Suppose that we start with two concepts (A and B) described by three descriptors (size, nationality, family) that can take the values (small, large) for the descriptor size, (French, German, Italian) for nationality, and (married, single) for family. Suppose that the concepts are illustrated by the examples in Table 8-1. ID3 will find an optimized decision tree to recognize concept A from concept B.

5.1.2 NUMERICAL OPTIMIZATION. For each descriptor that has not yet been used, the disorder left after applying a descriptor in the decision tree is computed. The one that leaves less disorder is chosen as the next node of the tree. The process stops when each leaf of the tree contains examples of one concept only. The disorder is computed after the classical entropy formula

$$E = - \sum f_s \log_2 f_s$$

where f_s is the relative frequency of class s.

In our example the initial disorder is given by

$$E0 = -(3/8 \log_2 3/8 + 5/8 \log_2 5/8) = 0.954$$

Let us now compute the information gain of applying each descriptor. For instance, the information value of *nationality* is obtained by analyzing the result of classifying the examples by *nationality* and computing the disorder after applying *nationality*. See Fig. 8-7.

The disorder in the cluster *Italian* is $-(3/3 \log_2 3/3) = 0$, in the cluster *French* it is $-(1/1 \log_2 1/1) = 0$, and in the cluster *German* it is $-(2/4 \log_2 2/4 + 2/4 \log_2 2/4) = 1$. Therefore, the disorder left after applying *nationality* is

$$E1 = (3/8 * 0) + (1/8 * 0) + (4/8 * 1) = 0.5$$

The gain in information is $E0 - E1 = 0.454$.

The same computation is done for, say, *family*. Applying *family*, we get Fig. 8-8.

The disorder after applying *family* in the cluster *single* is $-(2/5 \log_2 2/5 + 3/5 \log_2 3/5) = 0.971$ and in the cluster *married* is $-(3/3 \log_2 3/3) = 0$. The disorder left after applying *family* is therefore

$$E'1 = 5/8 * 0.971 + 3/8 * 0 = 0.607$$

TABLE 8-1

	Concept A			Concept B	
Size	Nationality	Family	Size	Nationality	Family
small	German	single	small	Italian	single
large	French	single	large	German	married
large	German	single	large	Italian	single
			large	Italian	married
			small	German	married

```
                        nationality
          ┌─────────────────┼─────────────────┐
       Italian            French            German
          │                  │                  │
          │       ┌──────────────────────┐     │
          │       │ large French single:A│     │
          │       └──────────────────────┘     │
┌──────────────────────┐         ┌──────────────────────┐
│ small Italian single:B│         │ small German single:A│
│ large Italian single:B│         │ large German married:B│
│ large Italian married:B│        │ large German single:A│
└──────────────────────┘         │ small German married:B│
                                 └──────────────────────┘
```

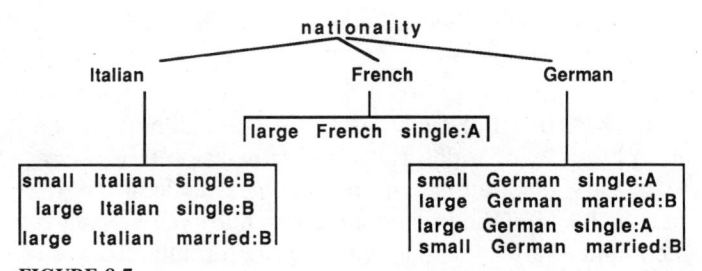

FIGURE 8-7
Result of the application of descriptor *nationality*.

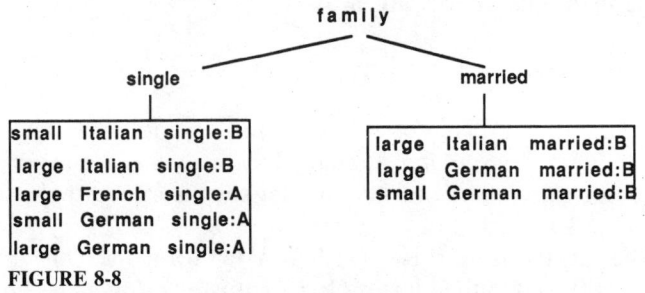

```
                        family
              ┌───────────────┴───────────────┐
            single                         married
              │                               │
┌──────────────────────┐          ┌──────────────────────┐
│ small Italian single:B│          │ large Italian married:B│
│ large Italian single:B│          │ large German married:B│
│ large French single:A │          │ small German married:B│
│ small German single:A │          └──────────────────────┘
│ large German single:A │
└──────────────────────┘
```

FIGURE 8-8
Result of the application of descriptor *family*.

and the associated gain in information is E0 − E'1 = 0.347. The gain is less than when *nationality* was applied, which therefore is preferred to *family*.

To complete our example and show a decision tree, let us know compute the information value of *nationality* followed by *family*. Applying them in order leads to Fig. 8-9, in which all leaves contain an example of a single concept. This is therefore a complete decision tree which allows us to decide whether an example belongs to concept *A* or to concept *B*.

From the initial examples, we knew that applying the three descriptors in any order was indeed a decision procedure. We have now learned that applying *nationality* and *family* is enough and that we optimize the search speed by applying them in that order.

5.1.3 IMPROVEMENTS ON ID3. All improvements performed are relative to a better understandability of the results. For instance, in some cases it seemed that preferring binary trees branching on their right only would improve understandability (Arbab and Michie, 1985). Other notable improvements include pruning the tree [see Bratko and Lavrac (1987)], transforming the tree to decision rules (Corlett, 1983), and using background knowledge (Kodratoff et al., 1987; Nunez, 1988).

```
                              nationality
          Italian             French              German
             |                   |                   |
         CONCEPT B           CONCEPT A               |
             |                   |                   |
    ┌────────────────────┐  ┌──────────────────┐   family
    │small  italian single│  │large French single│   ╱    ╲
    │small  italian single│  └──────────────────┘  single  married
    │large  italian single│                          |        |
    └────────────────────┘                       CONCEPT A  CONCEPT B
                                                     |        |
                                              ┌──────────────┐┌──────────────────┐
                                              │small German single││large German married│
                                              │large German single││small German married│
                                              └──────────────┘└──────────────────┘
```

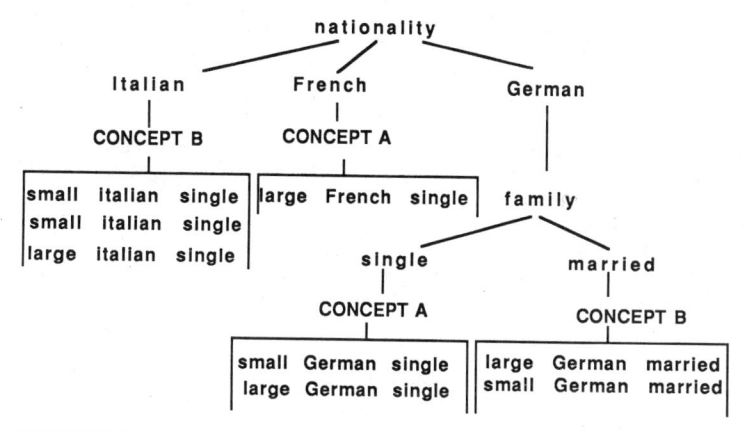

FIGURE 8-9
Result of the application of descriptor *nationality* followed by *family*.

5.2 INDUCE

5.2.1 METHODOLOGY. INDUCE builds a function R that recognizes positive examples and rejects negative ones, i.e., a complete and coherent function. To achieve this goal, it starts by building the most general recognition function R_i which recognizes the example e_i and rejects the sets of examples $\{e_j\}$. Details on how to build such a function are given later. Let POS be the set of positive examples and NEG be the set of negative examples; let e_i be the current member of POS. The algorithm to compute R is as follows. First, as we said, suppose that we have been able to compute R_i, the recognition function that recognizes e_i (and some others of POS) and rejects all instances of NEG. If it happens that R_i recognizes all of POS, we have obtained a complete and consistent recognition function $R = R_i$. If not, we choose another example e_j and compute R_j. Then $R_i \lor R_j$ (where $\lor$ is the logical disjunction) will improve on R_i; it is consistent by construction. Go on as long as $R_i \lor \cdots \lor R_n$ is not complete.

> **Example 2.** Suppose that we have 10 examples described by means of four descriptors and their values, as shown in Table 8-2. The background knowledge is expressed by the properties of the descriptors. They can be linear like X1, a linear descriptor describing the values of weight. This will be expressed by the formula {light < medium-weight < heavy}.
>
> Some descriptors can also be structured like X2 and X3.
>
> The background knowledge relative to X2 may be given by the taxonomies of generality in Fig. 8-10.
>
> It can also happen that no particular knowledge is available for a descriptor, such as X4 which is a nominal descriptor. This knowledge is expressed by describing the set of values as an unstructured set: {blond, red, chestnut}.

TABLE 8-2

	X1	X2	X3	X4
e1	light	boy	baby	red
e2	light	girl	baby	blond
e3	light	boy	child	chestnut
e4	medium	woman	teenager	chestnut
e5	medium	boy	child	red
e6	heavy	girl	child	blond
e7	heavy	man	teenager	red
e8	heavy	woman	teenager	chestnut
e9	heavy	man	adult	blond
e10	heavy	woman	adult	chestnut

5.2.2 BUILDING R_i

Definition 1. Let us define a recognition function R as being described by

$$R = (X_i = V_i) \, \& \cdots \& \, (X_k = V_k)$$

where the X_j's are descriptors and the V_j's are the values of these descriptors or disjunctions of possible values. For example, follow four different recognition functions:

$X2 = \text{man}$
$X4 = \text{man} \vee \text{woman} \vee \text{boy}$
$X3 = [\text{baby} \ldots \text{teenager}]$
$(X2 = \text{man}) \, \& \, (X3 = [\text{baby} \ldots \text{teenager}])$

Definition 2. A subset recognized by a recognition function is the subset of examples whose descriptor shows the same value as the recognition function. For

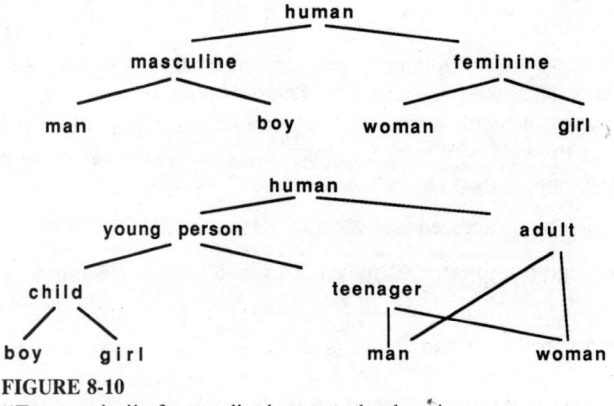

FIGURE 8-10
"Taxonomies" of generality between the descriptors.

instance, the recognition function R1: X2 = man$\vee$woman$\vee$boy recognizes the subset {e1, e3, e4, e5, e7, e8, e9, e10}. The function R2: (X1 = heavy$\vee$light) & (X4 = red$\vee$chestnut) recognizes {e1, e3, e7, e8, e10}.

Creation of a recognition function that recognizes e_i and rejects e_j: $G(e_i/e_j)$. In the beginning, $G(e_i/e_j)$ is empty. Then if the descriptor Xi has the same value in e_i and in e_j, then Xi does not occur in $G(e_i/e_j)$. If the descriptor Xi is different in e_i and e_j, let Vj be the value of Xi in e_j, then add the disjunct $\vee$(Xi $\neq$ Vj) to $G(e_i/e_j)$.

As an example, e1 and e2 differ by X2 and X4. In e2, X2 = girl and X4 = blond. It follows that G(e1/e2) = (X2 $\neq$ girl)$\vee$(X4 $\neq$ blond). Similarly, one sees that G(e1/e4) = (X1 $\neq$ medium)$\vee$(X2 $\neq$ woman)$\vee$(X3 $\neq$ teenager)$\vee$(X4 $\neq$ chestnut).

Comparison of an example and a set. We are now able to compute the most general recognition function which recognizes the example e_i and rejects the set of examples $\{e_j\}$. Let us denote it by $G(e_i/\{e_j\})$. This is what we have been calling R_i above.

For each e_k in $\{e_j\}$, calculate $G(e_i/e_k)$. The conjunction of all G obtained is $G(e_i/\{e_j\})$:

$$R_i = G(e_i/\{e_j\}) = \&_{e_k} G(e_i/e_k) \qquad e_k \in \{e_j\}$$

For example, G(e1/{e2, e4}) = G(e1/e2) & G(e1/e4) = ((X2 $\neq$ girl)$\vee$(X4 $\neq$ blond)) & ((X1 $\neq$ medium)$\vee$(X2 $\neq$ woman)$\vee$(X3 $\neq$ teenager)$\vee$(X4 $\neq$ chestnut)) which is then put in normal form.

5.2.3 IMPROVING THE RECOGNITION FUNCTIONS. These improvements are performed by using numerical information and background knowledge. Typical symbolic improvements minimize the number of disjuncts, improve the simplicity of the description, and replace disjunctions by interval when the descriptors are linear.

5.3 Structural Matching

5.3.1 DEFINITION OF STRUCTURAL MATCHING. Two formulas *structurally match* if they are identical except for the constants and the variables that instantiate their predicates. More formally, let E1 and E2 be two formulas. Then E1 structurally matches E2 if there exists a formula C and two substitutions s_1 and s_2 such that

1. $s_1 \circ C = E1$ and $s_2 \circ C = E2$.
2. Substitutions s_1 and s_2 never replace a variable by a formula or a function.

Here $\circ$ denotes the application of a substitution to a formula.

It must be understood that structural matching (SM) may be difficult up to undecidable. Nevertheless, in most cases, one can use the information coming from the other examples, to know how to orient the proofs necessary to the application of this definition (for details, see Vrain, 1987). Even if SM fails (which may often occur), the effects of the attempt to put into SM may still be interesting. We say that two formulas have been SMatched when every possible property has been used to put them into SM. If the SM is a success, then SMatching is identical to putting into SM. Otherwise, SMatching keeps the best possible result in the direction of matching formulas.

5.3.2 A SIMPLE EXAMPLE OF SUCCESSFUL SM. Let us consider the following two examples. E1 represents A piled on B, and E2 represents C and D, as shown in Fig. 8-11. The examples can be described by these formulas:

$$E1 = \text{SQUARE}(A) \ \& \ \text{SMALL}(A) \ \& \ \text{UPON}(A, B) \ \& \ \text{CIRCLE}(B)$$
$$\& \ \text{BIG}(B) \ \& \ \text{UNDER}(B, A)$$

$$E2 = \text{TRIANGLE}(C) \ \& \ \text{SMALL}(C) \ \& \ \text{LEFT\_SIDE}(C, D)$$
$$\& \ \text{SQUARE}(D) \ \& \ \text{SMALL}(D) \ \& \ \text{RIGHT\_SIDE}(D, C)$$

Let us suppose that the hierarchies in Fig. 8-12 are provided to the system. These taxonomies represent our semantic knowledge about the microworld in which learning is taking place. The SM of E1 and E2 proceeds by transforming them to equivalent formulas $E'1$ and $E'2$, such that $E'1$ is equivalent to E1 and $E'2$ is equivalent to E2 in the microworld (i.e., taking into account its semantics). When the process is completed, $E'1$ and $E'2$ are made of two parts: One is a version of E1 and E2 with variables called the *body* of the SMatched formulas. When SM succeeds, the bodies of $E'1$ and $E'2$ are identical. The other part, called the *bindings* (of the variables), gives all the conditions necessary for the body of each Ei' to be identical to the corresponding Ei.

The process starts as follows. Suppose that we know that the descriptors of the taxonomy FORM are "more interesting" than those of the taxonomy TOUCH. We therefore start attempting to match SQUARE and CIRCLE of E1 with SQUARE and TRIANGLE of E2. We replace A in E1 and D in E2 by the pseudovariable x, and we keep memory of this value in the bindings. This gives (note that the list of bindings is between brackets)

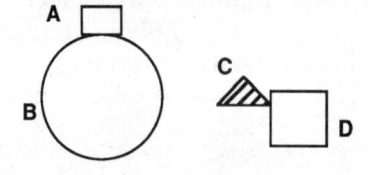

$$E'1 = \text{SQUARE}(x) \ \& \ \text{SMALL}(x) \ \& \ \text{UPON}(x, B) \ \& \ \text{CIRCLE}(B)$$
$$\& \ \text{BIG}(B) \ \& \ \text{UNDER}(B, x) \ \& \ [(x = A)]$$

FIGURE 8-11
E1 is the left part of the figure; E2 is its right part.

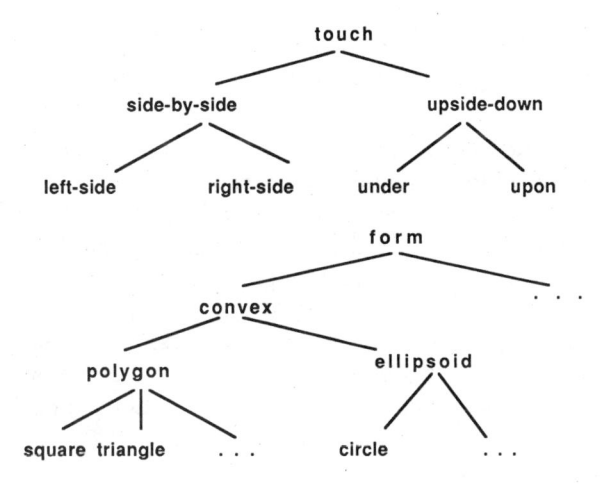

FIGURE 8-12 Taxonomies of generality relationships among the descriptors.

$$E'2 = \text{SQUARE}(x) \ \& \ \text{SMALL}(x) \ \& \ \text{RIGHT\_SIDE}(x, C) \ \& \\ \text{TRIANGLE}(C) \ \& \ \text{SMALL}(C) \ \& \ \text{LEFT\_SIDE}(C, x) \ \& \\ [(x = D)]$$

The next step uses the taxonomies by matching the CIRCLE of E1 to the TRIANGLE of E2. One obtains

$$E''1 = \text{SQUARE}(x) \ \& \ \text{SMALL}(x) \ \& \ \text{UPON}(x, y) \ \& \ \text{CONVEX}(y) \\ \& \ \text{BIG}(y) \ \& \ \text{UNDER}(y, x) \ \& \ [(x = A) \ \& \ (y = B)]$$

$$E''2 = \text{SQUARE}(x) \ \& \ \text{SMALL}(x) \ \& \ \text{RIGHT\_SIDE}(x, y) \ \& \\ \text{CONVEX}(y) \ \& \ \text{SMALL}(y) \ \& \ \text{LEFT\_SIDE}(y, x) \ \& \\ [(x = D) \ \& \ (y = C)]$$

This example shows well that once this SM step has been performed, the generalization step itself becomes trivial: we keep in the generalization all the bindings common to the SMatched formulas and drop all those not in common. In other words, this SM technique allows us to reduce the well-known generalization rules (Michalski, 1983, 1984) just to the "dropping-condition rule," which becomes legal on SMatched formulas. All the induction power is in the dropping-condition rule; all other rules are purely deductive. We must confess that formal proof of the above statement is still under research.

Since BIG and SMALL do not belong to a common taxonomy in our example, they will be dropped, and since x and y have always different values, this will be noted in the final generalization, which is

$$Eg = \text{SQUARE}(x) \ \& \ \text{SMALL}(x) \ \& \ \text{TOUCH}(x, y) \ \& \ \text{CONVEX}(y) \\ \& \ \text{TOUCH}(y, x) \ \& \ [(x \neq y)]$$

In this example, one feels that one could use theorems like

$$\forall x, y\ [\text{TOUCH}(x, y) \Leftrightarrow \text{TOUCH}(y, x)]$$

in order to improve the generalization.

5.3.3 USING THEOREMS TO IMPROVE GENERALIZATION.
It can be easily guessed that using theorems can lead to many difficulties, since one enters the realm of theorem proving, which is well known as a good source of unsolved problems. In the case of SM, one is driven by the need to put the examples into a similar form, and the usual difficulties of theorem proving are somewhat eased. We cannot formally prove this point, but the following example, taken from Vrain (1987), can illustrate our claim. Starting from two examples that have no common predicates, we show that they nevertheless have a common generalization, found by using theorems that link the predicates. Let the examples be

$$E1 = \text{MAMMALIAN}(A)\ \&\ \text{BRED\_ANIMAL}(A)$$
$$E2 = \text{TAME}(B)\ \&\ \text{VIVIPAROUS}(B)$$

to which the following theorems are joined

$$R1: \forall x\ [\text{MAMMALIAN}(x)\ \&\ \text{BRED\_ANIMAL}(x) \Rightarrow \text{TAME}(x)]$$
$$R2: \forall x\ [\text{TAME}(x)\ \&\ \text{VIVIPAROUS}(x) \Rightarrow \text{MAMMALIAN}(x)]$$
$$R3: \forall x\ [\text{TAME}(x) \Rightarrow \text{HARMLESS}(x)]$$

where $\forall$ is the universal quantification, & the logical conjunction, and $\Rightarrow$ the logical implication.

The first step of SM is here trivial. We replace the constants by a variable x and obtain the equivalent examples:

$$E'1 = \text{MAMMALIAN}(x)\ \&\ \text{BRED\_ANIMAL}(x)\ \&\ [(x = A)]$$
$$E'2 = \text{TAME}(x)\ \&\ \text{VIVIPAROUS}(x)\ [(x = B)]$$

Since the predicates have no common occurrence, we consider the first (this ordering is not significant and just follows the one in which the examples are given) predicate of E'1, MAMMALIAN. We see that we can deduce this predicate from E2, using rule R2. We get

$$E''1 = \text{MAMMALIAN}^*(x)\ \&\ \text{BRED\_ANIMAL}(x)\ \&\ [(x = A)]$$
$$E''2 = \text{TAME}(x)\ \&\ \text{VIVIPAROUS}(x)\ \&\ \text{MAMMALIAN}^{**}(x)\ \&$$
$$[(x = B)]$$

The MAMMALIAN of E'1 has been treated; this is why it is marked by an asterisk in E''1. The one of E''2 is derived from the use of theorems, which is why it is marked by double asterisks.

Using again the order in which the examples are given, we see that the next nonmarked predicate is BRED_ANIMAL. No rule can be applied to E''2 to make explicit the presence of BRED_ANIMAL in it. Nevertheless, we remark that applying rule R1 to E''1 uses the predicate concerned, BRED_ANIMAL. Checking the effect of this application, we see that it generates the atomic formula TAME(x) and that there is an occurrence of x in E''2 which matches this occurrence. Therefore, we conclude that we must apply R1 to E″1. We obtain

E‴1 = MAMMALIAN*(x) & BRED_ANIMAL*(x) & TAME**(x) & [(x = A)]

E‴2 = TAME*(x) & VIVIPAROUS(x) & MAMMALIAN**(x) & [(x = B)]

Now, the only unmatched predicate is VIVIPAROUS in E″2. No rules can be applied to E‴1 to make its presence explicit. The only rule which can be applied in E‴2 relative to VIVIPAROUS is R1. But it would introduce the atomic formula MAMMALIAN(x), which is already matched since its instances are starred.

No other rule can be applied, so we star the predicate VIVIPAROUS to remember that it has already been dealt with, obtaining

E″″1 = MAMMALIAN*(x) & BRED_ANIMAL*(x) & TAME**(x) & [(x = A)]

E″″2 = TAME*(x) & VIVIPAROUS*(x) & MAMMALIAN**(x) & [(x = B)]

All possible occurrences have been dealt with, a complete SM is not possible, therefore the SMatching operation stops here.

Now, the generalization step is trivial: one drops the noncommon occurrences, obtaining the generalization

G = TAME(x) & MAMMALIAN(x)

This example shows well how potential infinite proof loops can be easily avoided, simply because they do not improve the SMatching state of the examples. More generally, one can use theorem-proving techniques to improve the degree of similarity detected among the examples. Such a system is under development in our group (Vrain, 1987). It is not the concatenation of a classical theorem prover and of generalization algorithms, but instead is rigorously adapted to the kind of proofs required by machine learning. As an instance of its peculiarity (and of its incompleteness), it will not allow one to use the same theorem twice during a given derivation.

6 CONCLUSIONS

This chapter presents the main learning methodologies achieved by inductive, deductive, or analogical means. The deductive approach provides a collection of different types of explanations that can be obtained in domains having a strong and complete theory (EBG), a strong but incomplete theory, and a soft theory in which no formal proofs can be drawn. Our purpose is to show that all kinds of background knowledge have been studied as possible explanation generators that become, in turn, a kind of definition for understandability. More research needs to be done to be able to give a universal definition of understandability.

6.1 Enlarging the Knowledge Acquisition Bottleneck

All these have been or are currently being tested on real-life applications, with diverse success depending on the topic:

ID3 inspired several ML software packages that have been applied in the medical domain, usually performing better than the ES built by interrogation. ID3 has been also applied by insurance companies that thus saved large amounts of money (Michie, 1988).

INDUCE generated the first ES automatically obtained, the rules of which have been validated by experts.

Structural matching has been applied to air traffic control (Cannat and Vrain, 1988), and we say more about it at the end of this conclusion.

All the deductive methods have been applied to various problem solving. At any rate, this approach will be extremely useful for the refinement of existing ES, as we briefly describe it now. Given an existing ES, given a set of deep rules expressing the background knowledge of the expert, the rules will allow one to improve the ES through casual interaction with the expert. The expert uses the ES. If the expert agrees with the ES, then some generalization of the existing rules can be attempted. If the expert acts differently from the ES, then new efficient rules can be learned (Mitchell, Mahadevan, and Steinberg, 1985). If the expert contradicts the ES, new rules will be added to complete the knowledge of the ES.

All this shows that ML does help acquire or refine rules, thus helping to enlarge the knowledge acquisition bottleneck (Feigenbaum, 1981). Nevertheless, in the long run, we foresee an even more important application of ML to ES.

6.2 Enlarging the ES Certification Bottleneck

Existing expert systems tend to stay in the research and development department of the companies that developed them, instead of moving to application fields as they should. One of the main reasons for this lack of large application is that application fields need certified software, and there is presently no way to certify an ES, except through some kind of "examination" similar to those students undergo. We claim that ML techniques are the best tool available at present to improve this situation. This can be done in two ways: by helping to specify the knowledge acquisition problem of the ES and by providing some certification directly.

6.2.1 SPECIFYING THE KNOWLEDGE ACQUISITION PROBLEM.
ML programs themselves need to be characterized by a number of attributes, such as the change they make in their initial knowledge, their use of background knowledge, etc. Such benchmarks for ML programs are also specifications of the task to be performed by the output of the program. For instance, a clear

description of the changes performed in the knowledge base specifies the state of the intended output. Similarly, a description of the background knowledge of the ML program specifies the environment in which the ES rules are valid.

6.2.2 VALIDATING AN ES. Suppose that we have an ES built in a classical way (Human ES, HES) and an ML program whose output is a program that performs tasks, or at least parts of tasks, similar to those of the HES. This is an automatically built ES (MLES). When analyzing a rule of the HES, one can bias the ML system so that it generates a rule as similar as possible to the one under analysis. If one obtains the same rule, then the analysis as it has been automatically generated will tell which part of the knowledge, and which biasing, it relies on. This part of the knowledge is the validation of the rule if the knowledge itself has been validated, which brings us back to classical validation techniques. If one obtains a different rule, the same analysis can take place and one asks the human expert to react either by changing the rule or the background knowledge of the ML or by justifying the rule in the HES. If the expert justifies the rule, this is a validation; if the expert does not, one iterates as long as the rules are not identical. We had some experience for the latter case when building rules for an ES in aircraft control. The controller provided rules and examples illustrating their use. The ML system then infers rules from the examples (Cannat and Vrain, 1988). Our current practice is that the automatically generated rules are largely different from those expected by the expert. The explanations of the discrepancies are included in the background knowledge. When convergence is achieved, the obtained rule is validated by the background knowledge. We also have the whole history of the refinement that can be viewed as an elaborated validating argument.

REFERENCES

Addis, T. R. 1987. "A framework for knowledge elicitation," *Proc. 1st European Knowledge Acquisition Workshop (EKAW-87)* (T. R. Addis, J. Boose, and B. Gaines, eds.), Reading University.

Addis, T. R. 1988. "A knowledge organisation for abduction," unpublished draft.

Arbab, B., and D. Michie. 1985. "Generating rules from examples," *Proc. International Joint Congress on Artificial Intelligence*-85 (A. Joshi, ed.), Morgan Kaufmann, Los Altos, Calif., pp. 631–633.

Barr, A., and E. A. Feigenbaum. 1981. *The Handbook of Artificial Intelligence,* vol. 1, William Kaufmann Inc., Los Altos, Calif.

Bareiss, E. R., B. W. Porter, and C. C. Wier. 1989. "Protos: An exemplar-based learning apprentice," to appear in *Machine Learning: An Artificial Intelligence Approach,* vol. 3 (Y. Kodratoff and R. S. Michalski, eds.), Morgan Kaufmann, Los Altos, Calif.

Bratko, I. 1988. Unpublished set of lectures at European Summer School on machine learning, Les Arcs, France.

Bratko, I., and N. Lavrac, eds. 1987. *Progress in Machine Learning,* Sigma Press, Wilmslow, Great Britain.

Cannat, J. J., and C. Vrain. 1988. "Machine learning applied to air traffic control," *Proc. Human Machine Interaction, Artificial Intelligence, Aeronautics and Space,* CEPAD, Toulouse, France, September, pp. 265–274.

Chouraqui, E. 1985. "Construction of a model for reasoning by analogy." *Progress in Artificial Intelligence* (L. Steels and J. A. Campbell, eds.), Ellis Horwood, London, pp. 169–183.

Corlett, R. 1983. "Explaining induced decision tress," *Research and Development in Expert Systems* (M. A. Bramer, ed.), Cambridge University Press, London, pp. 136–142.

Cox, P. T., and T. Pietrzykowski. 1986. "Causes for events: Their computation and applications," *Proc. 8th International Conference on Automated Deduction,* Oxford.

Davies, T. R., and S. T. Russell. 1987. "A logical approach to reasoning by analogy," *Proc. IJCAI-87,* Milan, Morgan Kaufmann, Los Altos, Calif., pp. 264–269.

DeJong, G., and R. Mooney. 1986. "Explanation-based learning: An alternative view," *Machine Learning* 1:145–176.

Duval, B., and Y. Kodratoff. 1989. "A tool for the management of incomplete theories: Reasoning about explanations," to appear in *Machine Learning, Meta-Reasoning, Logic* (P. Brazdil, ed.), Pitman, London.

Feigenbaum, E. A., 1981. "Expert systems in the 1980s." In: *State of the Art Report on Machine Intelligence,* Pergamon-Infotech, Maidenhead, Great Britain.

Genesereth, M. R., and N. J. Nilsson. 1985. *Logical Foundations of Artificial Intelligence,* Morgan Kaufmann, Los Altos, Calif.

Kedar-Cabelli, S. T., and L. T. McCarty. 1987. "Explanation-based generalization as resolution theorem proving," *Proc. Fourth International Machine Learning Workshop,* Irvine, Calif., pp. 383–389.

Kodratoff, Y. 1988. *Introduction to Machine Learning,* Pitman, London.

Kodratoff, Y., and J. G. Ganascia. 1986. "Improving the generalization step in learning." In: *Machine Learning: An Artificial Intelligence Approach,* vol. 2 (R. S. Michalski, J. Carbonell, and T. Mitchell, eds.), Morgan Kaufmann, Los Altos, Calif., pp. 215–244.

Kodratoff, Y., J. G. Ganascia, B. Clavieras, T. Bollinger, and G. Tecuci. 1984. "Careful generalization for concept learning," *Proc. European Congress on Artificial Intelligence-84,* Pisa, pp. 483–492. Also in: *Advances in Artificial Intelligence* (T. O'Shea, ed.), North-Holland, Amsterdam, 1985, pp. 229–238.

Kodratoff, Y., M. Manago, and J. Blythe. 1987. "Generalization and noise," *International Journal of Man-Machine Studies* 27:181–204.

Kodratoff, Y., and G. Tecuci. 1987a. "DISCIPLE1: Interactive apprentice system in weak theory fields," *Proc. 10th International Joint Conference on Artificial Intelligence* (J. McDermott, ed.), Milan, Morgan Kaufmann, Los Altos, Calif., pp. 271–273.

Kodratoff, Y., and G. Tecuci. 1987b. "Techniques of design and DISCIPLE learning apprentice," *International Journal of Expert Systems* 1(1):39–66.

Laird, J. E., A. Newell, and P. L. Rosenbloom. 1986. *Universal Subgoaling and Chunking: The Automatic Generation and Learning of Goal Hierarchies,* Kluwer, Dordrecht.

Laird, J. E., P. L. Rosenbloom, and A. Newell. 1987. "SOAR: An architecture for general intelligence," *AI Journal* 33:1–64.

Michalski, R. S. 1983. "A theory and a methodology of inductive learning." In: *Machine Learning: An Artificial Intelligence Approach* (R. S. Michalski, J. G. Carbonell, and T. M. Mitchell, eds.), Morgan Kaufmann, Los Altos, Calif., pp. 83–134.

Michalski, R. 1984. "Inductive learning as rule-guided transformation of symbolic descriptions: A theory and implementation." In: *Automatic Program Construction Techniques* (A. Biermann, G. Guiho, and Y. Kodratoff, eds.), Macmillan, New York, pp. 517–552.

Michalski, R. S., J. G. Carbonell, and T. M. Mitchell, eds. 1983. *Machine Learning: An Artificial Intelligence Approach,* Morgan Kaufmann, Los Altos, Calif.

Michalski, R. S., J. G. Carbonell, and T. M. Mitchell, eds. 1986. *Machine Learning: An Artificial Intelligence Approach,* vol. 2, Morgan Kaufmann, Los Altos, Calif.

Michalski, R. S. and R. E. Stepp. 1983. "Learning from observation: Conceptual clustering." In: *Machine Learning: An Artificial Intelligence Approach* (R. S. Michalski, J. G. Carbonell, and T. M. Mitchell, eds.), Morgan Kaufmann, Los Altos, Calif., pp. 331–363.

Michie, D. 1988. Conference at European Summer School on ML, Les Arcs, France, July. Videotaped and distributed by PACE, Paris.

Mitchell, T. 1982. "Generalization as Search," *Artificial Intelligence* 18:203–226.
Mitchell, T., R. Keller, and S. Kedar-Cabelli. 1986. "Explanation-based generalization: A unifying view," *Machine Learning Journal* 1:47–80.
Mitchell, T. M., S. Mahadevan, and L. I. Steinberg. 1985. "LEAP: A learning apprentice for VLSI," *Proc. IJCAI-85* (A. Joshi, ed.), Morgan Kaufmann, Los Altos, Calif., pp. 573–580.
Nicolas, J. 1988. "Consistency and preference criteria for generalization languages handling negation and disjunction," *Proc. ECAI-88* (Y. Kodratoff, ed.), Pitman, London, pp. 402–407.
Nunez, M. 1988. "Economic induction: A case study," *Proc. 3d EWSL* (D. Sleeman, ed.), Glasgow, October, Pitman, London, pp. 139–145.
Priedetis, A., ed. 1988. *Analogica,* Pitman, London.
Puget, J.-F. 1987. "Apprentissage de plans à partir de preuves," *Proc. Association Française pour la Cybernétique Et la Technique*-87, Antibes, France.
Quinlan, J. R. 1983. "Learning efficient classification procedures and their application to chess end games." In: *Machine Learning: An Artificial Intelligence Approach* (R. S. Michalski, J. G. Carbonell, and T. M. Mitchell, eds.), Morgan Kaufmann, Los Altos, Calif., pp. 463–482.
Schank, R. C. 1986. Unpublished set of conferences during the winter of 1986, at Paris University 7.
Schank, R. C. 1987. *Explanation Patterns: Understanding Mechanically and Creatively,* Ablex Publishing Company, London.
Schank, R. C., and R. P. Abelson. 1977. *Scripts, Plans, Goals and Understanding,* Lawrence Erlbaum, Hillsdale, N.J.
Shortliffe, E. H. 1976. *Computer-Based Medical Consultations: MYCIN,* Elsevier, New York.
Siqueira, J., and J.-F. Puget. 1988. "Explanation-based generalization of failures," *Proc. ECAI-88* (Y. Kodratoff, ed.), Pitman, London.
Valiant, L. G. 1984. "A theory of the learnable," *Commn. ACM* 27:1134–1142.
Vrain, C. 1987. "Un Outil de Généralisation Utilisant Systématiquement les Théorèmes: le système OGUST," thèse de 3ème cycle, Univ. Paris-Sud.
Waldinger, R. 1977. "Achieving several goals simultaneously." In: *Machine Intelligence,* vol. 8 (E. Elcock and D. Michie, eds.), Ellis Horwood, London. A comprehensive presentation can also be found in section 7.4, pp. 287–297, of N. J. Nilsson's *Principles of Artificial Intelligence,* Tioga, Palo Alto, Calif.
Winston, P. H. 1982. "Learning new principles from precedents and exercises," *AI Journal* 19:321–350.

CHAPTER 9

PROPOSITIONAL LOGIC

RONALD R. YAGER

1 INTRODUCTION

In the construction of intelligent systems, a facility of primary importance is the ability to represent and reason with information and knowledge. In this chapter we present one such tool called the *binary propositional logic*. The fundamental building block of this system is the declarative sentence (datum), an object whose truth value can be determined as either true or false. We provide logical operations for connecting individual pieces of datum into more complex information in the form of propositions. A formal method is presented for using known information to deduce the validity of other facts. Two practical methods are presented for the inference process, the direct method and the resolution method.

    A major strength of this system is the straightforward nature of the process, making logical inferences with knowledge representable within its structure. A significant disadvantage is the limitation in its ability to represent some very important types of knowledge necessary for the construction of intelligent systems. This limitation becomes very apparent when we try to represent systems in which various forms of uncertainty and incompleteness are present. This lack has stimulated the development of various types of nonstandard logic (Smets et al., 1988) such as fuzzy logic (Yager et al., 1987) and nonmonotonic logic (Ginsberg, 1988). Nevertheless almost all these alternative logic forms are based on an extension of this fundamental tool.

2 REPRESENTING INFORMATION IN THE PROPOSITIONAL LOGIC

The basic building block of the propositional logic is the atomic proposition, or more simply the atom. It is a declarative sentence or datum whose truth is determinable as true or false. Examples of atoms would be:

John is a man.
Mary is 30 years old.
Bob likes Mary.
Ann is a man.

We use the symbols $A, B, C, \ldots$, uppercase letters, to represent these atoms. The atoms form the fundamental building blocks in the propositional logic. If A is an atom, we must be able to assign a value to the truth of statement A. In the standard propositional logic, the truth values are drawn from the set {true, false}. We sometimes use the equivalent representation of truth values as {T, F} or {1, 0}. The fact that there exist only two truth values leads people to call this a *binary logic*.

The problem of determining the truth of an atomic proposition is an important, and in some cases difficult, problem. Often it is not clear whether a truth value of true or false can be uniquely applied to a declarative sentence. To use the binary logic to represent knowledge, we must force this requirement on each datum. Consider the atom *John is tall*. Because of the imprecision in the word *tall*, it may not be easily decidable whether this statement is true or false. However, the binary logic requires that this choice be made.

In addition to the atomic statements and the acceptable truth values, another component of the binary logic is the logical operators (connectives). The significant feature of the logical operators is their ability to represent compound propositions in terms of the connection of atomic propositions. This ability allows us to represent in the formal system a knowledge of things in the world that are in forms more complex than just simple, one-fact declarative statements. If A and B are two atoms, we can represent compound propositions by combining these atoms through the logical connectives discussed below. Furthermore, within the framework of the propositional logic, we can calculate the truth of compound propositions from the truth values of the constituent atomic propositions. This property is called *truth functionality*.

In the following we let A indicate the atom *John is 30* and B indicate the atom *Mary is smart*.

1. *Conjunction* (and, $\wedge$). The statement

$$\text{John is 30 } and \text{ Mary is smart}$$

can be formally represented in this system as

$$A \wedge B$$

2. *Disjunction* (or, $\vee$). The statement

$$\text{John is 30 } or \text{ Mary is smart}$$

can be formally represented in this system as

$$A \lor B$$

3. *Conditional* (if...then, $\rightarrow$). The statement

If John is 30, *then* Mary is smart

can be formally represented in this system as

$$A \rightarrow B$$

In the above formula A is called the *antecedent* and B the *consequent*.

4. *Biconditional* (if and only if, $\leftrightarrow$). The statement

John is 30 *if and only if* Mary is smart

is represented as

$$A \leftrightarrow B$$

5. *Negation* (not, $^-$). The statement

John is *not* 30

can be represented as $\overline{A}$.

We shall use the forms $\overline{A}$ and $\neg A$ interchangeably. Our choice will depend on the clarity of presentation.

The following definition supplies the rules for the acceptable combination of atoms under the logical operations. The terms *sentences, formula, propositions,* and *well-formed formula* are used interchangeably to denote acceptable combinations.

> **Definition.** A proposition [well-formed formula (WFF), sentence, or formula] in the propositional logic is defined recursively by
>
> 1. An atom is a proposition.
> 2. If G is a proposition, then $\overline{G}$ is a proposition.
> 3. If G and H are propositions, then $G \land H$, $G \lor H$, $G \rightarrow H$, and $G \leftrightarrow H$ are propositions.
> 4. All propositions are formed from the above rules.

Essentially the structure of a well-formed formula in this logic is very similar to that of well-formed statements in arithmetic with parentheses being used in both systems for clarity. Thus with the aid of the propositional logic we can represent a large class of knowledge about the environment in a very formal manner.

We next turn to the question of determining the truth of a compound proposition from its constituents.

3 TRUTHS OF COMPOUND PROPOSITIONS

The binary propositional logic has a characteristic known as *truth functionality*. Essentially this means that the truth value of a complex formula can be uniquely determined from the truth values of its atomic constituents.

Before proceeding to the issue of compound truth values, we must comment on some notational convention.

Assume A is a proposition. Associated with A is a truth value, formally denoted Truth(A), where, as we previously noted in the binary logic, Truth(A) $\in$ {T, F}. A common practice in the propositional logic is to use the symbol A to indicate both the proposition A and the value Truth(A). Similarly, if $F(A, B, C, D)$ is some WFF, then we follow the standard convention of using $F(A, B, C, D)$ itself to indicate what we really mean as Truth($F(A, B, C, D)$). Inherent in this notational convention is the idea that the statement of some proposition P is effectively equivalent to Truth(P) = T.

Definition. Let G and H be propositions (hence either atomic or compound). Then the truth values associated with combinations of these propositions can be obtained by the following rules:

1. Conjunction:

G	H	$G \wedge H$
T	T	T
T	F	F
F	T	F
F	F	F

As we just pointed out, the headings of the above table are more correctly read as Truth(G), Truth(H), and Truth($G \wedge H$).

Note that a conjunction of two propositions is such that if any of the constituents is false, the conjunction is false.

If we represent T by 1 and F by 0, then the truth value of $G \wedge H$ can be easily obtained as min[G, H].

2. Disjunction:

G	H	$G \vee H$
T	T	T
T	F	T
F	T	T
F	F	F

The disjunction of two propositions is true if either one of the constituents is true.

Again, if we represent T by 1 and F by 0, the truth value of $G \vee H$ can be obtained as $\max[G, H]$.

3. Conditional:

G	H	$G \rightarrow H$
T	T	T
T	F	F
F	T	T
F	F	T

The conditional is true if either the consequent is true or the antecedent is false or both.

4. Biconditional:

G	H	$G \leftrightarrow H$
T	T	T
T	F	F
F	T	F
F	F	T

The biconditional is true if both elements have the same truth value.

5. Negation:

G	$\neg G$
T	F
F	T

Using 1 and 0 as the truth values, we see that $\neg G = 1 - G$.

The above defined rules can be used in tandem to evaluate the truth of compound propositions.

Example. Consider the formula

$$P \wedge (Q \rightarrow P) \quad \text{with } P = T, Q = F$$
$$T \wedge (F \rightarrow T)$$
$$T \wedge T$$
$$T$$

Definition. Assume G is a proposition. Let $A_1, A_2, \ldots, A_n$ be the atoms (or constituents) occurring in formula G. An *interpretation* of G is an assignment of truth values to $A_1, A_2, \ldots, A_n$ in which every A_i is assigned an element from the set $\{T, F\}$.

Definition. A proposition G is said to be *true under an interpretation* if and only if (abbreviated iff) G is evaluated as T in that interpretation by using the previous rules. Otherwise, G is said to be *false under the interpretation*.

Definition. If a formula G is true under an interpretation I, we say that I *satisfies* G. If a formula G is false under an interpretation I, we say that I *falsifies* G.

When an interpretation I satisfies a formula G, then I is also called a *model* of G.

A truth table provides an organized way to evaluate the truth of a proposition under all possible interpretations. We note that if a sentence G has n atoms, there are 2^n possible interpretations of G.

The following examples show how to use a truth table to evaluate the truth of a proposition under all possible interpretations.

Example. Consider $G = P \wedge (Q \to P)$.

P	Q	P	$\wedge$	(Q	$\to$	P)
T	T	T	T	T	T	T
T	F	T	T	F	T	T
F	T	F	F	T	F	F
F	F	F	F	F	T	F
			↑			

Here we first calculated the conditional $Q \to P$ and then conjuncted this with the truth value of P to obtain the overall truth value of G. The column corresponding to the final truth values is indicated by an up-pointing arrow.

Example. Consider $R \wedge \neg \overline{P} \vee (P \wedge Q)$. Since there are three constituents, there are 2^3, or 8, interpretations.

P	Q	R	(R	$\wedge$	$\neg \overline{P}$)	$\vee$	(P	$\wedge$	Q)
T	T	T	T	F	F	T	T	T	T
T	T	F	F	F	F	T	T	T	T
T	F	T	T	F	F	F	T	F	F
T	F	F	F	F	F	F	T	F	F
F	T	T	T	T	T	T	F	F	T
F	T	F	F	F	T	F	F	F	T
F	F	T	T	T	T	T	F	F	F
F	F	F	F	F	T	F	F	F	F
						↑			

We found the conjunction of R and $\neg \overline{P}$ and then of P and Q, and we disjuncted these values to get the enclosed overall truth values.

Situations in which a proposition is true under all possible interpretations as well as those in which a proposition is false under all possible interpretations play a significant role in the application of propositional logic to reasoning. Because of the significance of these situations, they are given special names.

Definition. A proposition is said to be a *tautology* (or is called *valid*) iff it is true under all interpretations of its constituents.

Definition. A proposition is said to be a *contradiction* (or *inconsistent* or *unsatisfiable*) iff it is false under all interpretations.

We note that a formula G is a tautology iff $\neg \overline{G}$ is a contradiction.

The following two propositions are the prototypical examples of tautology and contradiction.

Example. Consider the proposition $P \vee \neg \overline{P}$ from the following truth table. We are able to see that it is a tautology.

P	P	$\vee$	$\neg \overline{P}$
T	T	T	F
F	F	T	T
		↑	

Example. The following truth table shows that the sentence $P \wedge \neg P$ is a contradiction.

P	P	$\wedge$	$\neg \overline{P}$
T	T	F	F
F	F	F	T
		↑	

Actually we shall see shortly that $P \wedge \neg \overline{P}$ is equivalent to $\neg(P \vee \overline{P})$. In binary logic, the fact that $P \vee \neg \overline{P}$ is always true plays a central role and is often called the *law of the excluded middle*.

Another tautology is illustrated in the following example.

Example. $(P \wedge Q) \rightarrow Q$.

P	Q	$(P \wedge Q)$	$\rightarrow$	Q
T	T	T	T	T
T	F	F	T	F
F	T	F	T	T
F	F	F	T	F
			↑	

4 EQUIVALENCES OF PROPOSITIONS

A concept that plays an important role in the propositional logic is the relationship of equivalence of two propositions. Among other uses, this relationship will allow us to replace one sentence with another.

Definition. Two propositions G and H are said to be equivalent (or G is equivalent to H), denoted $H = G$, iff the truth values of F and G are the same under every consistent interpretation of the constituents in these propositions.

Example. Show that $P \wedge Q = Q \wedge P$.

P	Q	P	$\wedge$	Q	$=$	Q	$\wedge$	P
T	T	T	T	T		T	T	T
T	F	T	F	F		F	F	T
F	T	F	F	T		T	F	F
F	F	F	F	F		F	F	F
			↑				↑	

We note that both the indicated columns are identical.

An important ramification of the idea of equivalence is that within any proposition we can replace any subformula by an equivalent formula, and our resulting proposition will have the same truth values under all interpretations as the original proposition. This observation implies that if G is a proposition and G' is a new proposition obtained by replacing some subformula of G by an equivalent subformula, then $G = G'$.

Example. Consider the proposition

$$(P \wedge Q) \rightarrow R$$

Since we have shown that

$$(P \wedge Q) = (Q \wedge P)$$

then

$$(P \wedge Q) \rightarrow R = (Q \wedge P) \rightarrow R$$

A significant aspect of the logical system is that the equivalence of two propositions can be established within the framework of the logical operations.

Theorem. $G = H$ iff $G \leftrightarrow H$ is a tautology.

Proof

1. *Sufficiency:* Assume $G = H$. Under this assumption the only possible interpretations are those in which G and H have the same truth values. Hence

G	H	G	$\leftrightarrow$	H
T	T	T	T	T
F	F	F	T	F

Thus when $G = H$, $G \leftrightarrow H$ is a tautology.

2. *Necessity:* Assume $G \leftrightarrow H$ is a tautology, i.e., it always has truth value T. Assume G has truth value T. Then since $G \leftrightarrow H$ must be true, H must have truth value T. Assume G has truth value F. Again, since $G \leftrightarrow H$ must be true, H must have truth value F.

The relationship of equivalence between two propositions forms a classic example of a mathematical equivalence relationship in that

1. $P = P$ (reflexive)
2. $P = Q$ implies $Q = P$ (symmetric)
3. $P = Q$, $Q = R$, then $P = R$ (transitive)

Thus the set of all propositions in binary logic can be divided into a family of disjoint subsets covering the total set such that all the elements in a set are equivalent. These sets are called *equivalence classes*.

We note that if G is a tautology, then G = TRUE, where TRUE is a special proposition that can only take the truth value T. This equivalence follows since if G is a tautology, the only possible interpretation of G is T. Hence

G	G	$\leftrightarrow$	TRUE
T	T	T	T

Similarly, if G is a contradiction, then G = FALSE, where FALSE is the special proposition that can take only the truth value F. Since G is a contradiction, the only possible interpretation of G is F. Hence

G	G	=	FALSE
F	F	T	F

The following theorem follows naturally from our definition.

Theorem. If G and H are two tautologies, then

$$G = H$$

If G and H are two contradictions, then

$$G = H$$

If G is a tautology and H is a contradiction, then

$$G = \overline{H} \quad \overline{G} = H$$

We can provide two methods for proving that the two propositions G and H are equivalent. The first approach is to use a truth table to prove that $G \leftrightarrow H$ is a tautology. Using this method, we must show that $G \leftrightarrow H$ is true under all interpretations of the constituents of G and H. This effectively requires showing that G and H have the same truth tables. The second method allows us to use already proven equivalences and the fact that if we replace a subformula of a proposition by an equivalent subformula, we get a proposition equivalent to the original one. Thus, for example, starting with G and replacing subformulas by some equivalent subformulas, we obtain a string of equivalent propositions

$$G = G_1 = G_2 = G_3 = \cdots = G_n$$

We continue making substitutions until we obtain H.

The following set of examples illustrates the first approach.

Example. Show that $P \wedge Q = Q \wedge P$.

P	Q	$(P \wedge Q)$	$\leftrightarrow$	$(Q \wedge P)$
T	T	T	T	T
T	F	F	T	F
F	T	F	T	F
F	F	F	T	F
			↑	

Example. Show that $\neg(\neg P) = P$.

P	$\neg$	$(\neg$	$P)$	$\leftrightarrow$	P
T	T	F	T	T	T
F	F	T	F	T	F
				↑	

Example. Show that $\neg(P \wedge Q) = (\neg P \vee \neg Q)$.

P	Q	$(\neg$	$(P \wedge Q))$	$\leftrightarrow$	$(\neg P$	$\vee$	$\neg Q)$
T	T	F	T	T	F	F	F
T	F	T	F	T	F	T	T
F	T	T	F	T	T	T	F
F	F	T	F	T	T	T	T
				↑			

Example. Show that $P \to Q = \neg P \vee Q$.

P	Q	$(P \to Q)$	$\leftrightarrow$	$(\neg P$	$\vee$	$Q)$
T	T	T	T	F	T	T
T	F	F	T	F	F	F
F	T	T	T	T	T	T
F	F	T	T	T	T	F
			↑			

This last equivalence is particularly important and useful. It shows that the conditional operation in logic is the same as the negation of the antecedent disjuncted with the consequent.

The following examples illustrate the second approach.

Example. Show that

$$P \to Q = \neg(P \land \neg Q)$$
$$P \to Q = \neg P \lor Q = \neg P \lor \neg(\neg Q) = \neg(P \land \neg Q)$$

Example. Show that

$$\neg(P \lor Q) = \neg P \land \neg P$$
$$\neg(P \lor Q) = \neg(\neg P \to Q) = \neg(\neg(\neg P \land \neg Q))) = \neg P \land \neg Q$$

Table 9-1 summarizes some often used equivalences whose validities can be established by either of the above methods. In some cases these equivalences are given special names.

We show two other useful equivalences in the following examples.

Example. The rule of elimination is

$$(G \lor H) \land (\neg G \lor H) = H$$
$(G \lor H) \land (\neg P \lor H) = (G \land \neg G) \lor H$ distribution
$(G \land \neg G) \lor H = \text{FALSE} \lor H$ rule 5b
$\text{FALSE} \lor H = H$ rule 7a

Example

$G \to (H \land E) = (G \to H) \land (G \to E)$
$G \to (H \land E) = \neg G \lor (H \land E)$ conditional equivalence
$\neg G \lor (H \land E) = (\neg G \lor H) \land (\neg G \lor E)$ distribution
$= (G \to H) \land (G \to E)$ conditional equivalence

In passing we note that of the four binary operations $\land$, $\lor$, $\to$, and $\leftrightarrow$, all except $\to$ are commutative.

With the aid of equivalences we can replace the biconditional by terms involving conditions and conjunctions. Furthermore, the conditional can be replaced by terms involving negation and disjunction. In tandem these two equivalences allow us to represent any proposition simply in terms of negation, disjunction, and conjunction. In the next section we look at special forms which result from this type of substitution.

5 NORMAL FORMS

In this section we introduce some special forms called *normal forms*. We will see that any proposition can be written in normal form. These normal forms provide a convenient way for determining whether a proposition is a tautology

TABLE 9-1
Equivalent formulas

1. Commutativity
   (a) $G \vee H = H \vee G$
   (b) $G \wedge H = H \wedge G$

2. Idempotency
   (a) $G \vee G = G$
   (b) $G \wedge G = G$

3. Associativity
   (a) $(G \vee H) \vee E = G \vee (H \vee E)$
   (b) $(G \wedge H) \wedge E = G \wedge (H \wedge E)$

4. Distributivity
   (a) $E \vee (G \wedge H) = (E \vee G) \wedge (E \vee H)$
   (b) $E \wedge (G \vee H) = (E \wedge G) \vee (E \wedge H)$

5. (a) $G \vee \neg G = \text{TRUE}$
   (b) $G \wedge \neg G = \text{FALSE}$

6. Double negation
   $\neg(\neg G) = G$

7. (a) $G \vee \text{FALSE} = G$
   (b) $G \vee \text{TRUE} = \text{TRUE}$

8. (a) $G \wedge \text{FALSE} = \text{FALSE}$
   (b) $G \wedge \text{TRUE} = G$

9. DeMorgan's law
   (a) $\neg(G \vee H) = \neg G \wedge \neg H$
   (b) $\neg(G \wedge H) = \neg G \vee \neg H$

10. Conditional equivalence
    (a) $G \rightarrow H = \neg G \vee H$
    (b) $G \rightarrow H = \neg H \rightarrow \neg G$
    (c) $G \rightarrow H = \neg(G \wedge \neg H)$

11. Biconditional equivalence
    (a) $G \leftrightarrow H = (G \rightarrow H) \wedge (H \rightarrow G)$
    (b) $G \leftrightarrow H = H \leftrightarrow G$
    (c) $G \leftrightarrow H = \neg G \leftrightarrow \neg H$

or a contradiction. Later we will see that the conjunctive normal form plays a significant role in the resolution-based method of reasoning.

Definition. If $G_1, G_2, \ldots, G_n$ are propositions, then

$$G_1 \vee G_2 \vee G_3 \vee \ldots \vee G_n$$

is called the *disjunction* of $G_1, G_2, \ldots, G_n$, and

$$G_1 \wedge G_2 \wedge G_3 \wedge \ldots \wedge G_n$$

is called the *conjunction* of $G_1, G_2, \ldots, G_n$.

Definition. We use the term *literal* to indicate an atom or the negation of an atom. For example, if A is an atom, then A and $\neg A$ are literals.

Definition. We call a proposition or any portion of a proposition a *clause* if it is of the form of a disjunction of literals.

Example. If A, B, and C are atoms, then

$$C$$
$$A \vee B$$
$$A \vee \overline{B} \vee C$$
$$\overline{A} \vee \overline{B} \vee B \vee \overline{C}$$

are clauses.

Definition. A proposition G is said to be in *conjunctive normal form* (CNF) iff G has the form

$$G_1 \wedge G_2 \wedge G_3 \wedge \ldots \wedge G_n \quad n \geq 1$$

where each $G_1, G_2, \ldots, G_n$ is a clause, a disjunction of literals.

Example. If A, B, and C are atoms, then

$$(A \vee B \vee C) \wedge (\overline{A} \vee A \vee B) \wedge (\overline{B} \vee A \vee \overline{C})$$

is in CNF.

Definition. A proposition G is said to be in *disjunctive normal form* (DNF) if G has the form

$$G_1 \vee G_2 \vee \ldots \vee G_n \quad n \geq 1$$

where each G_i is a conjunction of literals.

The following observations provide the basis for our contention that the normal forms provide an easy method for determining whether a proposition is a tautology or a contradiction or neither.

Observation 1: A proposition in CNF evaluates to true iff all the constituent clauses are true. This observation follows from the property that the logical AND is only true if all the ANDed elements are true.

Observation 2: Since a proposition is a tautology iff it evaluates to true for all interpretations of the atoms, the CNF is a tautology iff all the constituent clauses are true for all interpretations. That is, each clause must itself be a tautology for the CNF to be a tautology.

Observation 3: A disjunction of literals, a clause, is a tautology iff there appears in it at least one pair of an atom and its negation.

Since $A \lor \neg A = \text{TRUE}$ and any proposition disjuncted with TRUE is true, the sufficiency is established. If P is a proposition which can be either true or false, such as an atom under all interpretations, it is not a tautology. Furthermore, $P \lor Q$ can both be true or false; thus, this is not a tautology.

We summarize the above three observations in the following theorem.

Theorem. A proposition in CNF is a tautology iff in each constituent clause there appears at least one atom together with its negation.

Example. Let A, B, and C be atoms. Consider the following CNFs:

1. $(A \lor B) \land (\neg A \lor A \lor C)$
2. $A \land B$
3. $A \land \neg A$
4. $(A \lor \neg A \lor B) \land (C \lor \neg C) \land (A \lor B \lor C \lor \neg B)$

The proposition given in item 4 is the only tautology.

In a manner analogous to that used for conjunctive normal forms, we can establish the following theorem for disjunctive normal forms.

Theorem. Any proposition in DNF is a contradiction iff in each of the constituent conjunctions of literals there appears one atom together with its negation.

Example. Assume A, B, and C are atoms. Consider the following DNFs:

1. $(A \land B) \lor (A \land B \land \neg B)$
2. $(A \land \neg A) \lor (B \land C \land \neg B)$
3. $A \lor \neg A)$

In the above only the second proposition is a contradiction.

The usefulness of the above theorems is enhanced by the fact that any proposition can be transformed by using appropriate equivalences into both a CNF and a DNF. Here is the procedure for obtaining these normal forms:

1. Use the equivalences

$$H \to G = \neg H \lor G$$

and

$$H \leftrightarrow G = (H \to G) \land (G \to H) = (\neg H \lor G) \land (\neg G \lor H)$$

to eliminate all conditionals and biconditionals.

2. Use the equivalences of double negation or DeMorgan's law to bring the negation sign immediately before atoms. As a result of steps 1 and 2, we are left with only literals and the operators $\land$ and $\lor$.
3. Use other laws, especially distributivity, to get the desired normal form.

Example. Consider $P \wedge (P \to Q)$.

$$P \wedge (P \to Q) = P \wedge (\neg P \vee Q) \quad \text{CNF}$$
$$P \wedge (\neg P \vee Q) = (P \wedge \neg P) \vee (P \wedge Q) \quad \text{DNF}$$

Note that $P \wedge (P \to Q)$ is neither a contradiction nor a tautology.

Example. Consider $(P \wedge (P \to Q)) \to Q$.

$$\begin{aligned}
(P \wedge (P \to Q)) \to Q &= (P \wedge (\neg P \vee Q)) \to Q \\
&= (P \wedge \neg(\neg P \vee Q)) \vee Q \\
&= \neg P \vee \neg(\neg P \vee Q) \vee Q \\
&= \neg P \vee (P \vee \neg Q) \vee Q \quad \text{DNF} \\
&= ((\neg P \vee P) \wedge (\neg P \vee \neg Q)) \vee Q \\
&= (\neg P \vee \neg P \vee Q) \wedge (\neg P \vee \neg Q \vee Q) \quad \text{CNF}
\end{aligned}$$

Note that this example is a tautology.

Before leaving this topic of normal forms, we state without proof two important related theorems.

Theorem. Given any proposition G which is formed solely from literals and which uses only conjunction ($\wedge$) and disjunction ($\vee$) as the connectives, we can obtain its negation $\neg G$ simply by interchanging the symbols $\wedge$ and $\vee$ and replacing each literal by its negation.

Example

$$G = (P \vee Q) \wedge (\neg P \vee R)$$
$$\neg G = (\neg P \wedge \neg Q) \vee (P \wedge \neg R)$$

Theorem. Principle of duality. From any equivalence $G = H$, where both sides are formed solely from literals, conjunctions, and disjunctions, there exists another equivalence obtained by simply interchanging the conjunctions and disjunctions.

Example. From the equivalence

$$P \vee Q = Q \vee P$$

we get, by using the principle of duality, the dual equivalence

$$P \wedge Q = Q \wedge P$$

Parenthetically we note that the principle of duality can be extended to allow the inclusion of the TRUE and FALSE proposition. In this case we interchange TRUE and FALSE.

Example. From $P \wedge \neg P = \text{FALSE}$ we get the dual equivalence $P \vee \neg P = \text{TRUE}$. From $P \vee \text{FALSE} = P$ we get $P \wedge \text{TRUE} = P$.

We will see in the next section that the ability to prove that a proposition is a tautology or a contradiction is an important aspect of the process of making inferences. At this point we summarize the methods available to us to show that a proposition is a tautology.

1. *Truth table method:* Create a truth table and evaluate the proposition under all possible interpretations of the constituent atoms. If under all interpretations the proposition evaluates to true, then it is a tautology.

    We note that although this method is rather straightforward and easily implemented, the table becomes rather large if the number of atoms is large, because there are 2^n interpretations if there are n atoms.

2. *Equivalence method:* Starting with the original proposition, we use equivalences to transform it into equivalent propositions until we get a known tautology such as TRUE, thus $G = G' = G'' = G''' = \ldots =$ tautology.

    While this method is more elegant than the truth table approach, it requires intelligence to not get detoured. In addition, the fact that one cannot show that $G =$ tautology may not necessarily mean that it is not a tautology. It may simply mean that one is not resourceful enough to do it.

3. *Normal form approach:* Convert the proposition to CNF. If each of the constituent clauses contains at least one pair of atom and negation, then the proposition is a tautology.

6 BASIC DEDUCTION PROCEDURE

With the aid of the formalism just developed, we can try to represent our knowledge about the environment by a collection of propositions, each stating some fact we know to be true. We call this collection of propositions our *knowledge base*. We call any proposition in our knowledge base a *premise*. Given a knowledge base, a central task of knowledge engineering involves using the knowledge base to generate other pieces of information not explicitly expressed in the knowledge base. The process just described is called *inference* or *deduction*. One of the most appealing features of a logic-based knowledge representation is the rigorous procedures available for making inferences. We now turn to the most significant feature of this chapter—the study of the inference mechanism of propositional logic.

Given a knowledge base and a set of valid propositions, a fundamental question becomes the specification of what constitutes a valid deduction or inference. The following definition provides the formal basis for determining valid deductions from knowledge bases.

> **Definition.** Assume $P_1, P_2, \ldots, P_n$ is a collection of propositions and G is also a proposition. Proposition G is said to be a *logical consequence* of $P_1, P_2, \ldots, P_n$ (or G logically follows from $P_1, P_2, \ldots, P_n$) iff for every interpretation I in which the compound proposition $P_1 \wedge P_2 \wedge \ldots \wedge P_n$ is true, G is also true. The property of G logically following from $P_1, P_2, \ldots, P_n$ is denoted
>
> $$(P_1, P_2, \ldots, P_n) \vdash G$$

In the above definition, the P_i's are called the *knowledge base* (or *premises*) and G is called the *conclusion*. We will find ourselves calling $(P_1, P_2, \ldots, P_n) \vdash G$ a *valid argument*.

Note that in order to deduce that G is *true* based on some premises, we need to do more than just show that G logically follows from the premises; we must also establish that the premises are true.

Theorem. Rule of reasoning. A sufficient condition to show that G is true based upon premises $P_1, P_2, \ldots, P_n$ requires us to establish that

1. G logically follows from $P_1, P_2, \ldots, P_n$.
2. $P_1, P_2, \ldots, P_n$ are all true.

Proof. First we note that $P_1 \wedge P_2 \wedge \ldots \wedge P_n$ is true if $P_1, P_2, \ldots, P_n$ are all true. If we have established first $(P_1, P_2, \ldots, P_n) \vdash G$ and then $P_1 \wedge P_2 \wedge \ldots \wedge P_n$ is true, i.e., all $P_1, P_2, \ldots, P_n$ are true, then from the definition of term *logically follows* we know that G is true.

We should point out that the establishment of a valid argument (logically follows) between a collection of premises (knowledge base) and a conclusion is purely a matter of structural relationship based on the forms of the premises and the conclusion. It is independent of the *content* of the premises. The connection between logic and the real world requires that the premises be true in order to deduce that the conclusion is true. Thus the logical consequence of G's following $P_1, \ldots, P_n$ is purely structural while the truth of G requires some connection with the world.

As we will subsequently show, the following is always a valid argument:

$$(P, P \rightarrow Q) \vdash Q$$

That is, Q logically follows from P and if P, then Q.

Now consider the statements

P_1: It is raining. P
P_2: If it is raining, then John is a monster. $P \rightarrow Q$

From the above we can conclude that it logically follows that

$$(P_1, P_2) \vdash \text{John is a monster.}$$

However, to establish the truth of the statement *John is a monster* via premises P_1 and P_2, we must establish that P_1 and P_2 are both true. In particular, we must establish that *If it is raining, then John is a monster* is true. The establishment of the truth of the premises from observations and data in the world is a question of epistemology. Our purpose here is to concentrate more on the issue of logical consequence, establishing valid formal arguments.

The following theorem provides a mechanism for moving the inference process into the truth function framework of the propositional logic.

Theorem. Given propositions $P_1, P_2, \ldots, P_n$, proposition G is a logical consequence of $P_1, P_2, \ldots, P_n$, $(P_1, \ldots, P_n) \vdash G$ iff the proposition

$$(P_1 \wedge P_2 \wedge \ldots \wedge P_n) \rightarrow G$$

is a tautology.

Proof. In the following we denote $P_1 \wedge \ldots \wedge P_n$ as P.

1. *Sufficiency:* Assume $P \rightarrow G$ is a tautology. Thus when P is true, since $P \rightarrow G$ is true by assumption from the definition of the conditional, G must be true.
2. *Necessity:* Assume $P \rightarrow G$ is not a tautology, i.e., that there exists some allocation of truth values such that $P \rightarrow G$ is false. Recalling the definition of $P \rightarrow G$, we see that the only way for $P \rightarrow G$ to be false is if P is true and G is false.

The significance of this theorem is that the question of logical validity can be brought into the framework of our truth function system.

We now offer an alternative characterization of the conditions necessary to make logical deductions. This is also within the truth function framework.

Theorem. $(P_1, P_2, \ldots, P_n) \vdash G$ is a valid argument if

$$P_1 \wedge P_2 \wedge \ldots \wedge P_n \wedge \neg G$$

is a contradiction (always false).

Proof. In the previous theorem we have shown that

$$(P_1, P_2, \ldots, P_n) \vdash G \quad \text{iff} \quad (P_1 \wedge P_2 \wedge \ldots \wedge P_n) \vdash G \text{ is a tautology.}$$

Since

$$(P_1 \wedge P_2 \wedge \ldots \wedge P_n) \rightarrow G = \neg(P_1 \wedge P_2 \wedge \ldots \wedge P_n) \vee G$$

then $(P_1, P_2, \ldots, P_n) \vdash G$ iff $\neg(P_1 \wedge P_2 \wedge \ldots \wedge P_n) \vee G = \text{TRUE}$. However,

$$(\neg P_1 \wedge P_2 \wedge \ldots \wedge P_n \vee G \; (\neg P_1 \wedge P_2 \wedge \ldots \wedge P_n \vee G = \text{TRUE}$$
$$\text{if} \quad \neg(\neg(P_1 \wedge P_2 \wedge \ldots \wedge P_n) \vee G) = \text{FALSE}$$

By applying DeMorgan's law, this becomes $(P_1 \wedge P_2 \wedge \ldots \wedge P_n \wedge \neg G) = \text{FALSE}$.

Using the results of the previous theorems, we can suggest two basic methods for showing that G logically follows from a knowledge base consisting of premises $P_1, P_2, P_3, \ldots, P_n$. The first is to set up the proposition

$$(P_1 \wedge P_2 \wedge \ldots \wedge P_n) \rightarrow G \tag{9.1}$$

and show it is a tautology. The second is to set up the proposition

$$P_1 \wedge P_2 \wedge \ldots \wedge P_n \wedge \neg G \tag{9.2}$$

and show that it is a contradiction.

Furthermore we recall that there are three basic approaches to determining the nature of a proposition.

1. Use the truth table approach and evaluate 9.1 or 9.2 under all possible interpretations of the atoms. Then show that 9.1 is always true (or 9.2 is always false).
2. Use the theorem approach of replacing elements in 9.1 or 9.2 with equivalent propositions and continue until you are left with a known tautology (contradiction).
3. Use the normal forms. Write 9.1 in CNF and see whether the constituent clauses all have at least one pair of atom and its negation; or write 9.2 in DNF and show that each constituent conjunction of literals has at least one pair of atoms and its negation.

The following example illustrates the use of these methods on the important logical form called *modus ponens*.

Example. Show that

$$(P, P \to Q) \vdash Q$$

is a valid argument.

1 $(P \land (P \to Q)) \to Q$ is a tautology.
  a Truth table approach

P	Q	$(P$	$\land$	$(P \to Q))$	$\to$	Q
T	T	T	T	T	T	T
T	F	T	F	F	T	F
F	T	F	F	T	T	T
F	F	F	F	T	T	F
					↑	

  b Equivalence method

$$
\begin{aligned}
(P \land (P \to Q)) \to Q &= (P \land (\neg P \lor Q)) \to Q & \text{equivalence 10}a \\
&= ((P \land \neg P) \lor (P \land Q)) \to Q & \text{equivalence 4}b \\
&= (\text{FALSE} \lor (P \land Q)) \to Q & \text{equivalence 5}b \\
&= (P \land Q) \to Q & \text{equivalence 7}a \\
&= \neg(P \land Q) \lor Q & \text{equivalence 10}a \\
&= \neg P \lor \neg Q \lor Q & \text{equivalence 9}b \\
&= P \lor \text{TRUE} & \text{equivalence 5}a \\
&= \text{TRUE} & \text{equivalence 7}b
\end{aligned}
$$

Since $(P \land (P \to Q)) \to Q$ is equivalent to TRUE, a tautology, it is true.

  c Use CNF.

$$\begin{aligned}
(P \wedge (P \to Q)) \to Q &= (P \wedge (\neg P \vee Q)) \to Q \\
&= \neg(P \wedge (\neg P \vee Q)) \vee Q \\
&= \neg P \vee \neg(\neg P \vee Q) \vee Q \\
&= \neg P \vee (P \wedge \neg Q) \vee Q \\
&= ((\neg P \vee P) \wedge (\neg P \vee \neg Q)) \vee Q \\
&= (\neg P \vee P \vee Q) \wedge (\neg P \vee \neg Q \vee Q)
\end{aligned}$$

Since this is in CNF with each clause having a pair of atom and its negation, the result is a tautology.

2 $P \wedge (P \to Q) \wedge \neg Q$ is a contradiction.
  a Truth table approach

P	Q	$(P$	$\wedge$	$(P \to Q))$	$\to$	$\neg Q$
T	T	T	T	T	F	F
T	F	T	F	F	F	T
F	T	F	F	T	F	F
F	F	F	F	T	F	T

  b Equivalence method

$$\begin{aligned}
P \wedge (P \to Q) \wedge \neg Q &= P \wedge (\neg P \wedge Q) \wedge \neg Q && \text{equivalence } 10a \\
&= P \wedge ((\neg P \wedge \neg Q) \vee (Q \wedge \neg Q)) && \text{equivalence } 4b \\
&= P \wedge ((\neg P \wedge \neg Q) \vee \text{FALSE}) && \text{equivalence } 5b \\
&= \neg P \wedge \neg P \wedge Q && \text{equivalence } 7a \\
&= \text{FALSE} \wedge Q && \text{equivalence } 5b \\
&= \text{FALSE} && \text{equivalence } 8a
\end{aligned}$$

  c Disjunctive normal form

$$\begin{aligned}
P \wedge (P \to Q) \wedge \neg Q &= P \wedge (\neg P \vee Q) \wedge \neg Q \\
&= ((P \wedge \neg P) \vee \neg P \wedge Q) \wedge \neg Q \\
&= (P \wedge \neg P \wedge \neg Q) \vee (\neg P \wedge Q \wedge \neg Q)
\end{aligned}$$

Since each term has a pair of an atom and its negation, the result follows.

7 RULES OF REASONING

As we have seen, the ability to show that a proposition is a tautology plays a key role in the determination of valid arguments. In this section we look at some basic examples of tautologies useful for the inference problem. Because of their central role, many of these are given special names, and as a group they are called the *rules of reasoning*.

Before providing these special rules, we present a general class of tautologies. In particular, all equivalences are tautologies.

Theorem. If $G = H$, then $G \to H$ is a tautology.

Proof. If we can show that $((G \leftrightarrow H) \to (G \to H)) = \text{TRUE}$, then for any $G = H$ which is valid, $G \to H$ is a tautology. Thus we need to show that $((G \leftrightarrow H) \to (G \to H) \leftrightarrow \text{TRUE})$ is a tautology.

G	H	$(G \leftrightarrow H)$	$\rightarrow$	$((G \rightarrow H)$	$\leftrightarrow$	TRUE
T	T	T	T	T	T	T
T	F	F	T	F	F	T
F	T	F	T	T	T	T
F	F	T	T	T	T	T
			↑			

Table 9-2 summarizes some tautologies which can be used in reasoning. These tautologies are sometimes called the *rules of inference* and in many cases are given special names.

These tautologies form the basis of the practical methods for providing reasoning to intelligent systems.

8 DIRECT METHOD OF INFERENCE

The *direct method* of inference is an organized methodology for proving that the knowledge base $P_1, P_2, \ldots, P_n$ logically implies a desired conclusion H. This approach is based on the fact that

TABLE 9-2
Useful tautologies

Rules of reasoning	
1. $P \vee$ TRUE	Law of the excluded middle
2. $P \vee \neg P$	
3. $P \wedge \neg P$	Law of contradiction
4. $P \rightarrow P$	
5. $(a)(P \wedge Q) \rightarrow P$	Law of simplification
$\quad (b)(P \wedge Q) \rightarrow Q$	
6. $P \rightarrow (P \vee Q)$	Law of addition
7. $((P \vee Q) \wedge (\neg P \vee R)) \rightarrow (Q \vee R)$	Resolution
8. $(P \wedge (P \rightarrow Q)) \rightarrow Q$	Law of detachment (modus ponens)
9. $(\neg Q \wedge (P \rightarrow Q)) \rightarrow \neg P$	Modus tollens
10. $(\neg P \wedge (P \vee Q)) \rightarrow Q$	Disjunctive syllogism
11. $(a)((P \rightarrow Q) \wedge (Q \rightarrow R)) \rightarrow (P \rightarrow R)$	Hypothetical syllogism
$\quad (b)$ More generally	
$\quad ((P \rightarrow S_1) \wedge (S_1 \rightarrow S_2) \wedge (S_2 \rightarrow S_3) \wedge \ldots \wedge (S_n \rightarrow R)) \rightarrow (P \rightarrow R)$	
12. $(P \wedge Q) \rightarrow (P \vee Q)$	
13. $(a)((P \rightarrow R) \wedge (Q \rightarrow R)) \rightarrow ((P \vee Q) \rightarrow R)$	Proof by cases
$\quad (b)$ More generally	
$\quad ((S_1 \rightarrow R) \wedge (S_2 \rightarrow R) \wedge (S_3 \rightarrow R) \wedge \ldots \wedge (S_n \rightarrow R)) \rightarrow ((S_1 \vee S_2 \vee \ldots \vee S_n) \rightarrow R)$	
14. $(P \rightarrow (Q \vee R)) \rightarrow ((P \wedge \neg Q) \rightarrow R)$	
15. $(P \leftrightarrow Q) \rightarrow (P \rightarrow Q)$	
16. $P \wedge Q \rightarrow (P \wedge Q)$	Law of adjunction

$$(P_1, P_2, \ldots, P_n) \vdash H \quad \text{if} \quad (P_1 \wedge P_2 \wedge P_3 \wedge \ldots \wedge P_n) \to H$$

is a tautology.

The direct method is based on the procedure of using the original premise along with the rules of reasoning to generate other true facts, forming an augmented set of premises, and continuing in this manner until we show that the desired hypothesis H is a true fact.

Before proceeding to the details of this method, we make some observations on the properties of the proof procedure which justify the method.

1. To prove

$$(P_1, P_2, \ldots, P_n) \vdash H$$

it is sufficient to prove that

$$(P_1, P_2, \ldots, P_j) \vdash H$$

where $P_1, P_2, \ldots, P_j$ is any subset of premises from the collection $P_1, P_2, \ldots, P_n$. This property is called the *monotonicity* of logic. The justification of this property is based on the following lemma.

Lemma

$$((P_1 \wedge P_2 \wedge \ldots \wedge P_j) \to H) \to ((P_1 \wedge P_2 \wedge P_j \wedge P_{j+1} \wedge \ldots \wedge P_n) \to H)$$

is a tautology.

Proof. Let $A = (P_1 \wedge P_2 \wedge \ldots \wedge P_j)$, and let $B = (P_{j+1} \wedge P_{j+2} \wedge \ldots \wedge P_n)$. We must show that $(A \to H) \to ((A \wedge B) \to H)$ is a tautology.

$$\begin{aligned}
(A \to H) \to ((A \wedge B) \to H) &= (\neg A \vee H) \to (\neg(A \wedge B) \vee H) \\
&= (\neg A \vee H) \to (\neg A \vee \neg B \vee H) \\
&= \neg(\neg A \vee H) \vee (\neg A \vee \neg B \vee H) \\
&= (A \wedge \neg H) \vee (\neg A \vee \neg B \vee H) \\
&= (A \vee \neg A \vee \neg B \vee H) \wedge (\neg H \vee \neg A \vee \neg B \vee H) \\
&= \text{TRUE}
\end{aligned}$$

To see that this lemma implies our observation, we proceed as follows. If we have shown that $(A) \vdash H$, then $A \to H$ is true. Since in the lemma we have shown that $(A \to H) \to ((A \wedge B) \to H)$ is a tautology, whenever $A \to H$ is true, $(A \wedge B) \to H$ must be true. Thus if $(A) \vdash H$, then $(A \wedge B) \to H$ is true, and hence whenever $A \wedge B$ is true, H must be true. So $(A \wedge B) \vdash H$.

Thus we conclude that we can use any subset of our premise to prove H and still be sure that H follows from all the premises. This property is the monotonicity of the proposition logic. In particular, adding premises does not negate any conclusions already reached. It just enables us to reach more conclusions.

2. Any proposition which is known to be always true, a tautology, can be added to our premises and used to prove H. The justification of this observation is based on the following lemma.

Lemma
$$((A \wedge \text{TRUE}) \to H) \to (A \to H)$$
is a tautology.

Proof

$$\begin{aligned}
((A \wedge \text{TRUE}) \to H) \to (A \to H) &= (\neg(A \wedge \text{TRUE}) \vee H) \to (\neg A \vee H) \\
&= (\neg A \vee \text{FALSE} \vee H) \to (\neg A \vee H) \\
&= (\neg A \vee H) \to \neg A \vee H \\
&= \text{TRUE}
\end{aligned}$$

Thus if $(A, \text{TRUE}) \vdash H$, then $(A \wedge \text{TRUE}) \to H$ is a tautology, true by definition. Since the lemma shows that $((A \wedge \text{TRUE}) \to H) \to (A \to H)$ is a tautology, then when $(A \wedge \text{TRUE}) \to H$ is true, $A \to H$ is true and since A is assumed true, H must be true. In essence, this fact says that

$(A \wedge ((A \wedge \text{TRUE}) \to H)) \to H$ is a tautology

$$\begin{aligned}
(A \wedge ((A \wedge \text{TRUE}) \to H)) \to H &= (A \wedge (\neg A \vee \text{FALSE} \vee H)) \to H \\
&= ((A \wedge \neg A) \vee (A \vee \text{FALSE}) \vee A \vee H) \to H \\
&= (\text{FALSE} \wedge \text{FALSE} \vee A \vee H) \to H \\
&= (A \wedge H) \to H \\
&= \text{TRUE}
\end{aligned}$$

The implication to this result is that we can add any tautology to our premises.

The following theorem tells us this: If from any subset of our knowledge base we can conclude some fact B as true and if from any other subset of premises we can conclude A, then if from A and B we can infer H, then our premises logically imply H.

Theorem. If $(P_1, P_2) \vdash A$ and $(P_3, P_4) \vdash B$ and $(A, B) \vdash H$, then $(P_1, P_2, \ldots, P_n) \vdash H$.

Proof. Formally we must show that

$$\begin{aligned}
(((P_1 \wedge P_2) \to A) \wedge ((P_3 \wedge P_4) \to B) \wedge \\
((A \wedge B) \to H))) \to ((P_1, P_2, \ldots, P_n) \to H)
\end{aligned}$$

is a tautology. Let us denote $P_1 \wedge P_2 = C$, $P_3 \wedge P_4 = D$, and $P_4 \wedge P_5 \wedge \ldots \wedge P_n = E$. Thus we get

$$\begin{aligned}
&((C \to A) \wedge (D \to B) \wedge ((A \wedge B) \to H)) \to ((C \wedge D \wedge E) \to H) \\
&= ((\neg C \wedge A) \vee (\neg D \vee B) \wedge (\neg A \vee \neg B \vee H)) \to (\neg C \vee \neg D \vee \neg E \vee H) \\
&= (C \wedge \neg A) \vee (D \wedge \neg B) \vee (A \wedge B \wedge \neg H) \vee \neg C \vee \neg D \vee H \vee \neg E
\end{aligned}$$

$$= (C \wedge \neg A) \vee \neg C \vee (D \wedge \neg B) \vee \neg D \vee (A \vee B \vee \neg H) \vee H \vee \neg E$$
$$= ((C \vee \neg C) \wedge (A \vee \neg C)) \vee ((D \vee \neg D) \wedge (\neg D \vee \neg B)) \vee (((A \vee B) \vee H)$$
$$\wedge (H \vee \neg H)) \vee \neg E$$
$$= \neg A \vee \neg C \vee \neg D \vee \neg B \vee (A \wedge B) \vee H \vee \neg E$$
$$= \neg A \vee \neg B \vee (A \wedge B) \vee \neg C \vee \neg D \vee H \vee \neg E$$
$$= \text{TRUE} \vee \neg C \vee \neg D \vee H \vee \neg E = \text{TRUE}$$

In essence, this theorem implies that if from any subset of premises we can deduce proposition B, then B can be added to our list of facts and treated as if it were a premise. This idea forms the basis of our direct method of proof, which is now presented.

The direct method of inference consists of the following procedures, where $P_1, P_2, \ldots, P_n$ are our initial premises and H is our desired conclusion.

1. We initialize our system by forming a list of current valid facts consisting of the collection of initial premises.
2. If H is a member of our current list of valid facts, then we stop and H is proved, as being valid from our premises; else we go to step 3.
3. If from any subset of our current list of valid facts we can deduce proposition B, then we add B to our current list of valid facts, forming a new list, and go to step 2.

The one remaining question regards implementation of step 3. The following rules provide the necessary tools.

A proposition B can be added to our list of premises, i.e., deduced as true from the list, under any of these conditions:

a. If B is a tautology, then we can add B to our current list. This follows from observation 2. However, we also note that ( )⊢ tautology, every tautology is inferable from the empty set of premises.
b. If for any subset $S_1, S_2, \ldots, S_q$ of our current valid premises $(S_1 \wedge S_2 \wedge \ldots \wedge S_q) \to B$ is a tautology, then B can be added.
c. If $B = P_i$, where P_i is any valid fact in our list, then B can be added.

This is justified by the fact that from tautology 15 we can add the fact that

$$(P_i \leftrightarrow B) \to (P_i \to B)$$

since $B = P_i$ then $B \leftrightarrow P_i$ is a tautology and hence can be added to our list. From modus pollens

$$((P_i \leftrightarrow B) \wedge ((P_i \leftrightarrow B) \to (P_i \to B))) \to (P_i \to B)$$

Thus using rule b above, we can add $P_i \to B$, again using modus pollens, with the facts $(P_i \to B)$ and P_i, to get

$$(P_i \wedge (P_i \to B)) \to B$$

and B is valid.

We note that of particular practical significance is rule b. This allows us to use any of our rules of reasoning as aids along the way. In practical examples the most important rules of reasoning are

$$(P \wedge (P \to Q)) \to Q \quad \text{modus pollens}$$

and

$$(P \wedge Q) \to P \quad \text{simplification}$$

Some examples will help in understanding this method.

Example. Show that $(A \vee B, B \to \neg C, C) \vdash A$. In this case our premises are

$$P_1: A \vee B \quad P_2: B \to \neg C \quad P_3: C$$

and our conclusion is A.

Proof

1. Using P_2 and P_3, via modus tollens we can deduce that $(B \to \neg C, C) \vdash \neg B$. (Denote $\neg B$ as P_4.)
2. From P_1 and the equivalence

$$A \vee B = \neg B \to A$$

we can add $\neg B \to A$ (call this P_5).

3. From P_4 and P_5, using modus pollens, we get our desired conclusion:

$$(\neg B, \neg B \to A) \vdash A$$

Example. Show that

$$(\neg A \vee B, \neg B \wedge C, (A \vee B) \to D) \vdash C \wedge D$$

Our premises are

$$P_1: \neg A \vee B \quad P_2: \neg B \wedge C \quad P_3: \neg(A \vee B) \to D$$

and our conclusion is $H: C \wedge D$.

Proof

1. From P_2, $\neg B \wedge C$. By the law of simplification we can deduce $P_4 = \neg B$ and $P_5 = C$.
2. Since $(\neg A \vee B) = (A \to B)$, then $(A \to B, \neg B) \vdash \neg A$ by modus tollens. (Denote $\neg A$ as P_6.)
3. Since $\neg(A \vee B) = \neg A \wedge \neg B$, proposition P_3 can be rewritten as $(\neg A \wedge \neg B) \to D$.
4. Using P_4 and P_6, from the law of adjunction we get $(\neg A, \neg B) \vdash \neg A \wedge \neg B$. Call this P_7.
5. Using P_7 and the equivalent form of P_3, by using modus pollens we get $((\neg A \wedge \neg B), (\neg A \wedge \neg B) \to D) \vdash D$. Call this P_8.
6. From P_5 and P_8 by the law of adjunction we get $(C, D) \vdash C \wedge D$, our desired conclusion.

Example. An American, French, or Chinese national was selected for a job. If the French person were selected, then the headquarters would be moved to Paris. If the Chinese citizen were selected, then rice would be served with every meal. The headquarters is not in Paris, and an American was not selected. Show that rice was served at every meal. Let

A: An American was selected.
B: A French national was selected.
C: A Chinese citizen was selected.
D: The headquarters was moved to Paris.
E: Rice was served with every meal.

Our proposition can be expressed formally as

$$P_1: A \lor B \lor C$$
$$P_2: B \to D$$
$$P_3: C \to E$$
$$P_4: \neg D$$
$$P_5: \neg A$$

and our conclusion is E.

Proof

1. Since P_1 is equivalent to $\neg A \to (B \lor C)$, using P_3 and modus pollens gives $(\neg A, \neg A \to (B \lor C)) \vdash B \lor C$.
2. Since $B \to D$ is equivalent to $\neg D \to \neg B$, we can use this along with P_4 to deduce, via modus pollens, $(\neg D \to \neg B, \neg D) \vdash \neg B$.
3. Since $B \lor C$ can be expressed as $\neg B \to C$, then by using $\neg B$ and modus pollens we get $(\neg B, \neg B \to C) \vdash C$.
4. From C and P_3 we get $(C, C \to E) \vdash E$.

9 RESOLUTION APPROACH TO INFERENCE

The resolution approach to inference, sometimes called *reductio ad absurdum*, is another organized way to prove that the knowledge base $P_1, P_2, \ldots, P_n$ logically implies a desired conclusion H. This approach is based on the fact that

$$(P_1, P_2, \ldots, P_n) \vdash H \quad \text{if} \quad P_1 \land P_2 \land \ldots \land P_n \land \neg H \text{ is a contradiction}$$

Before describing this method, we provide some useful observations and some justification for the methodology.

An integral part of this approach is the rule of reasoning called the *resolution law*. The resolution law says that the proposition

$$((P \lor Q) \land (\neg P \lor R)) \to (Q \lor R)$$

is a tautology, or

$$((P \lor Q), (\neg P \lor R)) \vdash (Q \lor R)$$

We first observe some tautologies which are special cases of this resolution rule.

1. The proposition $(P \wedge (\neg P \vee R)) \to R$ is a tautology.

Proof. Consider the proposition

$$(P \vee \text{FALSE}) \wedge (\neg P \vee R)) \to (R \vee \text{FALSE})$$

By the law of resolution this is a tautology; note $Q = \text{FALSE}$. Since $P \vee \text{FALSE} = P$ and $R \vee \text{FALSE} = R$, the above is equivalent to $(\neg P \wedge (\neg P \vee R)) \to R$. *Note:* Since $(\neg P \vee R) = (P \to R)$, we get $(P \wedge (P \to R)) \to R$ which is really modus pollens. Hence modus pollens is a special case of resolution.

2. The proposition $((P \vee Q) \wedge (\neg P \vee Q)) \to Q$ is a tautology.

Proof. Consider the resolution

$$((P \vee Q) \wedge (\neg P \vee Q)) \to (Q \vee Q)$$

which is a tautology. Since $Q \vee Q = Q$, the result holds.

3. The proposition $P \neg P \to \text{FALSE}$ is a tautology.

Proof. Consider

$$((P \vee \text{FALSE}) \wedge (\neg P \vee \text{FALSE})) \to (\text{FALSE} \vee \text{FALSE})$$

Since $(P \vee \text{FALSE}) = P$ and $(\neg P \vee \text{FALSE}) = \neg P$ and $(\text{FALSE} \vee \text{FALSE}) = \text{FALSE}$, the result follows.

The following theorem provides an important tool in the development of the resolution method.

Theorem

The proposition $(S_1 \to G) \to ((S_1 \wedge S_2 \wedge G) \leftrightarrow (S_1 \wedge S_2))$ is a tautology.

Proof

S_1	S_2	G	$(S_1 \to G)$	$\to$	$((S_1 \wedge S_2 \wedge G)$	$\leftrightarrow$	$(S_1 \wedge S_2))$
T	T	T	T	T	T	T	T
T	T	F	F	T	F	F	T
T	F	T	T	T	F	T	F
T	F	F	F	T	F	T	F
F	T	T	T	T	F	T	F
F	T	F	T	T	F	T	F
F	F	T	T	T	F	T	F
F	F	F	T	T	F	T	F
				↑			

Recall that

$$(P_1, P_2, \ldots, P_n) \vdash H \quad \text{if } (P_1 \wedge P_2 \wedge \ldots \wedge P_n \wedge \neg H) \text{ is a contradiction}$$

The implication of this theorem is as follows. If from any subset of $P_1, P_2, \ldots, P_n, \neg H$ we can infer G_1, then $P_1 \wedge P_2 \wedge \ldots \wedge P_n \wedge \neg H \wedge G_1 = P_1 \wedge P_2 \wedge \ldots \wedge P_n \wedge \neg H$. Then to show that $(P_1, \ldots, P_n) \vdash H$, it is sufficient to show that $P_1 \wedge P_2 \wedge \ldots \wedge P_n \wedge \neg H \wedge G_1 = \text{FALSE}$. Furthermore, if from any subset of $P_1, P_2, \ldots, P_n, \neg H, G_1$ we can infer G_2, then it is sufficient to show that $P_1 \wedge P_2 \wedge \ldots \wedge P_n \wedge \neg H \wedge G_1 \wedge G_2 = \text{FALSE}$. We can continue augmenting this proposition as much as we want. Furthermore, we note that a sufficient condition for a conjunction of propositions of the form $P_1 \wedge P_2 \wedge \ldots \wedge P_n \wedge \neg H \wedge G_1 \wedge \ldots \wedge G_p$ to be false is that any of the conjuncts is false. Thus if from any subset of propositions $P_1, P_2, \ldots, P_n, \neg H, G_1, \ldots, G_p$ we can generate the false propositions, then we have shown that $P_1 \wedge \ldots \wedge P_n \wedge \neg H = \text{FALSE}$ and thus $(P_1, \ldots, P_n) \leftrightarrow H$. With this understanding we can now present the resolution approach to inference.

We recall that we used the term *clause* to indicate a disjunction of literals, that is, for example, $A \vee B \vee \neg C$. Furthermore, we have shown that any proposition can be written in CNF, i.e., any $P = (\text{clause 1}) \wedge (\text{clause 2}) \wedge \ldots$. Parenthetically we note that a single element is a clause. In particular, we can express all our premises and the negation of the conclusion in CNF thus:

$$P_1 \wedge P_2 \wedge \ldots \wedge P_n \wedge \neg H = C_1 \wedge C_2 \wedge C_3 \wedge \ldots \wedge C_q$$

We denote $C_1, C_2, \ldots, C_q$ as our list of clauses. In the resolution method we work with the clauses as the basic elements instead of the propositions.

The resolution is as follows:

1. Write all premises and the negation of the conclusion in clause form. These clauses comprise our original list of clauses.
2. If one clause is the FALSE clause, stop because the theorem is proved; else, go to step 3.
3. Appropriately resolve two clauses in the list to obtain a new clause, called the *resolvent*.
4. If the resolvent is FALSE, stop; else, add the resolvent to the list of clauses and go to step 3.

To implement step 3, we note that any two clauses C_i and C_j can be resolved if there exists an atom such that the atom appears in one clause and its negation appears in the other clause. Note there are only three possible situations of interest.

1. $C_i = A \vee B \quad C_j = \neg A \vee D$

    In this case $((A \vee B), (\neg A \vee D)) \vdash (B \vee D)$. If $B = D$, then $B \vee D = B$. (We note that B and D may be compound disjunctions of literals, themselves clauses). Furthermore
2. $C_i = A \quad C_j = \neg A \vee B$

In this case $(A, (\neg A \lor B)) \vdash B$.
3. $C_i = A \quad C_j = \neg A$
In this case $(A, \neg A) \vdash \text{FALSE}$.

Here are some examples using the resolution method.

Example. Show that $(\neg Q, (P \to Q)) \vdash \neg P$. In this situation

$$P_1: \neg Q$$
$$P_2: (P \to Q) = (\neg P \lor Q)$$
$$\neg H = P$$

Since each of these are in clause form, our list of clauses is

$$C_1: \neg Q$$
$$C_2: \neg P \lor Q$$
$$C_3: P$$

Procedure:

1. Resolve C_1 and C_2 to obtain $C_4 = \neg P$.
2. Resolve C_3 and C_4 to obtain $C_5 = \text{FALSE}$.

Example. Show that $((P \lor Q), (\neg P \lor Q), (P \lor \neg Q)) \vdash (P \land Q)$. In this case

$$P_1: P \lor Q$$
$$P_2: \neg P \lor Q$$
$$P_3: P \lor \neg Q$$
$$\neg H = \neg P \lor \neg Q$$

Our list of clauses is

$$C_1: P \lor Q$$
$$C_2: \neg P \lor Q$$
$$C_3: P \lor \neg Q$$
$$C_4: \neg P \lor \neg Q$$

1. Resolve C_1 and C_2 to obtain $C_5 = Q$.
2. Resolve C_3 and C_4 to obtain $C_6 = \neg Q$.
3. Resolve C_5 and C_6 to obtain $C_7 = \text{FALSE}$.

Example. Show $(P \to Q, Q \to R, P) \vdash R$. In this case

$$P_1: (P \to Q) = (\neg P \lor Q)$$
$$P_2: (Q \to R) = (\neg Q \lor R)$$
$$P_3: P$$
$$\neg H = \neg R$$

Our clauses are

$$C_1: \neg P \lor Q$$
$$C_2: \neg Q \lor R$$
$$C_3: P$$
$$C_4: \neg R$$

1. Resolve C_1 and C_3 to obtain $C_5 = Q$.
2. Resolve C_2 and C_5 to obtain $C_6 = R$.
3. Resolve C_4 and C_6 to obtain $C_7 =$ FALSE.

Example. If the company has high profits, then its management is good or it was in a good industry and the year was a good year for business in general. The company has high profits. The year was not a good year for business in general. Prove that the company has good management.

Let $A \equiv$ the company has high profit, $B \equiv$ the company has good management, $C \equiv$ the company is in a good industry, and $D \equiv$ it was a good year for business in general. Using these as our atoms, we can formally express our information with the following propositions:

$P_1: A \to (B \lor (C \land D)) = \neg A \lor (B \lor (C \land D)) = (\neg A \lor B \lor C) \land (\neg A \lor B \lor D)$
$P_2: A$
$P_3: \neg D$
$\neg H: \neg B$

Our list of original clauses is

$C_1: \neg A \lor B \lor C$
$C_2: A \lor B \lor D$
$C_3: A$
$C_4: \neg D$
$C_5: \neg B$

1. Resolve C_3 and C_2 to obtain $C_6 = B \lor D$.
2. Resolve C_4 and C_6 to obtain $C_7 = B$.
3. Resolve C_7 and C_5 to obtain $C_8 =$ FALSE.

10 REFERENCES

Bittinger, M. L. 1982. *Logic, Proof and Sets*, Addison-Wesley, Reading, Mass.
Chang, C. L., and R. C. T. Lee. 1973. *Symbolic Logic and Mechanical Theorem Proving*, Academic, New York.
Enderton, H. B. 1972. *A Mathematical Introduction to Logic*, Academic, New York.
Generesreth, M. R., and N. J. Nilsson. 1987. *Logical Foundations of Artificial Intelligence*, Morgan Kaufmann, Los Altos, Calif.
Ginsberg, M. L. 1988. *Readings in Nonmonotonic Reasoning*, Morgan Kaufmann, Los Altos, Calif.
Hilbert, D., and W. Ackerman. 1950. *Principles of Mathematical Logic*, Chelsea Publishing, New York.
Smets, P., A. Mamdani, D. Dubois, and H. Prade. 1988. *Non-Standard Logics for Automated Reasoning*, Academic, London.
Suppes, P. 1957. *Introduction to Logic*, Van Nostrand, Princeton, N.J.
Wall, R. 1972. *Introduction to Mathematical Linguistics*, Prentice-Hall, Englewood Cliffs, N.J.
Yager, R. R., S. Ovchinnikov, R. M. Tong, and H. T. Nguyen. 1987. *Fuzzy Sets and Applications: Selected Papers of L. A. Zadeh*, Wiley, New York.

CHAPTER 10

NATURAL LANGUAGE PROCESSING: COMPUTER COMPREHENSION OF EDITORIAL TEXT

SERGIO J. ALVARADO
MICHAEL G. DYER
MARGOT FLOWERS

1 INTRODUCTION

Natural language processing (NLP) is an area of research within artificial intelligence (AI) concerned with the comprehension and generation of natural language text. Comprehension involves the dynamic construction of conceptual representations, linked by causal relationships and organized/indexed for subsequent retrieval. Once these conceptual representations have been created, comprehension can be tested by means of such tasks as paraphrasing, question answering, and summarization. Higher-level cognitive tasks are also modeled within the NLP paradigm and include translation, acquisition of word meanings and concepts through reading, analysis of goals and plans in multiagent environments (e.g., coalition and counterplanning behavior by narrative characters), invention of novel stories, recognition of abstract themes (such as irony and hypocrisy), extraction of the moral or point of a story, and justification/refutation of beliefs through argumentation.

This chapter describes a process model of argument comprehension. The model is implemented as a computer program, called OpEd (Alvarado, 1989;

Alvarado et al., 1985, 1986). The name is inspired by the op-ed page in newspapers and magazines, where readers send/receive opinions to/from the editor. The model is designed to take editorial text as input and answer subsequent questions concerning the beliefs/arguments of the editorial author and of others whom the author may criticize. The domain of knowledge is concerned with political/economic situations, focusing on arguments for/against economic protectionism.

1.1 Motivation

The ability to understand and engage in arguments is essential to reading editorials in magazines and newspapers, to following political debates, and to understanding religious, legal, and scientific reasoning. Humans use this ability to express and defend their opinions when discussing world affairs; when presenting their opinions on sex, religion or politics; when giving advice; and so on. Why should computers be able to comprehend and generate arguments? As computer systems become more widely used to aid in complex decision making and the generation of expert advice, they should exhibit the same cognitive skills possessed by their human-expert counterparts. That is, computers should be able not only to evaluate given situations and present their beliefs on possible courses of actions, but also to justify their beliefs, understand opposing beliefs, and argue persuasively against them. We would not accept advice from human experts who could not explain or defend their own points of view. Similarly, we should not accept advice from computer systems that lack these abilities.

Editorials are similar to argument dialogues: in both, argument participants present and justify their opinions. However, editorials lack the interactive elements of argument dialogues because the editorial writer is the only active argument participant. As a result, editorials can be viewed as *one-sided arguments,* where writers contrast their opinions against those of their implicit opponents (Bush, 1932; Stonecipher, 1979). By focusing on such editorials, we are able to concentrate on the representation and comprehension of argument structure, without having to be concerned with the many other processing and representational issues that arise when participating in open-ended discourse.

1.2 Sample Input/Output Behavior of OpEd

Editorial segments used as input to OpEd are in English and contain the essential wording and arguments of the original editorials. Here "essential" means that the original editorials have been edited to remove those parts which involve addressing issues that fall outside the scope of OpEd's process model, such as understanding references to historic events not encoded in OpEd's knowledge base and handling sarcastic or humorous statements. In the area of NLP, it is impossible to give a simple description of the kinds of texts that a model of conceptual analysis can handle. One cannot say, for example, that a given NLP system can understand "any text in length of 100 words or less." Such texts would include zen koans, mystery segments, jokes, dialog segments, poetry, religious statements, political satire, insults, advertisements, newspaper headlines, and so on. Each kind of text poses a uniquely different set of theoretical problems, and no NLP system yet exists that can handle literary text, where such disparate elements are often intertwined. At this stage in the development of NLP research, a system's capabilities must be described, in terms of not only the raw text handled, but also the knowledge structures and processing strategies embodied within the system and the ways in which such knowledge and processing interact.

Given the above caveat, let us consider an actual sample of OpEd's current input/output behavior. The input is labeled ED-JOBS and is a fragment of an editorial by Milton Friedman (1982). In this editorial fragment Friedman presents arguments against the Reagan administration's policies on international trade.

ED-JOBS

Recent protectionist measures by the Reagan administration have...disappointed ...us...[Voluntary]limits on Japanese...automobiles...[and]...voluntary limit[s] on steel...by the Common Market...are...bad for the nation....They do...[not]...promote the long-run health of the industries affected.... The... problem of the auto[mobile] and steel industries is:...in both industries, average wage rates are twice as high as the average....Far from saving jobs, the limitations on imports will cost jobs. If we import less, foreign countries will earn fewer dollars. They will have less to spend on [American] exports....The result will be fewer jobs in export industries.

An example of an editorial change made in ED-JOBS is the phrase "voluntary limits." In the original text, this phrase actually appears as "'voluntary' limits." The single quotation marks enclosing the world "voluntary" are a sarcastic reference to the fact that the limitations on imports are not really voluntary, but rather a response by foreign countries to economic pressure by the United States. Original editorials were selected so that OpEd could be tested on actual (versus fictitious) arguments. However, "raw" editorials from magazines and newspapers often contain numerous asides (i.e., concepts only tangential to the main argument) and references to issues and events outside of OpEd's knowledge and processing capabilities. Consequently, deletions had to be made and these are indicated by ellipses.

After processing an editorial fragment, OpEd is presented with a sequence of questions, also in natural language. Questions to OpEd regarding ED-JOBS are shown in bold lowercase letters. Output generated by OpEd is in uppercase.

Q1: **What has disappointed Milton Friedman?**
A1: PROTECTIONIST POLICIES BY THE REAGAN ADMINISTRATION.

Q2: **Who believes that the limitations on imports will save jobs?**
A2: THE REAGAN ADMINISTRATION.

Q3: **What is the result of the limitations on imports?**
A3: MILTON FRIEDMAN BELIEVES THAT PROTECTIONIST POLICIES BY THE REAGAN ADMINISTRATION DO NOT LEAD TO THE ACHIEVEMENT OF NORMAL PROFITS OF THE STEEL INDUSTRY AND THE AUTOMOBILE INDUSTRY.
THE REAGAN ADMINISTRATION BELIEVES THAT PROTECTIONIST POLICIES BY THE REAGAN ADMINISTRATION LEAD TO THE ACHIEVEMENT OF NORMAL PROFITS OF THE STEEL INDUSTRY AND THE AUTOMOBILE INDUSTRY.
MILTON FRIEDMAN BELIEVES THAT PROTECTIONIST POLICIES BY THE REAGAN ADMINISTRATION WILL THWART THE PRESERVATION OF JOBS FOR U.S.
THE REAGAN ADMINISTRATION BELIEVES THAT PROTECTIONIST POLICIES BY THE REAGAN ADMINISTRATION ACHIEVE THE PRESERVATION OF JOBS FOR U. S.

Q4: **Why does Milton Friedman believe that the limitations on imports will cost jobs?**
A4: MILTON FRIEDMAN BELIEVES THAT PROTECTIONIST POLICIES BY THE REAGAN ADMINISTRATION WILL THWART THE PRESERVATION OF JOBS FOR U.S. BECAUSE MILTON FRIEDMAN BELIEVES THAT AS A CONSEQUENCE OF PROTECTIONIST POLICIES BY THE REAGAN ADMINISTRATION, U.S IMPORTS FEWER PRODUCTS; AND IF U.S. IMPORTS FEWER PRODUCTS, THEN THERE IS A DECREASE IN PROFITS OF FOREIGN COUNTRIES; AND IF THERE IS A DECREASE IN PROFITS OF FOREIGN COUNTRIES, THEN FOREIGN COUNTRIES BUY FEWER AMERICAN EXPORTS; AND IF FOREIGN COUNTRIES BUY FEWER AMERICAN EXPORTS, THEN THERE IS A DECREASE IN PROFITS OF EXPORT INDUSTRIES; AND IF THERE IS A DECREASE IN PROFITS OF EXPORT INDUSTRIES, THEN THERE IS A DECREASE IN JOBS IN EXPORT INDUSTRIES; AND A DECREASE IN JOBS IN EXPORT INDUSTRIES THWARTS THE PRESERVATION OF JOBS FOR U.S.

OpEd's output is somewhat verbose because of our concentration on processes of comprehension in OpEd. Linguistic style in answer generation is not a major issue addressed in OpEd; as a result, its natural language generation strategies are not as well developed as its strategies for understanding.

1.3 Issues in Argument Comprehension

In OpEd, understanding editorial text involves six major issues: (1) applying domain-specific knowledge (i.e., politicoeconomic knowledge); (2) recognizing beliefs and belief relationships; (3) following causal chains of reasoning about goals and plans; (4) applying abstract knowledge of argument structure; (5)

mapping input text into conceptual structures which compose the internal representations of editorial arguments; and (6) indexing recognized concepts for later retrieval during question answering.

1.3.1 APPLYING DOMAIN-SPECIFIC KNOWLEDGE. OpEd must have a computational model of politicoeconomic knowledge in order to make sense of discussions about import restrictions. Consider three of the problems OpEd has to solve in order to process ED-JOBS:

CONCEPT REFERENCE

>...**protectionist measures by the Reagan administration...Voluntary limits on Japanese automobiles and voluntary limits on steel by the Common Market**...

The mention of "voluntary limits" does not constitute a sudden topic shift, but rather is a coherent continuation of Friedman's opinion concerning the Reagan administration's protectionist policies. To make such a connection possible, OpEd must understand that voluntary limits on imports are instances of protectionist policies.

CONCEPT INFERENCE

>...**the long-run health of the industries affected.**

What industries is Friedman referring to? Up to this point in ED-JOBS, there has not been an explicit reference to any industries. However, OpEd has been told that the limitations have been voluntarily imposed on exports of automobiles and steel to the United States. Therefore, OpEd must be able to infer that Friedman is referring to the U.S. automobile and steel industries.

CAUSAL COHERENCE

>**The problem of the...industries is...average wage rates**...

What do high wage rates have to do with industries? To establish the connection, OpEd must know that the profits of any industry are affected by the level of salaries it pays to its workers, i.e., the higher the salaries, the lower the profits.

OpEd's politicoeconomic knowledge need not include what an expert in politics and economics would know. Minimally, the model must encode what an average, well-informed adult reader would need to know to understand editorials of the type shown here. That is, OpEd's domain-specific knowledge need be only a "naive" model of politics and economics. Even so, such a model has to have representations for each of the following classes of concepts and their instances:

- *Authorities:* The Reagan administration
- *Institutions:* The Common Market, steel industry, and automobile industry

- *Countries:* United States, Japan, and countries from the Common Market
- *Products:* Imports, exports, steel, and automobiles
- *Economic quantities:* Earnings, spending, and cost
- *Occupations:* Jobs in export industries
- *Goals:* Saving jobs and attaining economic health of industries
- *Plans:* Protectionist policies
- *Events:* Importing and exporting

OpEd's politicoeconomic model must also include causal relationships that exist at the level of goals, plans, and events, such as how changes in one economic quantity affect other economic quantities and which plans can be used to achieve given goals. For example, understanding ED-JOBS requires recognizing the following relationships:

- Governments can impose or negotiate import restrictions to protect jobs in domestic industries and/or to help these industries become profitable.
- An industry's rate of employment depends on the industry's volume of sales.
- The capital available for importing goods depends on the capital produced from exporting goods.

This commonsense knowledge must be represented and formalized in OpEd so that it can be accessed and applied during editorial comprehension.

1.3.2 RECOGNIZING BELIEFS AND BELIEF RELATIONSHIPS. Another basic problem in editorial comprehension is to recognize the writer's explicitly or implicitly stated opinions along with their justifications. For example, after reading the sentence

> ...**protectionist measures by the Reagan administration have disappointed us.**

OpEd must infer that Friedman is against the Reagan administration's protectionist policies, although this opinion is not explicitly stated. Friedman's belief can be inferred from the affect description "disappointed," which indicates that "protectionist measures" cause goal or expectation failures. Friedman's position is later justified in the sentence

> ...**the limitations on imports will cost jobs.**

To recognize this support relationship, OpEd has to understand that losing jobs is a goal violation. Therefore, OpEd must use knowledge of goal/planning relationships to recognize belief justifications.

In addition to recognizing the editorial writer's beliefs and their justifications, OpEd has to be able to recognize other individuals' beliefs and how they support, are supported by, or are attacked by the editorial writer's beliefs. Recognizing these relationships is essential for comprehension because editorial arguments often present the writer's agreement or disagreement with respect to other individuals' beliefs, for instance, in the following sentence from ED-JOBS:

They do not promote the long-run health of the industries affected.

Here Friedman attacks the implicit belief by the Reagan administration that the limitations will help the automobile and steel industries recover from their economic slump. To recognize this attack relationship, OpEd must access its politicoeconomic model to realize that voluntary import restrictions are negotiated to provide the basis for long-term economic recovery of ailing industries. Thus, recognizing opposing beliefs relies on applying domain-specific knowledge.

1.3.3 FOLLOWING CAUSAL CHAINS OF REASONING. Understanding belief justifications also requires identifying and keeping track of causal chains of reasoning. These chains are sequences of cause-effect relationships which encode (1) why plans should or should not be selected, implemented, or terminated or (2) why goals should or should not be pursued. To follow such chains, OpEd must recognize explicit and implicit cause-effect relationships by applying politicoeconomic knowledge about goals and plans. For example, consider the causal chain which supports Friedman's belief that "the limitations on imports will cost jobs":

> **If we import less, foreign countries will earn fewer dollars. They will have less to spend on American exports. The results will be fewer jobs in export industries.**

Friedman's reasoning contains the following (implicit) relationships:

1. Import restrictions by the United States result in a decrease in imports to the United States.
2. The decrease in imports to the United States causes a decrease in foreign countries' export earnings.
3. The decrease in foreign countries' earnings causes a decrease in their spending on United States exports.
4. The decrease in spending on United States exports results in a decrease in earnings of United States export industries.
5. The decrease in earnings of United States export industries causes a decrease in the number of occupations in these industries.
6. The decrease in occupations thwarts the Reagan administration's goal of saving jobs.

When processing ED-JOBS, OpEd has to be able to infer relationships 1, 4, 5, and 6 along with the explicitly stated connection between 2 and 3. Therefore, OpEd must use domain-specific knowledge to infer missing steps in incomplete chains of reasoning in editorials.

1.3.4 REPRESENTING AND APPLYING ABSTRACT KNOWLEDGE OF ARGUMENTATION. Editorial comprehension implies applying abstract knowledge of argumentation. This knowledge is independent of domain-specific knowledge and thus fundamental to understanding and generating arguments in *any* domain. Abstract knowledge of argumentation is organized by memory structures called *argument units* (AUs) (Alvarado et al., 1985, 1986), which represent recurring patterns of support and attack relationships among beliefs. When combined with domain-specific knowledge, AUs can be used to argue about issues involving plans, goals, and beliefs in the particular domain. As a result, argument comprehension in OpEd is viewed as the process of recognizing, accessing, instantiating, and applying argument units.

The points of an editorial are organized by AUs recognized and instantiated during editorial comprehension. For example, in ED-JOBS, Friedman uses the following argument unit:

```
AU-ACTUAL-CAUSE: Although opponent O believes plan P should be used
to achieve goal G, SELF believes P does not achieve G because SELF
believes P does not affect situation S which thwarts G. Therefore,
SELF believes P should not be used.
```

Friedman uses AU-ACTUAL-CAUSE to argue that the Reagan administration's protectionist policies will not help the automobile and steel industries become profitable because such policies do not affect the actual cause of the industries' problems, namely, steel wage rates being much higher than average rates in other industries. To recognize this argument unit, OpEd needs to know that (1) import restrictions are intended to help struggling domestic industries become profitable again, (2) high salaries lower industries' profits, and (3) import restrictions are not wage control policies.

1.3.5 MAPPING LEXICAL ITEMS INTO CONCEPTUAL KNOWLEDGE. The process of recognizing argument units is also based on the capability of identifying specific linguistic constructs which signal opposition and expectation failures. For example, consider AU-OPPOSITE-EFFECT, another argument unit used in ED-JOBS:

```
AU-OPPOSITE-EFFECT: Although opponent O believes plan P should be
used to achieve goal G, SELF believes P does not achieve G because
SELF believes P thwarts G. Therefore, SELF believes P should not be
used.
```

Friedman uses AU-OPPOSITE-EFFECT to argue that he is against import restrictions because they will not save but will cost jobs. When processing ED-JOBS, OpEd must recognize AU-OPPOSITE-EFFECT after reading the sentence

Far from saving jobs, the limitations on imports will cost jobs.

This recognition process must be triggered by the explicit relationship of opposition between expected effects of import restrictions, namely, saving jobs and costing jobs. This relationship is signaled by the construct

```
''Far from'' X, Y
```

where X and Y are opposite (mutually exclusive) effects and the phrase "far from" indicates opposition. Therefore argument comprehension involves (1) recognizing specific linguistic constructs, accessing the specific conceptualizations they refer to, and (2) mapping from these conceptualizations into their appropriate argument units.

The process of mapping editorial text into conceptual representations includes other tasks, such as disambiguating words, resolving pronoun and concept references, and inferring implicit relationships in arguments. For example, the following problems must be solved when processing ED-JOBS:

- The phrase "protectionist measures" refers not to size measurements, but to economic protection policies.
- Although the United States is not explicitly mentioned in the sentence "voluntary limits...are bad for the nation," the word "nation" refers to the United States.
- In the phrase "the health of the industries," the word "health" means the economic well-being of the industries, as opposed to physical health.
- The phrase "far from" indicates not a spatial relationship but rather an opposition relationship.
- The phrase "limitations on imports" refers to protectionist policies and not to the greatest amount of imports allowed.
- In the phrase "If we import less," it is implicit that "we" refers to the United States.

As these examples show, editorial comprehension depends on the abilities to map verbatim text into conceptual representations and to represent and apply abstract concepts, such as protection, the health of an organization, and savings jobs.

1.3.6 MEMORY ORGANIZATION AND RETRIEVAL. Keeping track of the contents of the editorial involves building an internal conceptual model of editorial arguments. This model, known as an *argument graph* (Flowers et al., 1982), explicitly represents relationships of support and attack among beliefs as well as causal relationships among goals, plans, events, and states. The argument graph also aids the understanding process by representing and maintaining the current context of the editorial.

OpEd must parse input words and phrases into conceptual structures and integrate these structures into the editorial's argument graph. Every new belief has to be integrated into the graph by using links that indicate whether the belief supports, attacks, or is supported by other beliefs already existing in the graph

The argument graph of an editorial must also include indexing structures and access links which provide initial entry to the graph. These indices and access links need to be created during editorial comprehension and later used by search and retrieval processes when answering questions about the editorial. For instance, consider another question that OpEd can answer after reading ED-JOBS:

Q5: **What does Milton Friedman believe?**

To answer this question OpEd must retrieve the instantiations of the argument units used in ED-JOBS, namely AU-ACTUAL-CAUSE and AU-OPPOSITE-EFFECT. Retrieving these instantiated AUs requires (1) indices from argument participants to their professed beliefs (2) access links between beliefs and associated argument units, and (3) retrieval functions that take argument participants as input and retrieve argument units.

During question answering, the process of selecting appropriate retrieval strategies must depend on parsing the question and analyzing the conceptual content into one of a number of *conceptual question categories* (Lehnert, 1978). Different question categories lead to different search and retrieval processes. These processes must select indices according to the questions' input information. Once an index is selected, these processes must traverse access and memory links to locate an appropriate conceptualization. Answers retrieved must then be converted from memory representation to English. OpEd's verbatim answer to Q5 appears below:

A5: MILTON FRIEDMAN BELIEVES THAT PROTECTIONIST POLICIES BY THE REAGAN ADMINISTRATION ARE BAD BECAUSE MILTON FRIEDMAN BELIEVES THAT PROTECTIONIST POLICIES BY THE REAGAN ADMINISTRATION DO NOT LEAD TO THE ACHIEVEMENT OF NORMAL PROFITS OF THE STEEL INDUSTRY AND THE AUTOMOBILE INDUSTRY. MILTON FRIEDMAN BELIEVES THAT PROTECTIONIST POLICIES BY THE REAGAN ADMINISTRATION DO NOT LEAD TO THE ACHIEVEMENT OF NORMAL PROFITS OF THE STEEL INDUSTRY AND THE AUTOMOBILE INDUSTRY BECAUSE MILTON FRIEDMAN BELIEVES THAT NORMAL SALARY IN THE STEEL INDUSTRY AND THE AUTOMOBILE INDUSTRY HIGHER THAN THE NORM THWARTS THE ACHIEVEMENT OF NORMAL PROFITS OF THE STEEL INDUSTRY AND THE AUTOMOBILE INDUSTRY. MILTON FRIEDMAN BELIEVES THAT THE REAGAN ADMINISTRATION IS WRONG BECAUSE THE REAGAN ADMINISTRATION BELIEVES THAT PROTECTIONIST POLICIES BY THE REAGAN ADMINISTRATION LEAD TO THE ACHIEVEMENT OF NORMAL PROFITS OF THE STEEL INDUSTRY AND THE AUTOMOBILE INDUSTRY.

MILTON FRIEDMAN BELIEVES THAT PROTECTIONIST POLICIES BY THE REAGAN ADMINISTRATION ARE BAD BECAUSE MILTON FRIEDMAN BELIEVES THAT PROTECTIONIST POLICIES BY THE REAGAN ADMINISTRATION WILL THWART THE PRESERVATION OF JOBS FOR U.S. MILTON FRIEDMAN BELIEVES THAT THE REAGAN ADMINISTRATION IS WRONG BECAUSE THE REAGAN ADMINISTRATION BELIEVES THAT PROTECTIONIST POLICIES BY THE REAGAN ADMINISTRATION ACHIEVE THE PRESERVATION OF JOBS FOR U.S.

1.4 OpEd Architecture

OpEd consists of seven major interrelated components: semantic memory, lexicon, expectation-based conceptual parser, working memory, argument graph, memory search and retrieval processes, and English generator. These components along with their interactions are shown in Fig. 10-1 and described below.

INPUT: Editorial Text   Question Text

```
                EXPECTATION-BASED PARSER  ◄──────  LEXICON
                                                   Words
                                                   Phrases
                                                   Roots
                                                   Suffixes

  WORKING MEMORY       ARGUMENT GRAPH              SEMANTIC MEMORY
  Short-term Memory    Episodic Memory   Indexing  Argument Units
                                         Links     Belief Relationships
                                                   Beliefs
                                                   Reasoning Scripts
                                                   Goals and States
                                                   Plans and Events
        SEARCH AND RETRIEVAL PROCESSES   ◄──────   Economic Quantities
                                                   Authorities
                                                   Institutions
          Conceptual Answer                        Countries
                                                   Physical Objects
              ENGLISH GENERATOR  ◄─────────────────

             OUTPUT: Answer Text
```

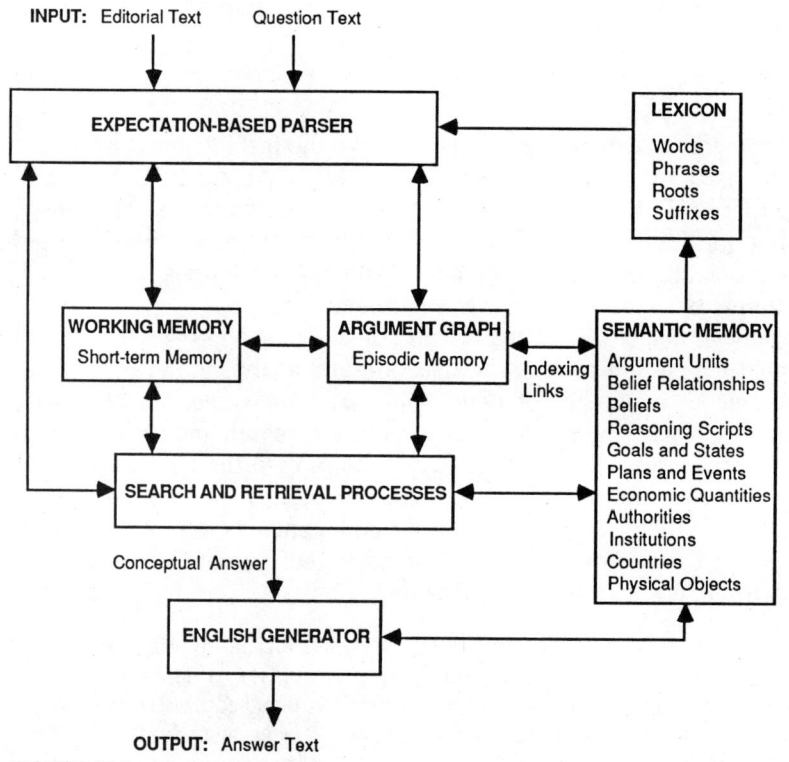

FIGURE 10-1
OpEd architecture.

SEMANTIC MEMORY. OpEd's semantic memory embodies a computational model of politicoeconomic knowledge and abstract knowledge of argumentation. OpEd's politicoeconomic model includes economic protection plans and their associated goals, causal relationships among economic quantities, and reasoning scripts containing prespecified causal chains in politics and economics. OpEd's model of argumentation involves representations of argument units which organize abstract patterns of support and attack relationships among beliefs.

LEXICON. OpEd has a lexicon in which words, phrases, roots, and suffixes are represented in terms of knowledge structures in its semantic memory. Each lexical item also has attached demons which perform such functions as disambiguating words, resolving pronoun and concept references, and role binding. For example, OpEd's lexicon contains the following words and phrases:

- *Lexical items for authorities, institutions, and countries:* "Reagan administration," "Common Market," "industry," "toolmaker," "manufacturer," "nation," "Japan," "Japanese," "American," "foreign," and "country."

- *Lexical items for products:* "Product," "steel," "import," "export," "automobile," "machine tool," and "machine-tool."
- *Lexical items for economic quantities:* "Wage rate," "dollar," "jobs," "limitation," "restriction," "less," "cheaper," and "expensive."
- *Lexical items that refer to goals:* "Affect," "hurt," "survive," "be bad for," "save," "cost," "promote," "lose," "disappoint," "problem," and "competition."
- *Lexical items for plans and events:* "Protectionist," "measure," "voluntary limit," "protection," "import," "seek," "impose," "demand," "produce," "make do," "sale," "earn," and "spend."
- *Connectives for cause-effect relationships:* "So that," "inevitably," "result," "if," and "then."
- *Connectives for opposition relationships:* "Far from" and "be wrongheaded."
- *Lexical items for beliefs:* "Argue that" and "argument."

DEMON-BASED PARSER. Each conceptual construct in semantic memory has attached processes called *demons* which perform knowledge application and knowledge interaction tasks, such as binding conceptualizations together, recognizing beliefs, recognizing support and attack relationships, recognizing argument units, and tracking causal chains of reasoning.

Input editorial text is mapped into conceptual representations by a demon-based parser which uses the techniques for conceptual parsing implemented in BORIS (Dyer, 1983a), an in-depth understanding of narrative text. Each input sentence is read from left to right on a word-by-word (or phrase-by-phrase) basis. When a lexical item is recognized, a copy of its associated conceptualization is placed into OpEd's working memory. Copies of the lexical item's demons and its conceptualization demons are placed into a demon agenda that contains all active demons. Then the parser tests all active demons and executes those whose test conditions are satisfied. After demons are executed, they are removed from the agenda.

WORKING MEMORY. When demons are executed, they bind together conceptualizations in working memory, and as a result, demons build the conceptual representation of the input sentence. Thus, working memory maintains the current context of the sentence being parsed. After the sentence has been read, its conceptual representation is removed from working memory, and the parsing process is repeated for the next sentence in the editorial.

ARGUMENT GRAPH. Also resulting from demon execution, the conceptualizations created in working memory get integrated with instantiated structures currently indexed by semantic memory's structures. These instantiations compose the editorial's argument graph which represents and maintains the current context of the editorial read so far. Thus, the argument graph can be viewed as OpEd's *episodic memory* (Tulving, 1972), as opposed to OpEd's semantic memory, which contains what OpEd knows before reading the editorial.

MEMORY SEARCH AND RETRIEVAL PROCESSES. During question answering, the argument graph also maintains the current context from which questions are understood. Each input question is parsed by the demon-based parser, and the question's conceptual representation is built in working memory. Question-answering demons attached to WH words (e.g., why, what, who, etc.) are activated whenever such words are encountered at the beginning of the question. These demons determine the question's conceptual category and activate appropriate search and retrieval demons that access the argument graph and return conceptual answers.

ENGLISH GENERATOR. Once an answer is found, it is generated in English by OpEd's recursive descent generator. This generator produces English sentences in a left-to-right manner by traversing instantiated knowledge structures and using generation patterns associated with knowledge structures in semantic memory. For example, instantiations of AU-OPPOSITE-EFFECT are generated by using the pattern

```
(BELIEF1-BY-SELF) ''because'' (BELIEF2-BY-SELF).
(SELF) ''believe that'' (OPPONENT) ''be wrong because''(BELIEF-
BY-OPPONENT).
```

where (1) SELF is the instance of the arguer using AU-OPPOSITE-EFFECT; (2) OPPONENT is the instance of the arguer's opponent; (3) BELIEF1-BY-SELF is the instance of the belief "P should not be used"; (4) BELIEF2-BY-SELF is the instance of the belief "P thwarts goal G"; (5)BELIEF-BY-OPPONENT is the instance of the belief "P achieves goal G"; and (6) the verbs "to believe" and "to be" are conjugated according to the contents of SELF, OPPONENT, and BELIEF-BY OPPONENT. Such patterns are used to generate English sentences, such as A5, during question answering.

2 REPRESENTING POLITICOECONOMIC KNOWLEDGE

During editorial comprehension, domain-specific knowledge must be used to recognize and represent the beliefs, belief relationships, and argument units that compose the editorial's argument graph. In the case of editorials such as ED-JOBS, the comprehension process involves dealing with one-sided arguments about protectionism. To process those editorials, OpEd must have a politicoeconomic model that represents knowledge of protectionism. Consider the following editorial fragment (Morrow, 1983) read by OpEd:

ED-RESTRICTIONS

The American machine-tool industry is seeking protection from foreign competition. The industry has been hurt by cheaper machine tools from Japan. The toolmakers argue that restrictions on imports must be imposed so that the industry can survive. It is a wrongheaded argument. Restrictions on imports would mean that American manufacturers would have to make do with more expensive American machine tools. Inevitably, those American manufacturers would produce more expensive products. They would lose sales. Then those manufacturers would demand protection against foreign competition.

When reading ED-RESTRICTIONS, OpEd must be able to infer the implicit connections between the U.S. manufacturers' act of "seeking protection," their current state of being "hurt by cheaper machine-tools from Japan," and their desire to "survive." Making these connections explicit requires modeling of the following relationships:

- *Politicoeconomic conflicts:* Domestic and foreign industries may get involved in disputes over low-priced imports sold by those foreign industries.
- *Conflict-resolution events:* To resolve trade disputes with foreign industries, domestic industries may petition the government of their country to implement economic protection plans.
- *Reasoning about economic goals:* Low-priced imports "hurt" domestic industries because (1) low-priced imports cause a decrease in sales of equivalent, high-priced domestic products; (2) a decrease in sales produces a decrease in earnings for domestic manufacturers; and (3) a decrease in earnings results in the goal by domestic manufacturers of preserving their level of earnings.
- *Economic-protection plans:* To preserve the level of earnings of domestic industries, a government can take two different courses of action: (1) unilaterally impose quotas or taxes on imports or (2) negotiate with foreign countries to limit imports from those countries.
- *Reasoning about economic protection plans:* Limits and taxes on imports help domestic industries to "survive" because (1) restrictions on imports result in a decrease in domestic spending on imports; (2) a decrease in spending on imports causes an increase in spending on domestic products; and (3) an increase in spending on domestic products achieves the goal by domestic manufacturers of preserving their level of earnings.

OpEd's model of politicoeconomic knowledge is composed of four major elements: *authority triangles* (Schank and Carbonell, 1979), which represent conflicts involving domestic and foreign industries; *planboxes* (Schank and Abelson, 1977), which represent protectionist plans as sets of actions for resolving trade disputes; a *graph of economic relations* (Riesbeck, 1984), which organizes cause-effect relationships among the economic quantities associated with economic actors; and *reasoning scripts* (Dyer, et al., 1987; Flowers and Dyer, 1984), which represent chains of reasoning about economic goals or about the effects of protectionist plans. This section describes OpEd's politicoeconomic constructs along with examples of their use within the framework of ED-JOBS and ED-RESTRICTIONS.

2.1 Social Acts and Authority Triangles for Politicoeconomic Conflicts

OpEd's representation of politicoeconomic conflicts is based on Schank's *basic social acts*, a representational system originally proposed by Schank (1978) and later developed and expanded by Schank and Carbonell (1979) and Carbonell (1981). To understand OpEd's model, it is necessary to have some exposure to the general issues addressed by those social acts. This section presents an overview of the basic social acts and describes how they are used in OpEd to encode knowledge of protectionism.

Social acts encode the basic concepts that organize knowledge in the domain of human social interactions. Seven basic social acts have been proposed by Schank and Carbonell (1979):

```
1. DISPUTE: The initiation or escalation of a conflict between two
   actors
2. PETITION: The request to an authority to settle a conflict
3. AUTHORIZE: The issuing of a decree by an authority
4. ORDER: The enforcement of a decree by an authority
5. INVOKE: The initiation of a direct course of action to settle a
   conflict
6. RESOLVE: The act of settling a conflict by means of a direct
   course of action
7. PRESSURE: The act by a third part of applying pressure to settle
   a conflict
```

For example, the following events in the domain of labor relationships can be represented by using basic social acts: (1) a violation of a worker's civil rights by her or his boss is a DISPUTE; (2) bringing a suit to a court of law to settle the conflict is a PETITION; (3) the jury's decision is an AUTHORIZE act; (4) the enforcement of such a decision by appropriate authorities is an ORDER; (5) a call for an out-of-court settlement is an INVOKE act; (6) the resolution of the case by means of direct negotiations is a RESOLVE act; and (7) the act by the worker's peers of forcing him or her to reconsider the court claim is an act of PRESSURE.

By using these social acts, any conflict situation can be represented in terms of a configuration composed of two basic elements: a DISPUTE between two actors and a resolution method for settling the conflict. Such a configuration is termed an *authority triangle* (Carbonell, 1981). According to the resolution method used, three basic types of authority triangles can be distinguished: AUTHORIZE triangle, RESOLVE triangle, and PRESSURE triangle:

```
1. AUTHORIZE triangle: DISPUTE → PETITION → AUTHORIZE + ORDER
2. RESOLVE triangle:   DISPUTE → INVOKE → RESOLVE
3. PRESSURE triangle:  DISPUTE → PETITION → PRESSURE
```

For instance, in the previous example on labor relationships, the DISPUTE between worker and boss can be settled by the worker PETITIONing to a court of law so that it AUTHORIZEs in his/her favor, the worker's boss INVOKing negotiations with the worker that directly RESOLVE the DISPUTE, and the boss PETITIONing to the worker's peers so that they PRESSURE that worker to accept the position of the boss.

2.2 Modeling Situations of Protectionism with Authority Triangles

Authority triangles provide a method for representing situations of protectionism, i.e., conflicts between domestic and foreign industries that are resolved through the application of economic protection plans. Those conflicts arise when domestic industries experience decreases in sales due to increases in sales of cheaper and better imports by foreign industries. Such decreases in sales motivate the three main goals underlying the conflicts between domestic and foreign industries: a short-term goal of preserving earnings by domestic industries, a long-term goal of attaining profitability by domestic industries, and a goal of preserving jobs by the workers in domestic industries. When these economic goals become active, domestic industries ask their government to implement trade policies that either decrease the amount of low-priced imports (i.e., an import quota) or increase their price (i.e., an import tax). Those restrictions can be unilaterally imposed through legislation or can be negotiated with foreign governments (Greenaway, 1983; Yoffie, 1983).

How are authority triangles used in OpEd's politicoeconomic model? Figure 10-2 shows how situations of protectionism are modeled with authority triangles. As Fig. 10-2 illustrates, two main situations of protectionism can be distinguished:

```
IMPOSED-LIMIT triangle: Industry I1 from country C1 has a DISPUTE
with industry I2 from country C2 over the sale price of product P2
by I2. To settle the conflict, I1 PETITIONs to government G1 of C1
for a limit (quota or tax) on P2 and G1 AUTHORIZEs such a limit.

NEGOTIATED-LIMIT triangle: Industry I1 from country C1 has a DISPUTE
with industry I2 from country C2 over the sale price of product P2
by I2 in C1. To settle the conflict, I1 PETITIONs to government G1
of C1 for a limit (quota or tax) on P2. G1 RESOLVEs the conflict by
INVOKing negotiations with government G2 of C2. The negotiations
result in a quota that is AUTHORIZEd by G2.
```

These two configurations can be used to represent the situations of protectionism referred to in the editorials processed by OpEd. For example consider the following excerpt from of ED-RESTRICTIONS:

> The American machine-tool industry is seeking protection from foreign competition. The industry has been hurt by cheaper machine tools from Japan. The toolmakers argue that restrictions on imports must be imposed...

302 KNOWLEDGE ENGINEERING: FUNDAMENTALS

IMPOSED-LIMIT TRIANGLE

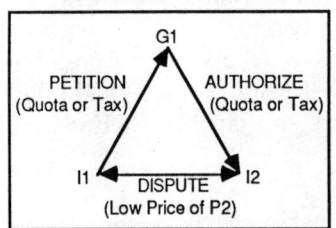

```
            G1
PETITION  /    \  AUTHORIZE
(Quota or Tax) (Quota or Tax)
I1 ←——DISPUTE——→ I2
     (Low Price of P2)
```

NOTATION FOR TRIANGLE ELEMENTS
G1: Government of importing country C1
G2: Government of exporting country C2
I1: Industry from C1
I2: Industry from C2
P2: Product by I2 equivalent to product P1 by I1
Quota: A decrease in amount of P2 in C1
Tax: An increase in price of P2 in C1
Negotiation: Meeting between G1 and G2 to set Quota

NEGOTIATED-LIMIT TRIANGLE

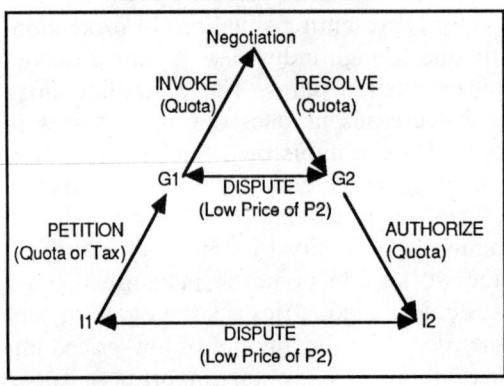

```
              Negotiation
         INVOKE    RESOLVE
         (Quota)   (Quota)
           G1 ←DISPUTE→ G2
              (Low Price of P2)
PETITION                AUTHORIZE
(Quota or Tax)           (Quota)
   I1 ←———DISPUTE———→ I2
         (Low Price of P2)
```

FIGURE 10-2
Situations of protectionism.

Here, the U.S. machine-tool industry has a DISPUTE with the Japanese machine-tool industry over low-priced Japanese machine tools sold in the United States. Due to this DISPUTE, the U.S. manufacturers have PETITIONed for economic protection from the U.S. government. Furthermore, the U.S. manufacturers believe that imports restrictions must be AUTHORIZEd (as opposed to negotiated) by the U.S. government. Thus, the conflict between the U.S. and Japanese manufacturers and its proposed resolution can be represented in terms of the IMPOSED-LIMIT triangle. This representation is illustrated in Fig. 10-3.

In contrast to ED-RESTRICTIONS, the conflict in international trade in ED-JOBS can be modeled as an instance of the NEGOTIATED-LIMIT triangle. To illustrate this fact, consider this fragment of ED-JOBS:

Recent protectionist measures by the Reagan administration... Voluntary limits on Japanese automobiles...

In ED-JOBS, the DISPUTE between the U.S. automobile industry and the Japanese automobile industry has been RESOLVEd through the negotiations INVOKEd by the Reagan administration. Those negotiations resulted in a limit on Japanese cars AUTHORIZEd by the Japanese government. Clearly,

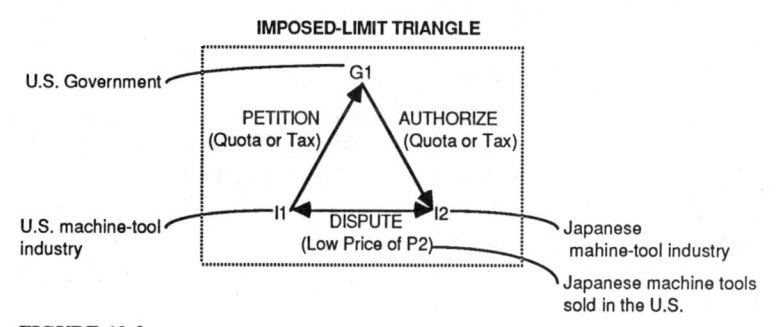

NATURAL LANGUAGE PROCESSING: COMPUTER COMPREHENSION OF EDITORIAL TEXT **303**

FIGURE 10-3
Representation of a trade conflict in ED-RESTRICTIONS.

the conflict settled through the use of "voluntary limits" can be modeled as an instance of the NEGOTIATED-LIMIT triangle, as shown in Fig. 10-4.

The main advantage of using authority triangles is that those constructs allow us to represent all the information associated with conflicts in international trade. As indicated in Fig. 10-2, both the IMPOSED-LIMIT triangle and the NEGOTIATED-LIMIT triangle organize general information about who may be the actors in a trade conflict, what may be the object of the conflict, and who may be the actors that settle the conflict. Therefore, if that information is implicit in an editorial, it will be made explicit in the triangle representations. For example the diagram in Fig. 10-4 represents explicitly the following missing relationships in ED-JOBS: (1) the actors of the DISPUTE are the

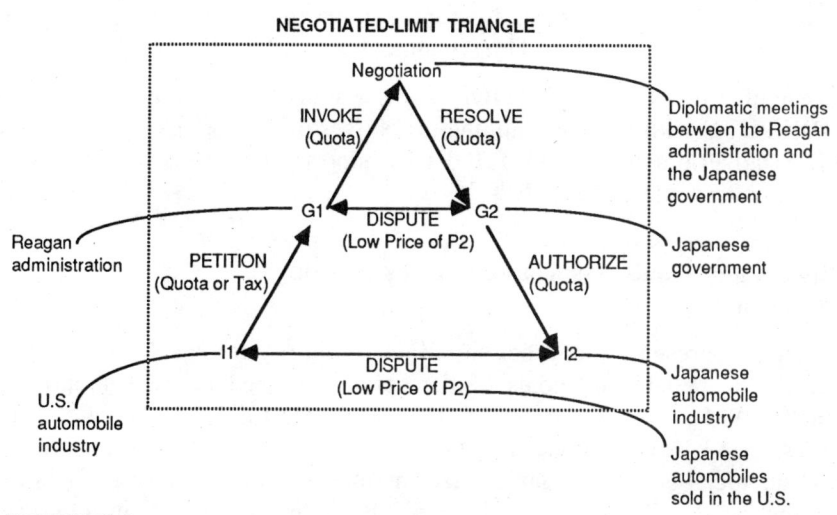

FIGURE 10-4
Representation of a trade conflict in ED-JOBS.

U.S. automobile industry and the Japanese automobile industry; (2) the object of the DISPUTE is the low price of Japanese automobiles sold in the United States; and (3) the actors of the negotiation that have RESOLVEd the DISPUTE are the Reagan administration and the Japanese government.

In addition to the above inferences, each basic social act in a triangle has a set of associated inferences that indicate the acts that may have happened or may happen according to that triangle configuration. As a result, if one of the acts in a triangle is known, it is possible to represent explicitly the other possible acts in that triangle. For instance, in the domain of situations of protectionism, the two inferences associated with the PETITION act are as follows:

```
PETITION-Antecedent Rule: IF industry I1 from country C1 has a
DISPUTE with industry I2 from country C2 over the price of a product
P2 by I2 which is equivalent to a product P1 by I1, THEN I1 will
PETITION for economic protection to government G1 of C1.

PETITION-Consequent Rule: IF industry I1 from country C1 PETITIONs
for economic protection to government G1 of C1, THEN G1 may either
(1) AUTHORIZE a quota or tax on P2 or (2) INVOKE negotiations on a
quota on P2 to RESOLVE the DISPUTE.
```

The above rules predict the DISPUTE and resolution methods encoded in the IMPOSED-LIMIT triangle and NEGOTIATED-LIMIT triangle. The application of these rules depends on the occurrence of the social act PETITION. For example, in the last sentence from ED-RESTRICTIONS:

> ...those [American] manufacturers would demand protection against foreign competition.

the above rules provide the basis for representing the events that may have caused and may result from the U.S. manufacturers' PETITION, namely:

- U.S. manufacturers will PETITION for economic protection because of their DISPUTE with foreign manufacturers over import prices.
- The U.S. government may AUTHORIZE import restrictions or INVOKE negotiations to settle the DISPUTE.

2.3 Beliefs and Goals Associated with Economic Protectionism

In addition to representing conflict situations, authority triangles can be integrated with the beliefs and goals of the actors involved in such situations (Carbonell, 1981). In the case of the IMPOSED-LIMIT triangle and NEGOTIATED-LIMIT triangle, knowledge of goals and beliefs is organized by the basic social acts that characterize the methods for settling conflicts in international trade. Two main rules encode such knowledge in OpEd's politicoeconomic model:

Economic-Protection Rule 1: IF government G1 of country C1 has the active goals of preserving jobs in C1 and helping industry I1 from C1 preserve earnings and attain profitability, AND G1 believes those goals will be achieved by implementing economic protection P, THEN G1 will AUTHORIZE P or INVOKE negotiations to implement P.

Economic-Protection Rule 2: IF industry I1 from country C1 has the active goals of preserving earnings and attaining profitability, AND I1 believes that those goals will be achieved by economic protection P from government G1 of C1, THEN I1 will PETITION G1 to implement P.

The above rules summarize the goals and beliefs of governments that impose or negotiate economic protection and industries that PETITION for economic protection. These rules can be applied to represent relationships implicitly stated in editorial text. For example, in ED-JOBS, the first rule can be used to infer the Reagan administration's goals and beliefs associated with the "voluntary limits on Japanese automobiles," namely, the administration's goals of preserving jobs in the United States and helping the U.S. automobile industry attain profitability and the administration's belief that negotiating "voluntary limits" will achieve its goals. Similarly, the second rule can be used in ED-RESTRICTIONS to represent explicitly the connections between the U.S. machine-tool industry's act of PETITIONing for "protection from foreign competition," the industry's active preservation goal of not being "hurt" by cheaper imports, and the industry's belief that "import restrictions must be imposed so that the industry can survive." The representation of those relationships is shown in Fig. 10-5.

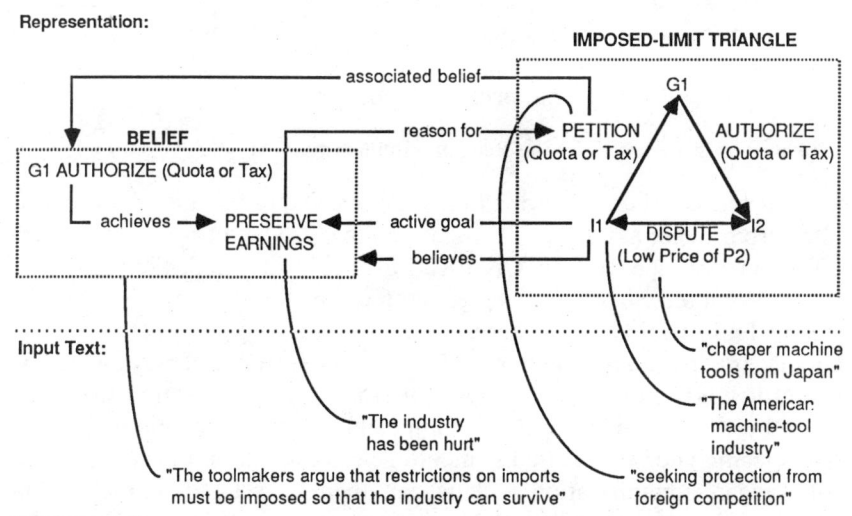

FIGURE 10-5
Goals and beliefs associated with a PETITION for economic protection.

As Fig. 10-5 indicates, representing a PETITION for economic protection requires representing the DISPUTE that precedes the PETITION, the course of action being PETITIONed to the government, the goals of the actor of the PETITION, and her or his belief that those goals will be achieved by the government's action. Thus, the representation of a conflict in international trade must include the conflict's triangle configuration and the beliefs and goals of the actors involved in that conflict.

2.4 Organizing Conflict-Resolution Events with Planboxes

From the perspective of goals and plans, the courses of action that a government can take to protect an industry can also be represented as *planboxes*. As originally defined by Schank and Abelson (1977), *planboxes* are possible sets of actions that can be executed to achieve delta goals, i.e., goals involving desires for a change in state. For example, three planboxes associated with the goal of gaining control of an object are asking, bargaining, and stealing. Similarly, the actions of imposing and negotiating limits on imports are the two planboxes associated with a government's economic goals of preserving jobs and helping domestic industries preserve earnings and attain profitability. In OpEd, those two planboxes are organized by a planning structure called P-ECON-PROTECTION (package of methods for achieving economic protection). P-ECON-PROTECTION is much like Schank and Abelson's PERSUADE package, which organizes planboxes associated with delta goals. The components of P-ECON-PROTECTION are illustrated in Fig. 10-6.

What do we gain by representing methods for achieving economic goals as planboxes organized by P-ECON-PROTECTION? One problem of representing verbatim arguments about protectionism involves dealing with descriptions of economic protection policies. Frequently, those descriptions do not mention the specific courses of action taken by a government. For example, the following phrase from the editorial ED-JOBS

...protectionist measures by the Reagan administration...

does not indicate whether the protectionist policies were unilaterally imposed or resulted from negotiations with foreign countries. This phrase can be represented as an instance of P-ECON-PROTECTION in which the importing country is the United States and the government of the importing country is the Reagan administration. That representation serves two purposes: it acts as a placeholder for the unknown course of action taken by the Reagan administration, and it holds expectations about possible courses of action that may have taken by the administration, i.e., imposing limits or negotiating limits. When the specific course of action is mentioned in the editorial, the representation of Reagan administration's plans can then be refined to include an instantiation of the corresponding planbox. For instance, after a reading of the following phrase in ED-JOBS:

P-ECON-PROTECTION

CONFLICT PARTICIPANTS C1: Importing country C2: Exporting country G1: Government of C1 G2: Government of C2 I1: Industry from C1 I2: Industry from C2
PRODUCTS P1: Product by I1 P2: Product by I2 equivalent to P1
GOALS G1 wants to help I1 PRESERVE EARNINGS G1 wants to help I1 ATTAIN PROFITABILITY G1 wants to PRESERVE JOBS in C1
CONFLICT SOLUTIONS Tax: An increase in price of P2 in C1 Quota: A decrease in amount of P2 in C1
IMPOSED-LIMIT PLANBOX G1 AUTHORIZE (Tax OR Quota) to I2
NEGOTIATED-LIMIT PLANBOX G1 INVOKE (Negotiations for Quota) —causes—> G1 and G2 RESOLVE (Quota) —causes—> G2 AUTHORIZE (Quota) to I2

FIGURE 10-6
Planboxes for economic protection.

Voluntary limits on Japanese automobiles...

it is clear that the administration's policies resulted from negotiations with Japan and, therefore, should be represented in terms of the negotiated-limit planbox. Thus, encoding conflict-resolution methods as planboxes in P-ECON-PROTECTION provides a system for dealing with unstated protectionist actions in editorials.

2.5 Economic Reasoning

A model of conflicts in international trade also requires a representation of the reasoning chains that show why economic goals become active as a result of changes in import prices and consumer spending and why economic protection plans result in changes in the level of earnings and employment in domestic industries. Those chains are sequences of cause-effect relationships among the economic quantities associated with the activity of trade. In OpEd, reasoning about goals and plans is represented by *reasoning scripts* (Dyer et al., 1987; Flowers and Dyer, 1984), memory structures that organize causal domain knowledge in the form of prespecified reasoning-chain sequences. OpEd also includes a model of the cause-effect relationships underlying reasoning scripts. As proposed by Riesbeck (1984), causal knowledge in the domain of economics can be modeled in terms of a network of economic quantities. In that network, nodes represent economic quantities, and links represent the effects quantities have on one another. Riesbeck's modeling approach has been

308 KNOWLEDGE ENGINEERING: FUNDAMENTALS

adopted in OpEd to represent the causal relationships that characterize activity of trade. OpEd's graph of trade relationships is shown in Fig. 10-7.

This graph organizes causal dependencies from the perspectives of producers and consumers. From a producer's point of view, trade can be characterized in terms of that producer's level of earnings, volume of sales, production costs, and product prices:

```
Trade Relationship 1: The level of earnings is directly proportional
to the volume of sales and inversely proportional to the level of
costs.

Trade Relationship 2: The level of costs is directly proportional to
the level of spending on basic machinery and production materials,
the level of salaries, and the level of employment.

Trade Relationship 3: The volume of sales is directly proportional
to the level of consumer spending on producer's products.

Trade Relationship 4: The level of prices is directly proportional
to the level of costs.

Trade Relationship 5: The levels of spending, salaries, and
employment are directly proportional to the level earnings.
```

In contrast, from a consumer's point of view, trade can be characterized in terms of product prices and level of consumer spending:

```
Trade Relationship 6: The level of spending on a product P1 is
inversely proportional to P1's price and to the level of spending on
an equivalent product P2 and directly proportional to P2's price.
```

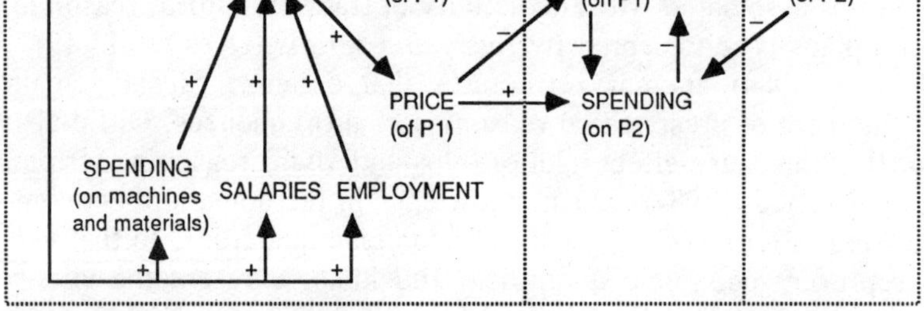

TRADE RELATIONSHIPS

```
PRODUCER-1          |  CONSUMER  |  PRODUCER-2
```

[Diagram showing causal graph with nodes: EARNINGS, COSTS, SALES (of P1), SPENDING (on P1), PRICE (of P2), PRICE (of P1), SPENDING (on P2), SPENDING (on machines and materials), SALARIES, EMPLOYMENT, connected by + and − signed arrows]

FIGURE 10-7
OpEd graph of trade relationships.

In OpEd, the trade relation graph provides the representational foundation for the causal chains of reasoning associated with economic goals. For instance, the graph shows that the price of produce P2 is connected to the earnings of PRODUCER1 by a positive sequence of links involving consumer spending on P2, consumer spending on P1, and sales of P1. According to that sequence, when the price of P2 is low, the level of earnings of PRODUCER1 is low. This causal relationship explains decreases in earnings experienced by producers whose prices are higher than their competitors'.

2.6 Comprehension of Causal Chains in Economics with Reasoning Scripts

The graph of trade relationships organizes chains of causality involving the effects of import pricing and other forms of economic protectionism on domestic industries, on their use of foreign materials and machinery, and on domestic industries that export their products. Often specific paths in the graph of trade relations will be well known by a given economist. As a result, paths in the chain are not explicitly mentioned by the editorial writer. In OpEd, a subset of the possible paths through the trade relation graph are preencoded in OpEd, in the form of a *reasoning script* (Dyer et al., 1987). Like a cultural script, a reasoning script holds a stereotypic chain of events. In this case, however, the events involve changes in economic quantities. In natural language understanding, scripts are assumed to have been learned through repeated interaction with stereotypic event sequences, such as sitting, ordering, eating, tipping, and paying in a restaurant. Scripts aid in the inference of missing events and reduce computation by avoiding a general search of all possible paths in the space of all possible inferences. For instance, an NLP system with $RESTAURANT (scripts are prefixed with a $) can immediately infer the events of READ-MENU and EAT-MEAL given only the input "John ordered a hamburger at the restaurant and afterward left a big tip."

A reasoning script (prefixed with $R-) is assumed to have been built up in the mind of the economist through repeated encounters with common sequences of economic effects. OpEd is currently not a learning system, so these reasoning scripts are encoded by hand. For example, in the domain of international trade conflicts, the reasoning script $R-LOW-IMPORT-PRICES → LOW-DOMESTIC-EARNINGS represents a causal chain of events that leads from low import prices to low earnings in domestic industries (see Fig. 10-8). In OpEd, this reasoning script is used to represent the meaning of the following sentence from ED-RESTRICTIONS:

> The [American machine-tool] industry has been hurt by cheaper machine tools from Japan.

In the above sentence, the word "hurt" does not refer to the physical state of U.S. manufacturers. Rather, it refers to the causal connection between cheap Japanese imports and the goal by U.S. manufacturers of preserving their earn-

```
$R-LOW-IMPORT-PRICES—>LOW-DOMESTIC-EARNINGS
┌─────────────────────────────────────────────────┐
│ ROLES                                           │
│   C1: Importing country                         │
│   I1: Industry from C1                          │
│   P1: Product by I1                             │
│   P2: Import equivalent to and less expensive than P1 │
├─────────────────────────────────────────────────┤
│ CAUSAL CHAIN                                    │
│   low PRICE of P2 —causes—>                     │
│    high SPENDING by C1 on P2 —causes—>          │
│     low SPENDING by C1 on P1 —causes—>          │
│      low SALES of P1 by I1 —causes—>            │
│       low EARNINGS of I1 —thwarts—>             │
│        G-PRESERVING-EARNINGS by I1              │
└─────────────────────────────────────────────────┘
```

FIGURE 10-8
$R-LOW-IMPORT PRICES → LOW-DOMESTIC-EARNINGS.

ings. This causal connection is an instantiation of $R-LOW-IMPORT-PRICES → LOW-DOMESTIC-EARNINGS:

```
low PRICE of Japanese machine tools —causes →
 high SPENDING by U.S. on Japanese machine tools —causes →
  low SPENDING by U.S. on U.S. machine tools —causes →
   low SALES of machine tools by U.S. machine-tool industry —causes →
    low EARNINGS by U.S. machine tool industry —thwarts →
     G-PRESERVING-EARNINGS by U.S. machine-tool industry
```

Thus, the use of reasoning scripts allows OpEd to represent abstract politicoeconomic concepts as sequences of cause-effect relationships associated with the activity of trade.

Why do economic protection plans help domestic industries? The immediate effect of such plans is to decrease domestic spending on cheap imports. This is the case because economic protections either reduce the amount of available imports through quotas or increase import prices through taxes. As indicated in the graph of trade relationships, decreasing domestic spending on imports causes an increase in domestic spending on more expensive domestic products. This increase in spending causes an increase in the volume of sales of domestic products and, consequently, an increase in the level of earnings of domestic manufacturers. This chain is represented by $R-ECON-PROTECTION → HIGHER-DOMESTIC-EARNINGS (not shown).

2.6.1 EFFECTS OF PROTECTIONISM ON INDUSTRIES USING IMPORTS. Although restrictions on imports benefit ailing industries, restrictions have negative side effects for other domestic industries. Due to the fact that restrictions shift spending from cheap imports to expensive domestic products, industries that use those imports experience an increase in their production costs. As indicated in the graph trade relationships, an increase in production costs results in a decrease in producer's earnings, because of a chain of effects involving an increase in product prices, a decrease in consumer spending, and a decrease in the volume of sales. This chain of reasoning is contained in $R-ECON-

PROTECTION → LOWER-DOMESTIC-EARNINGS and is instantiated when OpEd reads the following excerpt from ED-RESTRICTIONS:

> **Restrictions on [cheaper machine tools from Japan] would mean that American manufacturers would have to make do with more expensive American machine tools. Inevitably, those American manufacturers would produce more expensive products. They would lose sales.**

The representation of this cause-effect chain is depicted in Fig. 10-9. As indicated in the diagram, the instantiated reasoning script represents information that is implicitly stated in the editorial. Specifically, the script makes explicit why import restrictions result in an increase in spending on U.S. machine tools, why the increase in spending causes an increase in prices of U.S. products, why the increase in prices causes a decrease in the volume of sales by U.S. manufacturers, and why the decrease in sales results in a decrease in the manufacturers' level of earnings. Thus, reasoning scripts provide a representational system for dealing with missing steps in chains of reasoning in editorials.

2.6.2 EFFECTS OF PROTECTIONISM ON EXPORT INDUSTRIES. Another side effect of economic protection plans is that they may not preserve (or increase) the number of jobs in an importing country, but may decrease it. In politicoeconomic editorials, this side effect is frequently brought up in arguments against the use of import restrictions. For example, consider the following excerpt from ED-JOBS:

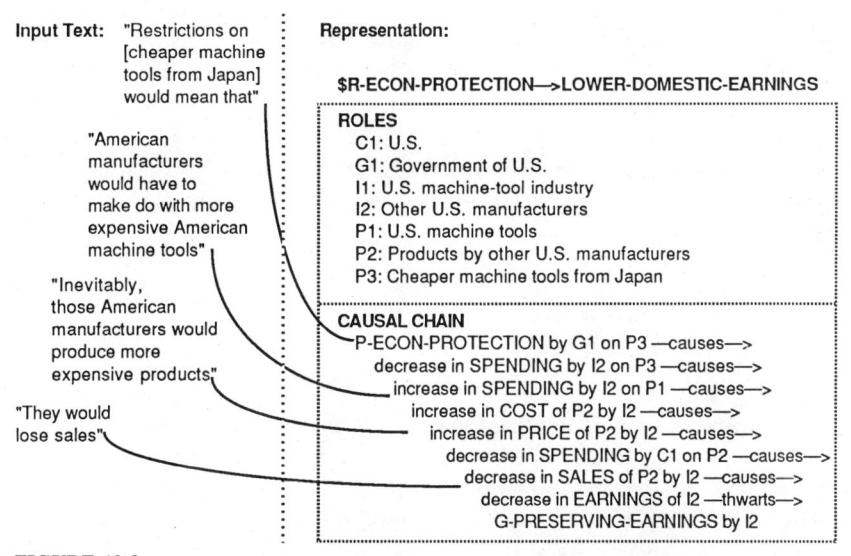

```
Input Text:   "Restrictions on        Representation:
              [cheaper machine
              tools from Japan]       $R-ECON-PROTECTION—>LOWER-DOMESTIC-EARNINGS
              would mean that"
                                      ROLES
              "American                  C1: U.S.
              manufacturers              G1: Government of U.S.
              would have to              I1: U.S. machine-tool industry
              make do with more          I2: Other U.S. manufacturers
              expensive American         P1: U.S. machine tools
              machine tools"             P2: Products by other U.S. manufacturers
                                         P3: Cheaper machine tools from Japan
              "Inevitably,
              those American          CAUSAL CHAIN
              manufacturers would       P-ECON-PROTECTION by G1 on P3 —causes—>
              produce more              decrease in SPENDING by I2 on P3 —causes—>
              expensive products"       increase in SPENDING by I2 on P1 —causes—>
                                        increase in COST of P2 by I2 —causes—>
              "They would               increase in PRICE of P2 by I2 —causes—>
              lose sales"               decrease in SPENDING by C1 on P2 —causes—>
                                        decrease in SALES of P2 by I2 —causes—>
                                        decrease in EARNINGS of I2 —thwarts—>
                                        G-PRESERVING-EARNINGS by I2
```

FIGURE 10-9
Representation of a causal chain of reasoning in ED-RESTRICTIONS.

> Far from saving jobs, the limitations on imports [by the Reagan administration] will cost jobs. If we import less, foreign countries will earn fewer dollars. They will have less to spend on American exports. The result will be fewer jobs in export industries.

The above excerpt contains a reasoning chain on how import restrictions cause a decrease in U.S. exports and, consequently, a decrease in jobs in U.S. export industries. In OpEd, this chain is represented in terms of $R-ECON-PROTECTION → LOWER-EXPORT-JOBS. This reasoning script describes the negative feedback that results from applying restrictions to international trade. There are four major reasons for this negative feedback:

1. Import restrictions cause a decrease in sales by exporting countries and, consequently, a decrease in their level of export earnings.
2. Countries play two different roles in international trade: As producers, they export their products to other countries; and as consumers, they import products from other countries. These roles depend on each other because the level of spending on imports is directly proportional to the level of export earnings.
3. The level of earnings of export industries is directly proportional to the industries' sales to importing countries.
4. The level of employment in export industries is directly proportional to the industries' level of earnings.

Therefore, import restrictions result in a decrease in the number of jobs in export industries of the countries that implement those restrictions.

Using $R-ECON-PROTECTION → FEWER-EXPORT-JOBS also allows OpEd to represent explicitly missing cause-effect relationships in the reasoning chain in ED-JOBS. OpEd representation of that chain is shown in Fig. 10-10. The diagram indicates that the following relationships are implicitly stated in ED-JOBS: the relationship between import restrictions and the level of U.S. spending on imports, between U.S. spending and the level of earnings by foreign countries, and between foreign earnings and the number of jobs in U.S. export industries.

3 BELIEFS, BELIEF RELATIONSHIPS, AND ARGUMENT UNITS

Computer comprehension of editorial arguments in OpEd is based on the capability of modeling beliefs. Beliefs can be set apart from other conceptual structures needed to understand narrative text. As pointed out by Abelson (1973, 1979), beliefs are not goals, plans, events, or states, but rather are predications about these structures and their relationships. Based on this view of beliefs, three types of predications have been characterized in OpEd:

```
Input Text:                    Representation:

                               $R-ECON-PROTECTION—>FEWER-EXPORT-JOBS
                               ┌─────────────────────────────────────────────┐
     "If we import less"       │ ROLES                                       │
                               │   C1: U.S.                                  │
      "foreign                 │   C2: Foreign countries                     │
      countries                │   G1: Reagan administration                 │
      will earn                │   I1: U.S. export industries                │
      fewer dollars"           │   P1: U.S. exports                          │
                               │   P2: Products from foreign countries       │
     "They will                ├─────────────────────────────────────────────┤
     have less                 │ CAUSAL CHAIN                                │
     to spend on               │   P-ECON-PROTECTION by G1 on P2 —causes—>   │
     American exports"         │   decrease in SPENDING by C1 on P2 —causes—>│
                               │   decrease in SALES of P2 by C2 —causes—>   │
     "The result               │   decrease in EARNINGS by C2 —causes—>      │
     will be fewer jobs        │   decrease in SPENDING by C2 on P1 —causes—>│
     in export industries"     │   decrease in SALES of P1 by I1 —causes—>   │
                               │   decrease in EARNINGS by I1 —causes—>      │
                               │   decrease in EMPLOYMENT in I1 —thwarts—>   │
                               │   G-PRESERVING-JOBS in C1 by G1             │
                               └─────────────────────────────────────────────┘
```

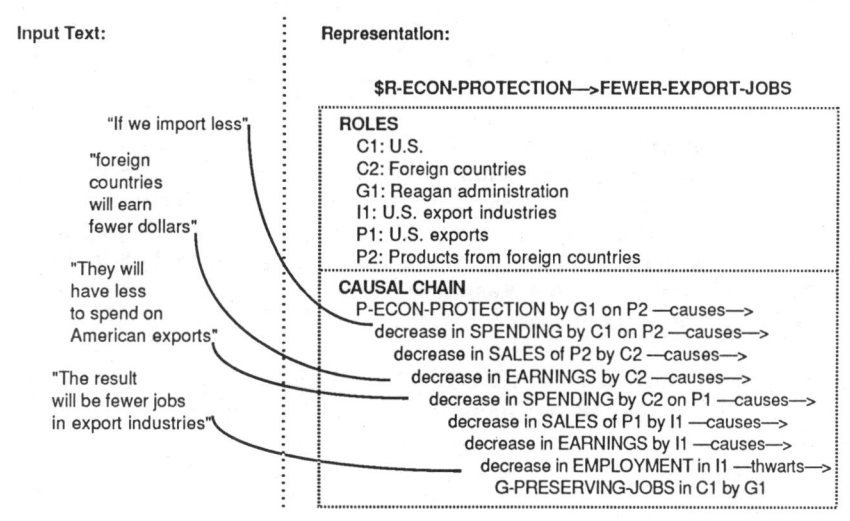

FIGURE 10-10
Representation of a causal chain of reasoning in ED-JOBS.

1. *Evaluative beliefs:* Judgments about the goodness or badness of domain-specific plans, such as "plan P is good/bad" and "plan P should/should not be implemented."
2. *Causal beliefs:* Expectations about the possible causes for the failure or achievement of domain-specific goals and the positive or negative effects that may result from implementing domain-specific plans.
3. *Beliefs about beliefs:* Predications about evaluative and causal beliefs, such as "belief B1 should not be held," "belief B1 does/does not provide evidence for belief B2," and "belief B1 contradicts belief B2."[1]

Why is it necessary to distinguish among these types of beliefs? A basic problem in editorial comprehension is to build an internal conceptual model of editorial arguments. This model, known as an *argument graph* (Flowers et al., 1982), represents explicitly whether beliefs in the editorial are involved in support relationships, because they provide evidence for one another, or attack relationships, because they contradict one another. For example, consider the following excerpt from an editorial by the Los Angeles Times (Dec. 9, 1984):

ED-CONTRADICTORY-POLICIES

American negotiators are pursuing a protectionist course in efforts to control steel imports, with restrictions under consideration that would place an extraordinary

---

[1] Other predications that fall within the category of beliefs about beliefs, such as "X believes that Y believes Z," are discussed in Wilks and Bien, 1983.

burden on consumers in what seems a vain effort to protect U.S. steel makers...This is the wrong way to go...The American steel industry...will be cushioned from the economic forces that alone...hold the hope of restoring productivity and competitiveness. And consumers will be forced to pay the cost through denial of the cheaper foreign products...This...protectionism comes at the very moment when the U.S. government has won international agreement...to liberalize trade in the service sector, where American companies compete so well. Washington is announcing to the world that a new wall is being built around the United States temporarily to bar the things that some foreigners do better than Americans, but that Washington wants others to pull down the walls that keep out things that U.S. industry does best.

Understanding ED-CONTRADICTORY-POLICIES requires representing the belief relationships that summarize the position of the Los Angeles Times, namely:

- *Support relationship between evaluative and causal beliefs:* The Los Angeles Times believes that protectionism by the United States is "wrong" because (1) import restrictions will not achieve the goals of "restoring productivity and competitiveness." for U.S. steel makers and (2) import restrictions will force U.S. consumers "to pay the cost" of protecting the steel makers.
- *Support relationship between causal beliefs:* The Los Angeles Times believes that (1) import restrictions, because they will result in a decrease in "cheaper foreign products," will force consumers "to pay the cost" in the United States, and (2) import restrictions will not achieve the goals of "restoring productivity and competitiveness" because they will block the "economic forces that alone" can achieve those goals.
- *Attack relationship between causal beliefs held by different arguers:* The U.S. government's belief that import restrictions will "protect U.S. steel makers" is contradicted by two of the Los Angeles Times' beliefs: (1) import restrictions will not achieve the goals of "restoring productivity and competitiveness" for steel makers and (2) import restrictions will force consumers "to pay the cost."
- *Attack relationship between evaluative beliefs held by the same arguer:* The Los Angeles Times believes that the U.S. government holds two contradictory positions with respect to international trade: (1) foreign countries should abolish the import restrictions "that keep out things that U.S. industry does best," and (2) the United States should implement import restrictions "to bar the things that some foreigners do better than Americans."

As the above relationships indicate, the way in which a belief is supported or attacked in editorial arguments depends on the nature of that belief. For instance, the evaluative belief that "plan P should not be implemented" can be supported by the causal belief that "implementing P will either fail to achieve or thwart a goal G." Similarly, the evaluative belief that "plan P should be implemented" can be supported by the causal belief that "implementing P will achieve a goal G." Clearly, editorial comprehension requires a taxonomy of the support and attack relationships that exist among different types of beliefs.

3.1 Belief Representation

How are beliefs represented in OpEd? The representation of a belief consists of three major components: the holder of the belief, the content of the belief, and links that indicate whether the belief attacks, supports, or is supported or attacked by other beliefs. For example, consider Fig. 10-11, which illustrates the representation of four beliefs from ED-CONTRADICTORY-POLICIES.

Figure 10-11 shows the attack and support relationships that contain BELIEF2, i.e., the Los Angeles Times' belief that import restrictions will

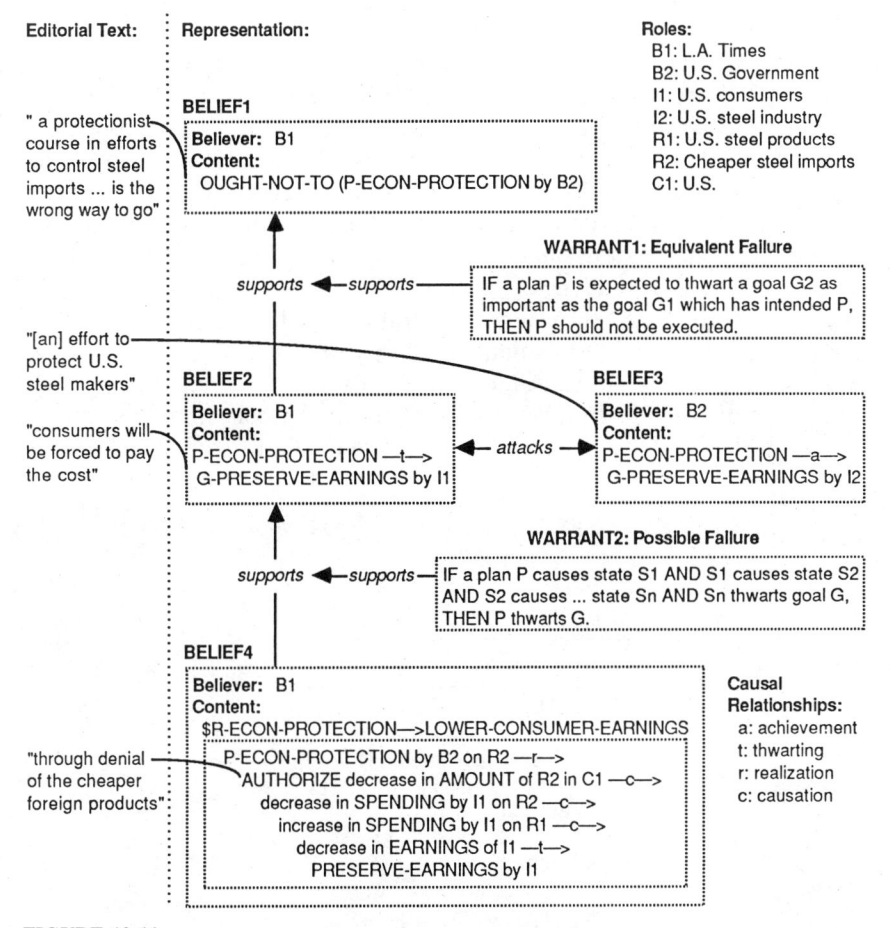

```
Editorial Text:        Representation:                                            Roles:
                                                                                  B1: L.A. Times
                                                                                  B2: U.S. Government
                       BELIEF1                                                    I1: U.S. consumers
" a protectionist                                                                 I2: U.S. steel industry
course in efforts      Believer:  B1                                              R1: U.S. steel products
to control steel       Content:                                                   R2: Cheaper steel imports
imports ... is the        OUGHT-NOT-TO (P-ECON-PROTECTION by B2)                  C1: U.S.
wrong way to go"

                                  ↑
                                                     WARRANT1: Equivalent Failure
                       supports ←—supports——  IF a plan P is expected to thwart a goal G2 as
                                              important as the goal G1 which has intended P,
"[an] effort to                               THEN P should not be executed.
protect U.S.
steel makers"          BELIEF2                             BELIEF3
                       Believer:  B1                       Believer:  B2
"consumers will        Content:                            Content:
be forced to pay       P-ECON-PROTECTION —t→   ← attacks → P-ECON-PROTECTION —a→
the cost"              G-PRESERVE-EARNINGS by I1           G-PRESERVE-EARNINGS by I2

                                  ↑                  WARRANT2: Possible Failure
                       supports ←—supports——  IF a plan P causes state S1 AND S1 causes state S2
                                              AND S2 causes ... state Sn AND Sn thwarts goal G,
                       BELIEF4                THEN P thwarts G.
                       Believer:  B1
                       Content:                                                   Causal
                       $R-ECON-PROTECTION—>LOWER-CONSUMER-EARNINGS                Relationships:
"through denial          P-ECON-PROTECTION by B2 on R2 —r→                         a: achievement
of the cheaper           AUTHORIZE decrease in AMOUNT of R2 in C1 —c→              t: thwarting
foreign products"        decrease in SPENDING by I1 on R2 —c→                      r: realization
                         increase in SPENDING by I1 on R1 —c→                      c: causation
                         decrease in EARNINGS of I1 —t→
                         PRESERVE-EARNINGS by I1
```

FIGURE 10-11
Representation of beliefs and belief relationships.

thwart the goal of preserving earnings for U.S. consumers. Each support relationship is itself supported by a more basic belief termed a *warrant* (Flowers et al., 1982; Toulmin, 1958; Toulmin et al., 1979). Warrants are inference rules that establish why conclusions can be drawn from supporting evidences (i.e., warrants are beliefs about beliefs). This explicit representation of warrants is needed because they can themselves be attacked in arguments about the use of support strategies. For instance, the support relationship between BELIEF2 and the instantiation of the reasoning script $R-ECON-PROTECTION → LOWER-CONSUMER-EARNINGS in BELIEF4 is based on the following warrant:

```
Possible-Failure Warrant: IF plan P causes state S1 AND S1 causes
state S2 AND S2...causes state Sn AND Sn thwarts goal G, THEN P
thwarts G.
```

Figure 10-11 also shows that the representation of *contents* of beliefs involve (1) a causal dependency between a plan and a goal, (2) a chain of causal dependencies organized by a reasoning script, or (3) an evaluative component. Causal dependencies include intentional relationships among goals, plans, events, and states, such as goal achievement, goal failure, goal motivation, goal suspension, plan intention, plan enablement, plan disablement, event realization, and forced events. These dependencies are represented by means of *intentional links* (I-links), a representational system developed in (Dyer, 1983a) that encodes the motivations and intentions of narrative characters. Other nonintentional causal dependencies, such as those among states of economic quantities, are represented by using a general causal link. The major causal dependencies used in OpEd are summarized in Table 10-1.

The representation of causal dependencies in OpEd expands Dyer's work on intentional links. In Dyer's model, states are considered as part of events. In contrast, states and events are separated in OpEd due to the need to represent explicitly the causal relationships that exist among conflict-

TABLE 10-1
Causal dependencies in OpEd

Relationship name	Representation
Goal achievement	STATE —achieves → GOAL
Goal failure	STATE —thwarts → GOAL
Goal motivation	STATE —motivates → GOAL
Goal suspension	GOAL1 —suspends → GOAL2
Plan intention	GOAL —intends → PLAN
Plan enablement	STATE —enables → PLAN
Plan disablement	STATE —disables → PLAN
Event realization	PLAN —realizes → EVENT
Forced event	STATE —forces → EVENT
Consequent state	EVENT —causes → STATE
	STATE1 —causes → STATE 2

resolution events, states of economic quantities, economic goals, and protectionist plans. As Table 10-1 indicates, in OpEd goals are *motivated* by desires to attain, change, or maintain given states, and plans are *intended* to achieve active goals. Further interactions among plans and goals are mediated by chains of causal effects among events and states. That is, plans may *achieve* or *thwart* goals because (1) once plans are executed, they *cause* events to be *realized* and (2) those events result in states that may achieve or thwart goals. For example, the Los Angeles Times can argue that P-ECON-PROTECTION—thwarts → G-PRESERVING-EARNINGS because of the following chain of effects:

```
P-ECON-PROTECTION by U.S. government on cheap steel imports —realizes →
   AUTHORIZE decrease in AMOUNT of cheap steel imports —causes →
     decrease in SPENDING by U.S. consumers on cheap steel imports —causes →
       increase in SPENDING by consumers on expensive U.S. products —causes →
         decrease in EARNINGS of U.S. consumers —thwarts →
           G-PRESERVING-EARNINGS by U.S. consumers
```

Therefore, dependencies of the form PLAN—achieves → GOAL and PLAN —thwarts → GOAL can be viewed as *condensed* causal chains of reasoning.

These plan-goal dependencies also form the representational foundation underlying evaluative components of beliefs. Evaluative components are high-level abstractions that categorize and organize concepts in terms of being "good" or "bad," or leading to "good" or "bad" (Abelson, 1979). In OpEd, evaluative components are used to represent the main standpoints that argument participants hold with respect to a given plan P, i.e., whether they support or oppose the use of P. These plan evaluations are captured by the following constructs:

```
OUGHT-TO (P): A plan P should be executed IF the following
situations can be expected: (1) P will achieve the goal G1 which has
intended P; AND (2) P will not thwart a goal G2 which is more
important than or as important as G1.

OUGHT-NOT-TO (P): A plan P should not be executed IF any of the
following situations can be expected: (1) P will not achieve the
goal G1 which has intended P; OR (2) P will thwart a goal G2 which
is more important than or as important as G1.
```

For example, in the following excerpt from ED-CONTRADICTORY-POLICIES

> ...**a protectionist course in efforts to control steel imports...is the wrong way to go**...

the phrase "wrong way to go" indicates that the Los Angeles Times opposes restrictions on steel imports. This sentence is represented as an instantiation of OUGHT-NOT-TO, as indicated in Fig. 10-11.

OpEd's evaluative components categorize plans in terms of the possible positive or negative effects of implementing those plans.[2] Frequently, these outcomes are explicitly mentioned in editorial arguments to justify evaluative beliefs about plans. For instance, consider the Los Angeles Times' argument in ED-CONTRADICTORY-POLICIES:

> **American negotiators are pursuing a protectionist course in efforts...to protect U.S. steel makers...And consumers will be forced to pay the cost through denial of the cheaper foreign products...**

Here the Los Angeles Times contrasts the following plan-goal relationships: (1) import restrictions are intended to achieve the goal of preserving earnings for U.S. steel makers, and (2) import restrictions will thwart the goal of preserving earnings for U.S. consumers. These relationships refer to opposite effects on two equivalent (i.e., equally important) goals and consequently justify the Los Angeles Times' belief that the United States OUGHT-NOT-TO negotiate import restrictions. Thus, evaluative components not only provide a representational system for contents of beliefs, but also organize belief justifications in terms of goal achievements and goal failures.

3.2 Attack Relationships

Contents of beliefs serve as the basis for establishing whether those beliefs attack one another. In OpEd, an *attack* (A) is modeled as a bidirectional relationship between two contradictory beliefs; i.e., if belief B1 attacks belief B2, then belief B2 attacks belief B1. Two beliefs are considered contradictory if their contents involve either planning situations that cannot occur at the same time (i.e., mutually exclusive planning situations) or opposite effects of a plan P on two interrelated goals. These relationships are summarized in Table 10-2 and described below.

3.2.1 ATTACKS BASED ON MUTUALLY EXCLUSIVE PLANNING SITUATIONS. An evaluative or causal belief B1 about a plan P can be contradicted by stating a belief B2 which negates the content of B1.[3] This type of contradiction, termed *contradiction by negation* (Flowers, 1982), is the basis for three different attack structures developed within the framework of OpEd:

---

[2] The notion of evaluative components in OpEd is similar in nature to the deontic notion of the *"ought" of reasons* (Harman, 1986), which characterizes judgments that use the term "ought" to indicate reasons for doing or not doing something.

[3] At this point it is important to notice that the plan-goal relationship P—thwarts → G is not the same as the relationship P—not-achieves → G. Instead, the goal-thwarting relationship indicates one of the reasons why P cannot achieve G. Similarly, the relationship P—achieves → G is not equal to the relationship P—not-thwarts → G, but rather a reason why P cannot result in G's failure.

TABLE 10-2
Attack relationships

Type of belief contents	Content of belief B1	Content of belief B2	Attack relationship between B1 and B2
Mutually exclusive planning situations	OUGHT-TO (P)	OUGHT-NOT-TO (P)	A-OBJECTIONABLE-PLAN
	P —achieves → G	P —not-achieves → G	A-UNREALIZED-SUCCESS
	P —thwarts → G	P —not-thwarts → G	A-UNREALIZED-FAILURE
	P —thwarts → G1	P —achieves → G2 G1 less important than G2	A-GREATER-SUCCESS
Opposite effects of a plan P on interrelated goals	P —achieves → G1	P —thwarts → G2 G1 less important than G2	A-GREATER-FAILURE
	P —achieves → G1	P —thwarts → G2 G1 as important as G2	A-EQUIVALENT-FAILURE
	P1 —achieves → G1	P1 —thwarts → G2, at time T G2 —initiates → P2, at time T1 > T G1 as important as G2 P1 instance of P P2 instance of P	A-SPIRAL-FAILURE

```
1. A-OBJECTIONABLE-PLAN: Although arguer A1 believes that plan P
   should be executed, arguer A2 believes that P should not be
   executed.
2. A-UNREALIZED-SUCCESS: Although arguer A1 believes that plan P
   achieves goal G, arguer A2 believes that P does not achieve G.
3. A-UNREALIZED-FAILURE: Although arguer A1 believes that plan P
   thwarts goal G, arguer A2 believes that P does not thwart G.
```

Both A-OBJECTIONABLE-PLAN and A-UNREALIZED-SUCCESS can be used to represent attacks on two beliefs associated with the execution of a plan P, namely, (1) the actor of P believes that P OUGHT-TO be implemented and (2) the actor of P believes that P will achieve the goal G which has intended P. Frequently, these beliefs are implicitly stated in editorial arguments. For example, consider again the following fragment from ED-CONTRADICTORY-POLICIES:

> American negotiators are pursuing a protectionist course...in what seems a vain effort to protect U.S. steel makers...This is the wrong way to go...

Here, the word "vain" stands for the negative-achievement relationship between a plan and a goal. Within the context of ED-CONTRADICTORY-POLICIES, this relationship refers to the Los Angeles Times' belief that import restrictions will not achieve the U.S. government's goals of helping steel makers preserve earnings and attain profitability. This belief contradicts the implicitly stated belief by the U.S. government that import restrictions will achieve its goal of helping steel makers. Similarly, the Los Angeles Times' belief that import restrictions are "wrong" (i.e., OUGHT-NOT-TO be executed) attacks the implicitly stated belief by the U.S. government that import restrictions OUGHT-TO be executed. These two attacks correspond to instances of A-UNREALIZED-SUCCESS and A-OBJECTION-ABLE-PLAN, respectively.

In contrast to these two attack relationships, A-UNREALIZED-FAILURE is used to represent attacks on a belief often professed by opponents of a plan P, that is, the belief that P thwarts the goal G which has intended P. For instance, consider the following excerpt from an editorial by Lee Iacocca (1986):

ED-TOUGH-POLICY

It's time to quiet down all...[free-trade purists] who keep telling us that getting tough on [international] trade will cost us jobs. It won't.

In ED-TOUGH-POLICY, Iacocca argues against the free trader's belief that imposing restrictions on international trade thwarts the goal of preserving jobs. Since this goal is one of the goals that import restrictions are intended to achieve, Iacocca's argument can be represented in terms of A-UNREALIZED-FAILURE. This representation is shown in Fig. 10-12.

3.2.2 ATTACKS BASED ON OPPOSITE EFFECTS ON INTERRELATED GOALS. Another way to contradict a belief about the effect a plan P has on a goal G1 is by stating that P has the opposite effect on G2, a goal more important than or equally important to G1. According to this type of contradiction by opposite effects, four attack relationships can be distinguished:

```
1. A-GREATER-SUCCESS: Although arguer A1 believes that plan P
   thwarts goal G1, arguer A2 believes that P achieves a more
   important goal G2.
2. A-GREATER-FAILURE: Although arguer A1 believes that plan P
   achieves goal G1, arguer A2 believes that P thwarts a more
   important goal G2.
3. A-EQUIVALENT-FAILURE: Although arguer A1 believes that the plan P
   achieves goal G1, arguer A2 believes that P thwarts an equally
   important goal G2.
4. A-SPIRAL-FAILURE: Although arguer A1 believes that the instance
   P1 of plan P achieves goal G1, arguer A2 believes that P1 thwarts
   an equally important goal G2 AND G2's failure will require using
   P2, another instance of P.
```

Representation:

A-UNREALIZED-FAILURE

```
BELIEF1                              BELIEF2
Believer: Freetraders                Believer: Lee Iacocca
Content:                  ← attacks → Content:
P-ECON-PROTECTION —thwarts→          P-ECON-PROTECTION —not-thwarts→
  G-PRESERVING-JOBS in U.S.            G-PRESERVING-JOBS in U.S.
```

EditorialText: —"[free-trade purists] ... keep telling us that getting tough on trade will cost us jobs" —"It won't"

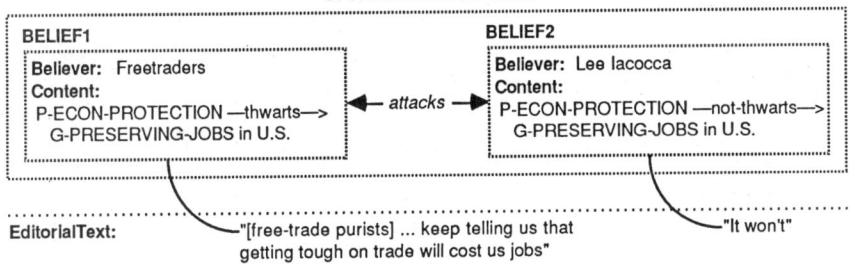

FIGURE 10-12
Attack relationship in ED-TOUGH-POLICY.

These four attack relationships are used to represent arguments that contrast the negative and positive effects of a plan P in order to show that P should be favored or opposed. For example, A-GREATER-SUCCESS shows that the negative side-effects of a plan P are a small price to pay for P's positive effects. To illustrate this type of attack, consider the following passage taken from Greenaway and Milner (1979, pp. 18–19):

ED-REVENUE-TARIFF

> If the government feels it requires additional revenue to finance higher state expenditure, tariff imposition may be viewed as a suitable source... Since the tariff must be paid by... domestic consumers, the government is guaranteed a yield.

The above excerpt presents an argument which Greenaway and Milner call "the argument for the revenue tariff." In this argument, the position of the government is that (1) imposing a tariff achieves the government's goal of attaining a higher level of spending and (2) this goal is more important than the goal of preserving earnings by domestic consumers. Clearly, the argument for the revenue tariff amounts to an implicit attack on the belief (by protectionism opponents) that imposing tariffs will thwart the goal of preserving earnings by domestic consumers. Such an attack can be represented in terms of A-GREATER-SUCCESS, as illustrated in Fig. 10-13.

In contrast to A-GREATER-SUCCESS, the other attack relationships involving opposite effects show that the negative side effects of a plan P do not grant the implementation of P. For example, the following excerpt from an editorial by the Los Angeles Times (Oct. 5, 1985)

> ...legislation to limit textile and apparel imports...will do more harm than good...

322 KNOWLEDGE ENGINEERING: FUNDAMENTALS

is an instance of A-GREATER-FAILURE. This instance contrasts the implicitly stated belief by legislators that import restrictions will lead to goal achievements and the Los Angeles Times' belief that those restrictions cause major goal failures. Similarly, the following excerpt from ED-CONTRADICTORY-POLICIES

> American negotiators are pursuing a protectionist course in efforts...to protect U.S. steel makers...And consumers will be forced to pay the cost through denial of the cheaper foreign products...

is an instance of A-EQUIVALENT-FAILURE. The contradictory beliefs in the above excerpt are the government's belief that import restrictions will achieve the goal of preserving earnings for steel makers and the Los Angeles Times' belief that those restrictions will thwart the same type of goal for consumers. (This attack relationship has already been illustrated in Fig. 10-11.)

Finally, A-SPIRAL-FAILURE is used to represent arguments about goal failures triggered by repeated applications of the same plan P. For instance, consider a fragment of Lance Morrow's (1983) argument in ED-RESTRICTIONS:

> The toolmakers argue that import restrictions must be imposed so that the industry can survive...Restrictions on imports would mean that [other] American manufacturers would...loose sales. Then those manufacturers would demand protection against foreign competition.

The above editorial segment contains an attack relationship between the toolmakers' belief that import restrictions will achieve their goal of preserving the industry's level of earnings and Morrow's belief that those restrictions will thwart the same goal for other manufacturers and, consequently, motivate the use of more import restrictions. The representation of this attack relationship is shown in Fig. 10-14.

3.3 Support Relationships

Beliefs can also relate to one another via relationships of support. In OpEd, a *support* (S) is a construct composed of three major elements: a supported be-

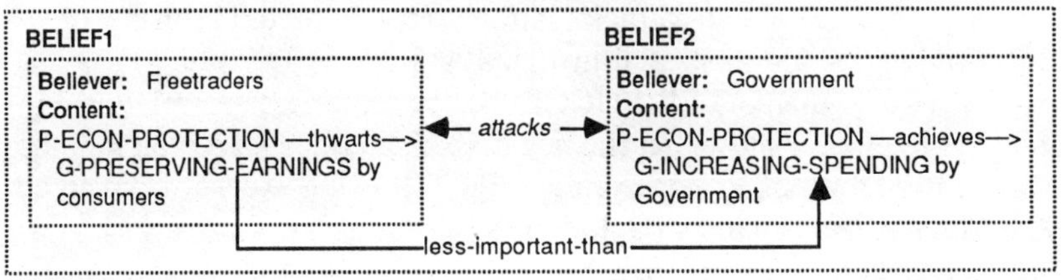

```
                          A-GREATER-SUCCESS
┌─────────────────────────────────────────────────────────────────┐
│ BELIEF1                          BELIEF2                        │
│ Believer: Freetraders            Believer: Government           │
│ Content:                         Content:                       │
│ P-ECON-PROTECTION —thwarts—→  ← attacks → P-ECON-PROTECTION —achieves—→ │
│ G-PRESERVING-EARNINGS by         G-INCREASING-SPENDING by       │
│ consumers                        Government                     │
│           └──────────less-important-than──────────┘             │
└─────────────────────────────────────────────────────────────────┘
```

FIGURE 10-13
Attack relationship in ED-REVENUE-TARIFF.

Representation:

A-SPIRAL-FAILURE

```
┌─────────────────────────────────────┐     ┌──────────────────────────────────────────┐
│ BELIEF1                             │     │ BELIEF2                                  │
│ Believer: U.S. toolmakers           │     │ Believer: Lance Morrow                   │
│ Content:                            │     │ Content:                                 │
│ P-ECON-PROTECTION1 —achieve→        │← attacks →│ P-ECON-PROTECTION1 —thwart→       │
│   G-PRESERVE-EARNINGS1              │     │   G-PRESERVE-EARNINGS2 —intend→          │
│                                     │     │   P-ECON-PROTECTION2                     │
└─────────────────────────────────────┘     └──────────────────────────────────────────┘
                    └──────── equally-important ────────┘
```

Editorial Text: "The toolmakers argue that restrictions on imports must be imposed so that the industry can survive" — "Restrictions on imports would mean that [other] American manufacturers would ... lose sales. Then they would demand protection against foreign competition"

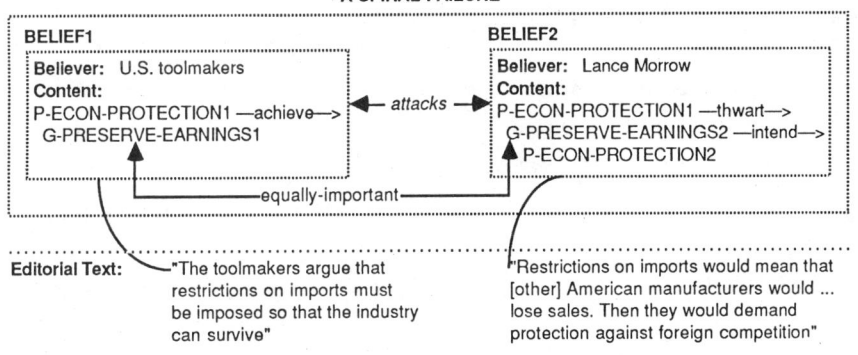

FIGURE 10-14
Attack relationship in ED-RESTRICTIONS.

lief B, a justification J that contains a single supporting belief or a conjunction of supporting beliefs, and a warrant W that grants the existence of the support relationship from J to B. Support structures are used in OpEd to represent instances of plan-based reasoning in editorial arguments, i.e., the reasoning used by arguers to justify why plans should or should not be implemented or why plans will or will not cause goal achievements or failures. According to the nature of these reasoning instances, four basic types of support relationships have been characterized: refinements of plan evaluations, refinements of plan-goal relationships, analogies, and examples.

3.3.1 SUPPORTS BASED ON REFINEMENTS OF PLAN EVALUATIONS. An evaluative belief about a plan P can be justified by stating the goal failures or achievements that result from implementing P. This type of reasoning, termed *refinement of plan evaluations,* is the basis for the support relationships summarized in Table 10-3.

The five support structures in Table 10-3 represent arguments in favor of or against the use of a plan P. For example, S-REALIZED-SUCCESS embodies the following reasoning:

1. S-REALIZED-SUCCESS: Arguer A believes that plan P should be executed because A believes that P will achieve the goal G which has intended P.

To illustrate this support structure, consider the following excerpt from an editorial by the Los Angeles Times (Feb. 16, 1984):

ED-JOB-SAVING-QUOTAS

The Japanese quotas were pushed hardest by the United Auto Workers union, which touted them... as a means of restoring American jobs...

TABLE 10-3
Supports based on refinements of plan evaluations

Content of supported belief B	Content of justification J	Support relationship between B and J
OUGHT-TO (P)	G —intends → P, at time T P —achieves → G, at time T1 > T	S-REALIZED-SUCCESS
OUGHT-NOT-TO (P)	G —intends → P, at time T P —not-achieves → G, at time T1 > T	S-UNREALIZED-SUCCESS
OUGHT-NOT-TO (P)	G1 —intends → P, at time T P —thwarts → G2, at time T1 > T G1 less important than G2	S-GREATER-FAILURE
OUGHT-NOT-TO (P)	G1 —intends → P, at time T P —thwarts → G2, at time T1 > T G1 as important as G2	S-EQUIVALENT-FAILURE
OUGHT-NOT-TO (P1)	G1 —intends → P1, at time T P1 —thwarts → G2, at time T1 > T G2 —intends → P2, at time T2 > T1 G1 as important as G2 P1 instance of P P2 instance of P	S-SPIRAL-FAILURE

In ED-JOB-SAVING-QUOTAS, the position of the U.A.W. is that restrictions on Japanese automobiles should be imposed and import restrictions will achieve the goal of preserving jobs in the United States. Clearly, ED-JOB-SAVING-QUOTAS can be represented in terms of S-REALIZED-SUCCESS, as shown in Fig. 10-15.

In contrast to the above support relationship, S-UNREALIZED-SUCCESS involves the use of a negative-achievement relationship to justify the belief that a plan P should be opposed:

2. S-UNREALIZED-SUCCESS: Arguer A believes that plan P should not be executed because A believes that P will not achieve the goal G which has intended P.

For instance, S-UNREALIZED-SUCCESS can be used to represent the following fragment of Milton Friedman's argument in ED-JOBS:

> Recent protectionist measures...have disappointed us...They do not promote the long-run health of the [automobile and steel] industries...

Here, the affect description "disappointed" indicates that Friedman believes that import restrictions should not be implemented. This belief is justified by

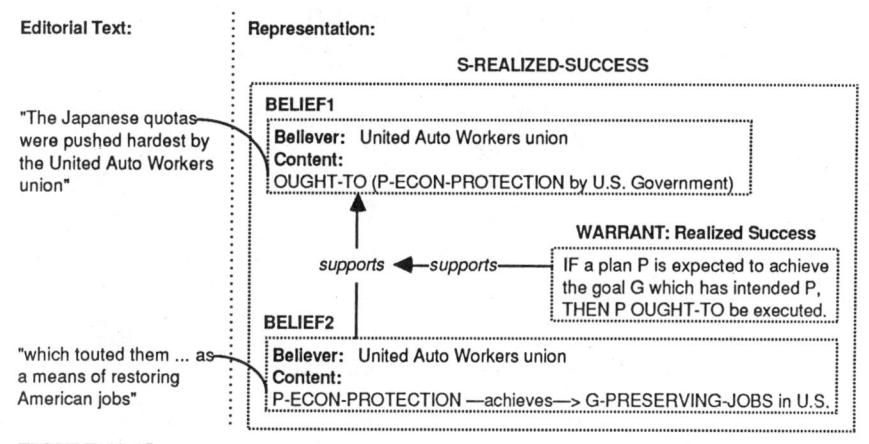

FIGURE 10-15
Support relationship in ED-JOB-SAVING-QUOTAS.

the negative achievement relationship stated in the second sentence of the above editorial segment. In that sentence, the phrase "long-run health" refers to the industries' goal of attaining profitability. Since this goal is one of the goals that import restrictions are intended to achieve, Friedman's argument can be represented as an instantiation of S-UNREALIZED-SUCCESS. This representation is shown in Fig. 10-16.

In addition to S-UNREALIZED-SUCCESS, three more support structures can be used to justify the belief that a plan P should not be implemented:

3. S-GREATER-FAILURE: Arguer A believes that plan P should not be executed because A believes that P will thwart a goal G2 more important than the goal G1 which has intended P.

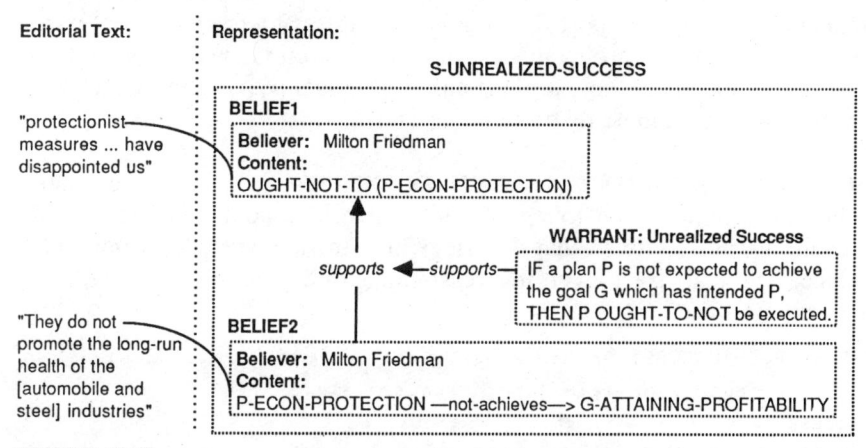

FIGURE 10-16
Support relationship in ED-JOBS.

4. S-EQUIVALENT-FAILURE: Arguer A believes that plan P should not be executed because A believes that P will thwart a goal G2 as important as the goal G1 which has intended P.
5. S-SPIRAL-FAILURE: Arguer A believes that the instance P1 of plan P should not be executed because A believes that P1 will thwart a goal G2 as important as the goal G1 (which has intended P) and G2's failure will require using P2, another instance of P.

For example, in the following excerpt from an editorial by Timothy Bresnahan (1984)

> ...I think the import quotas are terrible public policy...[P]rotecting domestic industries from foreign competition does more harm than good.

Bresnahan's argument is an instance of S-GREATER-FAILURE. Similarly, in the following excerpt from an editorial by Robert Samuelson (1984)

ED-PROTECTION-OPPONENTS

> ...In the last months...major retailers and farm groups have...vigorously protested proposed new trade restrictions...Major retailers have formed a group to resist...restrictions on apparel, which the retailers said would have cost them...millions of dollars...Farm groups have joined coalitions opposing...textile...restrictions. About two-thirds of America's wheat, half the soybeans and a third of the corn is exported. Farm groups fear that U.S. import restrictions will cause some countries to retaliate...

The retailers' argument and farmers' argument correspond to instances of S-EQUIVALENT-FAILURE and S-SPIRAL-FAILURE, respectively. In ED-PROTECTION-OPPONENTS, the negative-spiral effect caused by import restrictions is indicated by the word "retaliate." These instances are based on the following relationships: (1) import restrictions are intended to achieve the goal of preserving earnings for the U.S. textile and apparel industries; (2) major retailers believe that import restrictions will thwart their goal of preserving earnings; and (3) farmer groups believe import restrictions will thwart foreign countries' goal of preserving earnings and, consequently, will motivate those countries to impose import restrictions on U.S. grains.

3.3.2 SUPPORTS BASED ON CAUSAL-CHAIN EXPANSION. One way to justify the belief that a plan P leads to a goal achievement or a goal failure is by making a chain of causal effects that describes how those goal relationships may take place. At the abstract level, this reasoning strategy is captured by the following support structures:

1. S-POSSIBLE-SUCCESS: Arguer A believes that plan P will achieve goal G because A believes that P causes state S1 AND S1 causes...state Sn AND Sn achieves G.
2. S-POSSIBLE-FAILURE: Arguer A believes that plan P will thwart

```
goal G because A believes that P causes state S1 AND S1
causes...state Sn AND Sn thwarts G.
```

At the level of domain knowledge, the chains of causal effects organized by instances of the above support structures in editorials correspond to instances of reasoning scripts. For example, consider the following passage from Greenaway and Milner (1979, pp. 40–41):

ED-BENEFICIAL-TARIFF

Suppose...policy-makers impose a tariff on low-price textiles from abroad...Because the post-tariff price of imports is higher than their free-trade price, domestic textile producers...can now supply more of the [diminished] market. Thus,...domestic producers...benefit from tariffs.

Here, the belief that tariffs achieve the goal of attaining profitability for domestic producers is supported by a causal chain on how tariffs switch domestic spending from imports to domestic products and, consequently, increase the level of earnings of domestic producers. This causal chain is an instance of $R-ECON-PROTECTION → HIGHER-DOMESTIC-EARNINGS. Similarly, consider the following fragment of Milton Friedman's argument in ED-JOBS:

...the limitations on imports [by the Reagan administration] will cost jobs. If we import less, foreign countries will earn fewer dollars. They will have less to spend on American exports. The result will be fewer jobs in export industries.

In ED-JOBS, Friedman's belief that "limitations on imports will cost jobs" is supported by a reasoning chain which describes how import restrictions by the United States trigger a negative feedback on U.S. export jobs. This chain is an instance of the reasoning script $R-ECON-PROTECTION → FEWER-EXPORT-JOBS.

Instances of reasoning scripts can also be used to justify belief about the negative-spiral effects resulting from implementing a plan P. At the abstract level, those justifications are characterized by the following support structure:

```
3. S-POSSIBLE-SPIRAL-FAILURE: Arguer A believes that plan P1 (an
   instance of P) will thwart goal G AND G will intend plan P2
   (another instance of P) because A believes that P1 causes state
   S1 AND S1 causes...state Sn AND Sn thwarts G AND G's failure
   requires using P2.
```

For example, consider the following paragraph, which summarizes an argument presented in Cuddington and McKinnon (1979, pp. 4–6):

ED-COUNTERATTACK

Free-trade economists believe that import restrictions by the U.S. will cause foreign countries to retaliate. They argue that import quotas will cause trade losses for foreign countries. To recover from those losses, foreign countries will impose tariffs on products they import from the U.S.

The above argument involves the use of S-POSSIBLE-SPIRAL-FAILURE and the reasoning script $R-ECON-PROTECTION → ECON-RETALIATION to justify the belief that import restrictions lead to retaliations. These constructs are illustrated in Fig. 10-17.

3.3.3 SUPPORTS BASED ON ANALOGIES. Another strategy used to justify causal beliefs in editorial arguments is reasoning by analogy. The approach taken here to represent the use of analogies in editorial arguments is similar to the one proposed in August and Dyer (1985a, 1985b). For example, consider the following argument by Lester Thurow (1983):

ED-MOTORCYCLES

...[A] tariff on large motorcycles...will not give America a world-class motorcycle industry...The American steel industry has been protected since the late 1960s and is less competitive today than it was then...

The argument in ED-MOTORCYCLES is an instance of the following support structure:

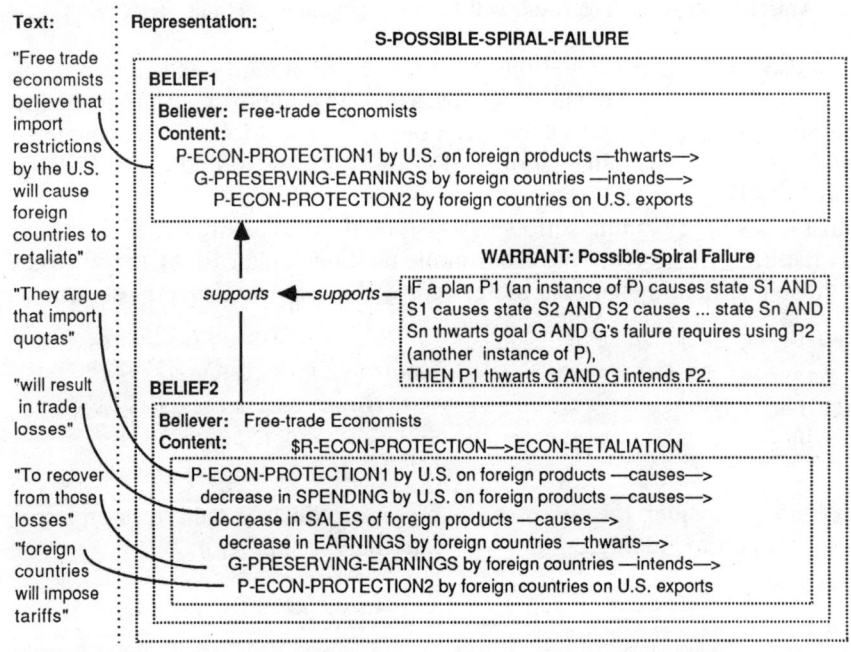

```
Text:                Representation:
                                        S-POSSIBLE-SPIRAL-FAILURE
"Free trade
economists           BELIEF1
believe that         Believer:  Free-trade Economists
import               Content:
restrictions             P-ECON-PROTECTION1 by U.S. on foreign products —thwarts—>
by the U.S.              G-PRESERVING-EARNINGS by foreign countries —intends—>
will cause               P-ECON-PROTECTION2 by foreign countries on U.S. exports
foreign
countries to
retaliate"
                                        WARRANT: Possible-Spiral Failure
"They argue          supports  <—supports— IF a plan P1 (an instance of P) causes state S1 AND
that import                              S1 causes state S2 AND S2 causes ... state Sn AND
quotas"                                  Sn thwarts goal G AND G's failure requires using P2
                                         (another instance of P),
"will result                             THEN P1 thwarts G AND G intends P2.
in trade
losses"              BELIEF2
                     Believer:  Free-trade Economists
                     Content:      $R-ECON-PROTECTION—>ECON-RETALIATION
"To recover              P-ECON-PROTECTION1 by U.S. on foreign products —causes—>
from those               decrease in SPENDING by U.S. on foreign products —causes—>
losses"                  decrease in SALES of foreign products —causes—>
                         decrease in EARNINGS by foreign countries —thwarts—>
"foreign                 G-PRESERVING-EARNINGS by foreign countries —intends—>
countries                P-ECON-PROTECTION2 by foreign countries on U.S. exports
will impose
tariffs"
```

FIGURE 10-17
Supporting causal chain in ED-COUNTERATTACK.

S-SIMILAR-UNREALIZED-SUCCESS: Arguer A believes that plan P1 will
not achieve goal G1 because A believes that a plan P2 (similar to
P1) has not achieved goal G2 (similar to G1) in the past.

Thurow uses S-SIMILAR-UNREALIZED-SUCCESS to argue that an import tax will not achieve the U.S. motorcycle industry's goal of becoming competitive because similar protectionist measures have not achieved that type of goal for the U.S. steel industry. This instance of S-SIMILAR-UNREALIZED-SUCCESS is illustrated in Fig. 10-18.

Reasoning by analogy also serves as the basis for showing why a plan P (1) can not achieve a goal G, (2) may achieve or thwart a goal G, or (3) may result in a negative-spiral effect. These uses of analogy are organized by the support structures shown in Table 10-4. For example, consider S-SIMILAR-SPIRAL-FAILURE. This structure embodies the following reasoning:

S-SIMILAR-SPIRAL-FAILURE: Arguer A believes that plan P1 (an
instance of P) will thwart goal G1 AND G1's failure will require
using P2 (another instance of P) because A believes that plan P3
(similar to P1) has thwarted goal G2 (similar to G1) in the past AND
G2's failure has required using plan P4 (similar to P2).

S-SIMILAR-SPIRAL-FAILURE is used in the following segment from an editorial by Feldstein and Feldstein (1985):

ED-1930-RETALIATION

...[A] 20% tax on imports...could easily provoke retaliation by foreign governments...The last major trade war [was] precipitated by our 1930 Hawley-Smoot tariff...

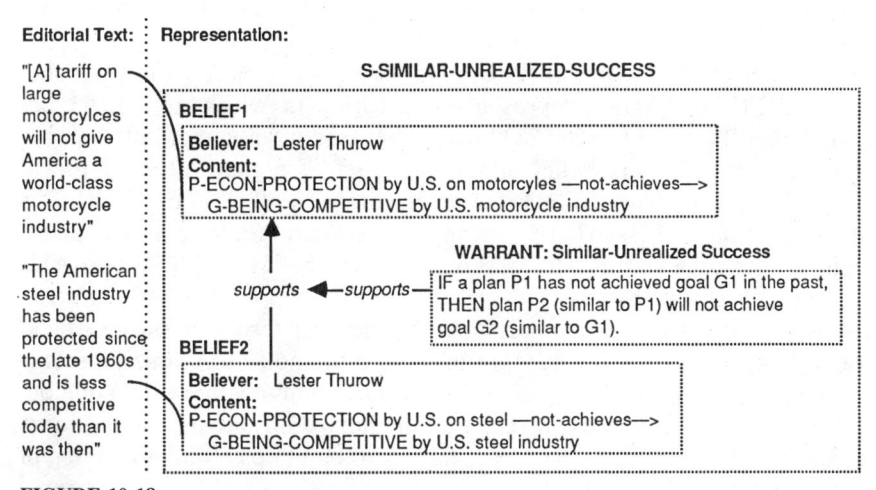

FIGURE 10-18
Support relationship in ED-MOTORCYCLES.

Here Feldstein and Feldstein predict the outcome of the 20 percent tax on imports by using an historical precedent. Specifically, their argument contains the following relationships: (1) the proposed tax on imports will thwart the goal by foreign countries of preserving earnings, (2) this goal failure will cause foreign countries to impose import restrictions on U.S. products, (3) the proposed tax is similar to the Hawley-Smoot tariff which caused trade losses for foreign countries in the 1930s, and (4) those losses caused foreign countries to impose tariffs on U.S. products. Clearly, ED-1930-RETALIATION is an instance of S-SIMILAR-SPIRAL-FAILURE.

3.4 Argument Units

Support and attack relationships can be combined with beliefs held by two or more debaters to form arguments. Domain-independent, abstract representations of such two-party belief structures are termed *argument units* (AUs). An example of an AU is AU-GREATER-SIDE-EFFECT, where one arguer A1 grants to A2 that plan P1 will achieve goal G1 (so A1 and A2 share a belief). However, A1 holds that A2 OUGHT-NOT-TO execute P1, since A1 believes that P1, while achieving G1, as a side effect will cause the failure of G2. Furthermore, A1 believes that G2 is a more important goal than G1.

This AU appears in many different domains, and whenever an arguer can recognize (or generate) it, the AU provides an organizing structure for argument interpretation. We believe that most arguments consist of a relatively small number of argument units and that more complex arguments are built up by elaboration and composition operations over a more primitive set of simple argument units. Argument units thus provide a foundation for understanding arguments across domains.

For example, Fig. 10-19 shows a simplified version of the argument graph of ED-JOBS. This graph contains two overlapping argument units, AU-ACTUAL-CAUSE and AU-OPPOSITE EFFECT. For readability, the conceptual content of each belief in the argument graph is described in English.

AU-ACTUAL-CAUSE represents a situation in which arguer A1 believes that plan P OUGHT-NOT-TO be used because it will not affect situation S, which thwarts G. This belief attacks a belief of arguer A2, that P will achieve G and therefore OUGHT-TO be used. AU-OPPOSITE-EFFECT is similar to AU-ACTUAL-CAUSE, but in this case one arguer believes that a plan will do the opposite of what the other arguer believes. These two AUs overlap through shared beliefs.

It is possible to represent arguments solely in terms of basic attack/support relationships. However, AUs provide a method of organizing recurring configurations of attack/support relationships involving more than one believer. By having AUs already in semantic memory, OpEd's task of argument analysis is greatly simplified. Instead of having to build up an understanding of each argument at the belief level, OpEd can use fragments of belief structures to infer an argument unit and thus instantiate and infer implicit beliefs and unstated relationships in an argument. It is likely that the more skilled and ex-

TABLE 10-4
Supports based on analogies

Content of supported belief B	Content of justification J	Support relationship between B and J
P1 —not-achieves → G1	P2 —not-achieves → G2 P1 similar to P2 G1 similar to G2	S-SIMILAR-UNREALIZED-SUCCESS
P1 —not-thwarts → G1	P2 —not-thwarts → G2 P1 similar to P2 G1 similar to G2	S-SIMILAR-UNREALIZED-FAILURE
P1 —achieves → G1	P2 —achieves → G2 P1 similar to P2 G1 similar to G2	S-SIMILAR-SUCCESS
P1 —thwarts → G1	P2 —thwarts → G2 P1 similar to P2 G1 similar to G2	S-SIMILAR-FAILURE
P1 —thwarts → G1 G1 —intends → P2 P1 and P2 instances of P	P3 —thwarts → G2 G2 —intends → P4 P1 similar to P3 P2 similar to P4 G1 similar to G2	S-SIMILAR-SPIRAL-FAILURE

perienced human arguers, along with remembering specific argument instances, have acquired, abstracted, and generalized a repertoire of abstract AUs and use these AUs for generating arguments in new domains.

4 DEMON-BASED PROCESSING

All dynamic, process knowledge in OpEd is embodied in demons. A *demon* is a delayed procedure that implements test/action rules, where the tests and actions may involve memory search, pattern matching, inferencing, disambiguation of word senses, reference resolution, role bindings, and dynamic instantiations of knowledge constructs.

As each word and/or phrase in the input is read, lexical memory is accessed. Attached to each word or phrase in the lexicon are one or more knowledge structures, and associated with each structure are one or more demons. Demons are *spawned;* i.e., an instance of the demon is created with its own binding environment, and the demon is placed in a demon agenda, where its test portion is polled on regular intervals. Whenever the test portion is satisfied, the action portion is then executed.

All knowledge structures (goals, social acts, argument units, etc.) are implemented as *frames* (Minsky, 1977). A frame consists of *roles* and associated *slots,* where the roles specify semantic cases and the slots describe default values. For example, the word "disappointed" in ED-JOBS is an AFFECT (Dyer, 1983b) and represents a situation in which the one feeling the disappointment believes that a goal would succeed, followed by a belief that it will

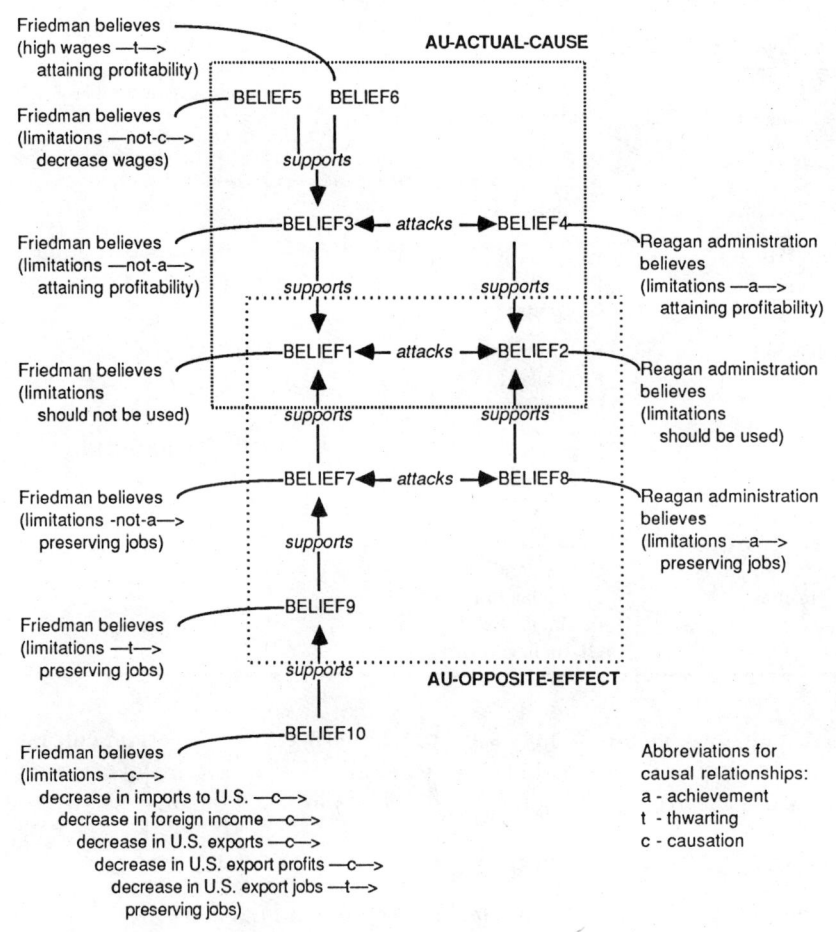

```
Friedman believes
(high wages —t—>                      AU-ACTUAL-CAUSE
 attaining profitability)
                          ┌─ BELIEF5    BELIEF6
Friedman believes  ◄──────┘
(limitations —not-c—>           supports
 decrease wages)
                                    │
                          BELIEF3◄─ attacks ─►BELIEF4
Friedman believes  ◄──────┘                         └──► Reagan administration
(limitations —not-a—>           supports    supports       believes
 attaining profitability)                                 (limitations —a—>
                          BELIEF1◄─ attacks ─►BELIEF2       attaining profitability)
Friedman believes  ◄──────┘                         └──► Reagan administration
(limitations                                               believes
 should not be used)            supports    supports      (limitations
                                                           should be used)
                          BELIEF7◄─ attacks ─►BELIEF8
Friedman believes  ◄──────┘                         └──► Reagan administration
(limitations -not-a—>           supports                   believes
 preserving jobs)                                         (limitations —a—>
                          BELIEF9                          preserving jobs)
Friedman believes  ◄──────┘
(limitations —t—>               supports           AU-OPPOSITE-EFFECT
 preserving jobs)
                          BELIEF10
Friedman believes  ◄──────┘
(limitations —c—>                              Abbreviations for
 decrease in imports to U.S. —c—>              causal relationships:
 decrease in foreign income —c—>               a - achievement
 decrease in U.S. exports —c—>                 t - thwarting
 decrease in U.S. export profits —c—>          c - causation
 decrease in U.S. export jobs —t—>
 preserving jobs)
```

FIGURE 10-19
Argument graph of ED-JOBS.

now fail. In ED-JOBS, Friedman is disappointed because he believed that Reagan would behave in one way, but now Reagan is behaving in a way that violates Friedman's beliefs and/or goals. A simplified representation of the lexical entry for "disappointment" is shown in Fig. 10-20.

The world "disappointed" is represented in terms of several related frames, involving AFFECTs, BELIEFs, and GOALs. Variables are indicated by single letters, and attached demons appear to the right of each arrow. The FIND-WM-OBJECT and FIND-WM-ACTOR demons each search working memory for ANIMATEs who are the OBJECT and ACTOR (the one feeling "disappointed"). These demons know about active and passive voice. For example, in the case of ⟨"Friedman" "was" "disappointed" "by" "Reagan"⟩, Friedman would be the ACTOR. Since the verb is passive and "Reagan" is the object of a "by," Reagan is the cause of Friedman's AFFECT.

NATURAL LANGUAGE PROCESSING: COMPUTER COMPREHENSION OF EDITORIAL TEXT   **333**

patterns: <y "disappointed" x>
              <x "was" "disappointed">
              <x "was" disappointed" "by" y>
              <y "has" "disappointed" x>
frame:

```
(AFFECT
   ACTOR X <== (FIND-WM-OBJECT 'ANIMATE)
   STATE (NEGATIVE)
   G-SITU (BELIEF
             ACTOR X
             CONTENT
              (GOAL
                  IS G <== (FIND-DISAPPOINT-GOAL-OF 'ACTOR)
                  STATUS (SUCCESS)
                  FOLLOWED-BY
                      (G   (STATUS (FAILURE)
                           CAUSED-BY
                              (PLAN/BELIEF
                                  IS    Z <== ((FIND-PLAN-BELIEF)
                                              (INFER-BELIEF-ACTOR))
                                  IS    ACTOR Y <== ((FIND-WM-ACTOR
                                                      'ANIMATE)))))
```

FIGURE 10-20
Simplified lexical entry for "disappointed."

In ED-JOBS, Friedman does not ever explicitly say that he is against Reagan's plan of "protectionist measures." Friedman simply states that these measures have "disappointed" him. OpEd must actually *infer* that Friedman therefore does not believe in this plan. Once the ACTOR of the AFFECT (Friedman) and the ACTOR of the BELIEF/PLAN (Reagan) are bound, then both Friedman's and Reagan's beliefs concerning this plan can be inferred. Once the demon FIND-PLAN-BELIEF finds the protectionist plan, the demon INFER-BELIEF-ACTOR can infer that the ACTOR of the AFFECT does not hold this BELIEF.

Below is a simplified, annotated trace of the major inferences being made by OpEd during the reading of ED-JOBS.

...measures...disappointed us

```
======> NEGATIVE-AFFECT by Friedman
==infer ==> B1: Friedman believes (measures OUGHT-NOT-TO be executed)
```

...limits...bad for the nation

```
======> limits —cause → GOAL FAILURE U.S.
==infer ==> B2: Friedman believes (limits —cause → GOAL FAILURE U.S.)
            B2 supports B1
```

The text does not explicitly state what the connection is between "limits" and "protectionist measures." The inference that belief B2 supports

B1 is made by examining the relationship between "limits on Japanese automobiles" and "protectionist measures" and recognizing that limitations in product imports is an instance of the general plan of protectionism.

...not promote health of industries...

```
=====> limits —not-achieve → G-ATTAIN-PROFITABILITY
=infer ==> B3: Friedman believes (limits —not-achieve →
G-ATTAIN-PROFITABILITY)
          B3 supports B1
```

...the problem...is...wage rates...

```
=====> wage-rates —thwart → G-ATTAIN PROFITABILITY
=infer ==> B4: Friedman believes (wage-rates —thwart →
G-ATTAIN-PROFITABILITY)
      activate AU-ACTUAL-CAUSE (Arguer: Friedman)
      = infer ==> B5: Reagan adm. believes (limits —achieve →
GAIN-ATTAIN-PROFITABILITY)
```

As each belief content is constructed, the goal/plan relationships are compared against argument units that have similar I-links. Since wage rates thwart the same goal that limits achieve, the argument unit AU-ACTUAL-CAUSE is recognized.

...Far from saving...limitations cost...

```
=====> limits —achieve → G-PRESERVE-JOBS  limits —thwart → G-
PRESERVE JOBS
                      |                          |
                      +———————— opposite ————————+
=refine ==> B2: Friedman believes (limits —thwart → G-PRESERVE-
JOBS U.S.)
      activate AU-OPPOSITE-EFFECT (Arguer: Friedman)
      = infer ==> B6: Reagan adm. believes (limits
—achieve → G-PRESERVE-JOBS U.S.)
```

Here the same plan both achieves and thwarts the same goal. This opposition is built by a demon attached to the pattern ⟨"far from" x, y⟩. This demon examines the conceptual structures bound to x and y. If they are conceptual opposites and believed by different individuals, then AU-OPPOSITE-EFFECT is instantiated.

...import less...earn fewer dollars...less to spend...fewer jobs

```
=====> CAUSE-EFFECT-CHAIN0
          decrease imports —cause →
             decrease foreign-income —cause →
                decrease export —cause →
                   decrease jobs
```

```
=infer ==> B7: Friedman believes (CAUSE-EFFECT-CHAIN0)
          B7 supports B2
```

The causal chain is inferred by a combination of traversing the trade relationship graph and application of reasoning scripts.

Once the argument graph is completed, OpEd parses natural language questions in a manner similar to parsing the text. When a question word is encountered (e.g., "what," "why," "how"), a question-answering demon is spawned which then examines the conceptual content of the question. For example, if a demon attached to "why" encounters a belief of x (e.g., "Why does Reagan believe in limitations?"), then it will search the argument graph for this belief and return its supporting beliefs. If a demon attached to "what" encounters a belief of x with the CONTENT field not specified (e.g., "What does Friedman believe?"), then it will return all beliefs held by x.

5 COMPARISON TO RELATED WORK

The theory of argument comprehension implemented in OpEd has been developed from the perspective of natural language understanding. That is, argument comprehension in OpEd is not considered as an isolated process but rather an an integral aspect of natural language understanding. As a result, OpEd builds upon knowledge constructs and processing strategies developed originally for computer comprehension of natural language texts, including conceptual dependency theory (Schank, 1973, 1975), goals and plans (Wilensky, 1983; Carbonell 1981), social acts (Schank and Carbonell, 1979), scripts (Schank and Abelson, 1977), affects (Dyer, 1983b); intention/causal links (Dyer and Lehnert, 1982; Lehnert et al., 1983); themes (Schank, 1982; Dyer, 1983a); belief relationships (Toulmin, 1958; Flowers et al., 1982), reasoning scripts (Dyer et al., 1987; Flowers and Dyer, 1984), and conceptual question types (Lehnert, 1978).

5.1 Formal/Logical Approaches to Belief

Belief research in AI has concentrated on truth maintenance (Doyle, 1979; de Kleer, 1986) and evidential reasoning (Pearl, 1986; Dechter and Dechter, 1988). The goal of these systems is to maintain and/or propagate evidence for or against beliefs, once these beliefs and their relationships are already resident in a knowledge base. A major task of such systems is to maintain logical consistency among a set of beliefs by dynamically altering the truth value of tentative beliefs (i.e., assumptions). Logicians have a long history of examining the notions of knowledge and belief (Hintikka, 1962; Chellas, 1980), but their approach has been to establish axiomatic systems for deducing consistent believes, without regard to how humans understand the beliefs of others and relate them to their own beliefs. None of these groups of researchers has addressed the natural language comprehension task, which involves dynamically constructing a knowledge base of the beliefs and belief justifications of an arguer "on the fly," from textual input. In OpEd, the vast majority of the processing and knowledge structures are not involved in logical operations *per se*. Major tasks involve commonsense inferencing, memory search, planning, application of world knowledge, and recognition of AUs.

5.2 A Structural Approach to Argument Comprehension

Cohen (1983) has postulated a *structural model* for argument understanding, in contrast to the *conceptual model* presented in OpEd. In Cohen's model, understanding an argument requires building a tree where argument propositions are connected by a single evidence link. The root of the tree contains the major claim made in an argument. Relations between propositions are determined by using (1) a proposition analyzer that produces a proposition from the input and integrates it into the tree built so far, (2) a clue interpreter that analyzes the role of special linguistic connectives (e.g., "as a result," "similarly"), and (3) an evidence oracle that accesses a knowledge base and model of the speaker in order to determine whether any evidence relation exists between two propositions. Unfortunately, Cohen's model was not implemented, so it is difficult to assess its potential. Our analysis, however, indicates that arguments do not conform to a simple tree structure, but consists of a directed graph. In addition, since Cohen's tree lacks conceptual content, it cannot indicate how either explicit or implicit conceptualizations (contained in a given proposition) relate to and/or provide evidence for conceptualizations contained in other propositions. Moreover, by using an "oracle," the model avoids having to deal with the critical problems of how lexical items are mapped from natural language to conceptual structures, how world knowledge is represented and applied during the comprehension process, and how argument strategies are represented and applied within a given domain.

In contrast to Cohen's strictly structural approach, OpEd analyzes editorial arguments by building a conceptual graph which captures interactions between goals, plans, events, emotional states, beliefs, and argument units. This conceptual graph results from recognizing and instantiating those knowledge structures along with causal, intentional, and belief relationships. In addition, OpEd's comprehension process also results in building indices that are subsequently used during question answering.

5.3 Psycholinguistic Analysis of Editorial Text

While there has been an extensive examination of expository, narrative, and discourse text (Britton and Black, 1985; van Dijk, 1985; Spiro et al., 1980) in the psychological literature, there appears to be little direct analysis of argument-based editorial text. Van Dijk and Kintsch (1983, p. 101) did perform an extensive analysis of an editorial from *Newsweek* that begins with the sentence:

Compared with the relative shades of gray in El Salvador, Guatemala is a study in black and white. On the left is a collection of extreme Marxist-Leninist groups led by what one diplomat calls "a pretty faceless bunch of people."

The editorial is quite complex and goes on to discuss the effect of Reagan's election on Guatemala, military aid to Guatemala from Israel, mass murders committed by the Guatemalan "regime," and so on. Van Dijk and Kintsch apply a very general theory of discourse and schema coherency to the text, which includes coherency structures at the local (sentence), macrostructure, and schematic levels. Much of the discussion is concerned with the discourse structure of what is actually expository text. Unfortunately, van Dijk and Kintsch do not have any computer implementation of their model, so the discussion is at a very general descriptive level. However, the work is useful, to the extent that it gives one an idea of a number of discourse issues (e.g., topic shifts) involved in complex, expository text—issues that have been largely ignored in OpEd.

6 CURRENT SCOPE OF OpEd

OpEd is currently able to read ED-JOBS, an editorial fragment by Milton Friedman (1982) and ED-RESTRICTIONS, an editorial segment taken from an editorial by Lance Morrow (1983). ED-RESTRICTIONS presents Morrow's arguments against implementing import restrictions to protect the U.S. machine-tool industry. A portion of OpEd's input/output behavior on this editorial is shown below.

ED-RESTRICTIONS

The American machine-tool industry is seeking protection from foreign competition. The industry has been hurt by cheaper machine tools from Japan. The toolmakers argue that restrictions on imports must be imposed so that the industry can survive. It is a wrongheaded argument. Restrictions on imports would mean that American manufacturers would have to make do with more expensive American machine tools. Inevitably, those American manufacturers would produce more expensive products. They would lose sales. Then those manufacturers would demand protection against foreign competition.

Q1: **What does the American machine-tool industry believe?**
A1: THE AMERICAN MACHINE TOOL INDUSTRY BELIEVES THAT PROTECTIONIST POLICY BY THE AMERICAN GOVERNMENT ACHIEVES THE PRESERVATION OF NORMAL PROFITS OF THE AMERICAN MACHINE TOOL INDUSTRY.

Q2: **Why does Lance Morrow believe that import restrictions would mean fewer sales?**
A2: LANCE MORROW BELIEVES THAT PROTECTIONIST POLICY BY THE AMERICAN GOVERNMENT MOTIVATES THE PRESERVATION OF NORMAL PROFITS OF AMERICAN INDUSTRIES BECAUSE LANCE MORROW BELIEVES THAT AS A CONSEQUENCE OF PROTECTIONIST POLICY BY THE AMERICAN GOVERNMENT, AMERICAN INDUSTRY PRODUCES WITH HIGH COST AMERICAN MACHINE TOOL; AND IF AMERICAN INDUSTRY PRODUCES WITH HIGH COST AMERICAN MACHINE TOOL, THEN AMERICAN INDUSTRIES PRODUCE HIGH COST PRODUCTS, AND IF AMERICAN INDUSTRIES PRODUCE HIGH COST

PRODUCTS, THEN AMERICAN INDUSTRIES SELL FEWER PRODUCTS; AND IF AMERICAN INDUSTRIES SELL FEWER PRODUCTS, THEN THERE IS A DECREASE IN PROFITS OF AMERICAN INDUSTRIES; AND A DECREASE IN PROFITS OF AMERICAN INDUSTRIES MOTIVATES THE PRESERVATION OF NORMAL PROFITS OF AMERICAN INDUSTRIES.

Although both ED-RESTRICTIONS and ED-JOBS argue against protectionism by the United States, Morrow's argument in ED-RESTRICTIONS is conceptually different from that of Friedman in ED-JOBS. This difference stems from the fact that Morrow uses argument units different from those used by Friedman. The argument units used in ED-RESTRICTIONS are

```
AU-EQUIVALENCE: Although opponent O believes plan P should be used
to achieve goal G1, SELF believes P should not be used to achieve G1
because SELF believes P thwarts goal G2 which is equivalent to G1.

AU-SPIRAL-EFFECT: Although opponent O believes plan P1 should be
used to achieve goal G1, SELF believes P1 should not be used to
achieve G1 because SELF believes P1 thwarts goal G2 and this goal
failure will require using plan P2 which is equivalent to P1.
```

Morrow uses AU-EQUIVALENCE to argue that

- The U.S. machine-tool industry believes that import restrictions on Japanese machine tools should be implemented because they will achieve their goal of preserving the industry's finances.
- However, implementing these import restrictions will thwart other U.S. manufacturers' goal of preserving their finances.
- Therefore, the import restrictions on machine tools should not be implemented.

AU-SPIRAL-EFFECT is used to argue further that these import restrictions should not be implemented because the goal violation resulting from these policies will motivate the use of more protectionist policies. That is, Morrow believes import restrictions will trigger a protectionist spiral.

To recognize the argument units in ED-RESTRICTIONS, OpEd must keep track of a causal chain of reasoning that contains the following relationships: (1) Import restrictions will result in an increase in production costs for other U.S. manufacturers that use costly U.S. machine tools. (2) The increase in production costs will cause an increase in the price of products made by those manufacturers. (3) The increase in the products' prices will lead to a decrease in their sales. (4) The decrease in products' sales will result in a decrease in earnings of U.S. manufacturers. (5) The decrease in earnings thwarts those manufacturers' goal of preserving their finances. (6) Because of this goal violation, U.S. manufacturers will petition for import restrictions.

When processing ED-RESTRICTIONS, OpEd has to be able to infer relationships 3, 4, 5, and 6 along with the implicitly stated connection between 1 and 2. In addition, mapping ED-RESTRICTIONS into conceptual representations requires solving the following problems:

- In the phrases "seeking protection from foreign competition" and "demand protection against foreign competition," it is implicit that petitions for protection are directed to the U.S. government.
- It is also implicit that in the phrase "import restrictions must be imposed" the U.S. government would be the one imposing the restrictions.
- In the phrases "the industry has been hurt" and "the industry can survive," the words "hurt" and "survive" refer not to the physical state of the industry but rather to its economic well-being.
- When referring to activities of industries, the phrase "make do with" has to be understood as "manufacture products using."

Getting OpEd to read ED-RESTRICTIONS in addition to handling ED-JOBS did not require modifying its process model of argument comprehension, but rather augmenting its lexicon and politicoeconomic knowledge, augmenting its knowledge of argument units to include AU-EQUIVALENCE and AU-SPIRAL-EFFECT, and specifying the processing strategies needed to manipulate the conceptual structures added. In addition, OpEd's search and retrieval processes did not need any modifications to retrieve answers to questions about ED-RESTRICTIONS.

7 LIMITATIONS OF OpEd AND FUTURE DIRECTIONS

Editorial comprehension in OpEd does not account for the processes responsible for (1) establishing the intention of an editorial, i.e., whether the editorial is intended to explain or persuade; (2) recognizing what it means to be persuaded by, in agreement with, or in disagreement with an argument; (3) reorganizing and updating beliefs once a persuasive argument has been recognized; and (4) recognizing whether an argument is sound or contains errors in reasoning. We must first model the process of parsing editorial text into conceptual representations before we can ever attempt to model such processes as persuasion, agreement, belief revision, and argument-error recognition.

OpEd currently has no background knowledge concerning historical events in the domain of economics. For instance, the Reagan administration has always professed its belief in free international markets and never viewed the voluntary limits as restrictions on international trade to protect domestic industries. However, OpEd does not know these facts because OpEd does not support a historical memory for politicoeconomic events and beliefs. The model of beliefs which OpEd currently constructs during editorial comprehension is based *only* on those beliefs explicitly or implicitly stated in the editorial. Modeling and applying historical memories during argument comprehension currently fall outside the scope of OpEd.

7.1 Fragility of Rules and Knowledge Engineering Bottleneck

The design of OpEd is a labor-intensive process, involving the hand coding of many knowledge constructs, along with the processing rules that integrate these constructs during comprehension. This knowledge engineering bottleneck could be overcome if OpEd had a general learning capability. Currently, OpEd's argument comprehension ability does not improve with experience in reading editorials. To improve argument comprehension would require a more sophisticated theory of representation, episodic memory organization, and generalization.

In addition, OpEd suffers from the same problem that plagues all rule-based systems—a lack of robustness in handling exceptions to rules and unanticipated rule interactions and an inability to recover from the incorrect application of rules, due to changes in context. We are currently exploring a variety of "subsymbol" (i.e., connectionist, PDP, and artificial neural systems) processing and knowledge representation methods, in order to address both fragility and learning bottlenecks (Dyer, 1989).

7.2 Metaarguments

The theory of processing and argument units in OpEd is designed to recognize and instantiate attack/support relations among beliefs. However, these relationships are themselves beliefs and are termed *warrants* (Toulmin, 1958). Arguments that attack AUs and/or warrants (i.e., the rules that allow one belief to justify another) may be termed *metaarguments*. Such arguments involve disputes concerning the strategies by which one forms arguments. Metaarguments arise whenever arguer A1 tells arguer A2 that A2 is not allowed to use a given argument structure to support or attack a belief. Metaarguments do not occur very often, but can be very effective when they are applicable, since it is the underlying logic of an argument that is being attacked by a metaargument. A metaargument often attempts to show a logical inconsistency or circularity within the argument unit being used by another arguer.

Metaarguments occur in debates over the nature of valid scientific reasoning and in the nature of rational evidence within religious contexts. Consider the following two argument fragments from *The Atheist Debater's Handbook* (Johnson, 1981, p. 14) in which the author refutes arguments for the existence of God:

> ...Many theists insist that it is the responsibility of the atheist to offer evidence justifying his lack of belief in God. But is the theist's demand rational?...
> Imagine that I just claimed there to be a gigantic man-eating frog in a local lake, and a friend denies it. It would be incumbent upon me to prove that such a thing exists, and not my friend's responsibility to disprove it....
> The theist claims that the atheist must disprove God's existence. The atheist could reply that there is conclusive evidence to suggest that God does not exist and thus it is the theist who must disprove the existence of such evidence. The

demand for disproof inevitably leads to an inconclusive farce. The demand for proof, on the other hand, can have conclusive results. Therefore, the only sensible procedure would be to demand proof, not disproof.

Here, the argument is not about whether God exists, but about *what kinds of strategies for proof or refutation are acceptable or unacceptable* when one is justifying God's existence. Johnson argues that the theist must offer proof for God's existence and cannot claim that God exists simply because God's existence has not been disproved. The justification for this metaargument is that disproof leads to attempts to disprove evidence for prior disproof—resulting in an infinite regress of disproved disproofs.

> There is a tendency among theists to offer, as evidence for the existence of God, phenomena which "science cannot explain." For example, neo-Darwinism has thus far been unsuccessful in explaining the development of the turtle's shell... God, the theist claims, must therefore be the explanation of the turtle's shell.
>   God certainly could be used to explain puzzling phenomena. However, the issue is not whether a particular explanation can be proved but, instead, whether the explanation presented is in fact correct....
>   The theist...contends that since God is an *adequate* explanation of a puzzling phenomenon, and no other adequate explanation is known, then God must be the correct explanation...but...the adequacy of an explanation is no guarantee of its correctness. (pp. 17–18)

In this fragment, Johnson argues that one cannot argue for God's existence based solely on the ability of the notion of God's existence to explain puzzling phenomena. (For example, the existence of life is puzzling. God could have created life; therefore, God must exist, since God's existence can then be used to "explain" the existence of life.) Johnson argues that many past (and currently unexplained) phenomena are now (or will be) explained without need for postulating a God; therefore a current *lack* of a correct explanation does not lend support for any proposed explanation.

The representational constructs used for beliefs in OpEd appear to be adequate for *representing* metaarguments, since attack and support relationships could be constructed among warrants. However, as can be seen from the above examples, metaarguments are highly abstract and complex in nature and currently beyond OpEd's *processing* (i.e., recognition and comprehension) capabilities. Any program capable of comprehending a wide range of such metaarguments would indeed be exhibiting a very high level of cognitive skill in abstract reasoning. We are currently exploring comprehension issues posed by metaargument texts.

8 CONCLUSIONS

This chapter has presented a highly knowledge-based approach to natural language comprehension of a class of editorials labeled one-sided arguments. The theory of editorial comprehension discussed here has been implemented in a computer program, OpEd, currently capable of reading two short editorial fragments in the domain of economic protectionism and answering questions about the argument content. The process of editorial comprehension is viewed as one of managing a wide variety of disparate knowledge sources, including scripts, social acts, affects, causal chains of reasoning, economic entities, plans, actions, characters, beliefs, and belief relationships. The theory of argument comprehension postulates the existence of *argument units* as a central, organizing construct of argument knowledge, which consist of configurations of belief attack and support relationships, where the content of each belief refers to abstract goal/plan situations.

Argument units represent knowledge of argumentation which is highly abstract and independent of any particular domain. As a result, knowledge of argument units allows a comprehension system to recognize and interpret arguments in disparate domains, as long as the system has sufficient planning skill and domain knowledge to build up instances of the particular goals, plans, and beliefs occurring within that domain.

The knowledge-intensive nature of editorial comprehension indicates that major advances in handling complicated editorial text will require breakthroughs in the areas of learning and reasoning. Learning is required to overcome the labor-intensive task of hand coding the vast amounts of general world knowledge needed for language comprehension. Current rule-based approaches appear to be too fragile to adequately model the recovery and reinterpretation processes that arise as the context of the argument changes. The major benefit derived from constructing OpEd is that it forces us to make explicit the knowledge and processing required for comprehension of complex editorial texts.

ACKNOWLEDGMENTS

This research was supported in part by a contract to the second two authors with the JFT Program of the DoD, monitored by the Jet Propulsion Lab, Pasadena. The research was also supported in part by an IBM Faculty Development Award and an ITA Foundation Grant to the second author. The OpEd process model was implemented on Apollo workstations acquired through a grant from the W. M. Keck Foundation.

REFERENCES

Abelson, R. P. (1973). The Structure of Belief Systems. In: R. C. Schank and K. M. Colby (Eds.), *Computer Models of Thought and Language,* San Francisco: Freeman.
Abelson, R. P. (1979). Differences between beliefs and knowledge systems. *Cognitive Science* 3:335–366.
Alvarado, S. J. (1989). Understanding editorial text: A computer model of argument comprehension. UCLA Ph.d. thesis.
Alvarado, S. J., M. G. Dyer, and M. Flowers (1985). Memory representation and retrieval for editorial comprehension. *Proceedings of the Seventh Annual Conference of the Cognitive Science Society,* Irvine, California, 228–235.

Alvarado, S. J., M. G. Dyer, and M. Flowers (1986). Editorial comprehension in OpEd through argument units. *Proceedings of the Fifth National Conference on Artificial Intelligence.* Philadelphia, pp. 250–256 (distributed by Morgan Kaufmann, Los Altos, Calif.).
August, S. E., and M. G. Dyer (1985a). Analogy recognition and comprehension in editorials. *Proceedings of the Seventh Annual Conference of the Cognitive Science Society.* Irvine, California, pp. 228–235 (distributed by Lawrence Erlbaum Assoc., Hillsdale, NJ).
August, S. E., and M. G. Dyer (1985b). Understanding analogies in editorials. *Proceedings of the Ninth International Joint Conference on Artificial Intelligence.* Los Angeles, pp. 845–847 (distributed by Morgan Kaufmann, Los Altos, Calif.).
Bresnaham, T. F. (1984). Quotas, Not Bonuses, Hurt Auto industry (Editorial). *Los Angeles Times*, Part II, p. 7, May 11.
Britton, B. K., and J. B. Black (1985). *Understanding Expository Text.* Hillsdale, NJ: Lawrence Erlbaum Assoc.
Bush, C. R. (1932). *Editorial Thinking and Writing.* New York: D. Appelton.
Carbonell, J. G. (1981). *Subjective Understanding: Computer Models of Belief Systems.* Ann Arbor, MI: UMI Research Press.
Chellas, B. F. (1980). *Model Logic.* New York: Cambridge University Press.
Cohen, R. (1983). *A Computational Model for the Analysis of Arguments.* Tech. Rep. CSRG-151, Computer Science Department, University of Toronto, Canada.
Cuddington, J. T., and R. I. McKinnon (1979). Free trade versus protectionism: A perspective. In: *Tariffs, Quotas, and Trade: The Politics of Protectionism.* San Francisco: Institute of Contemporary Studies.
Dechter, R., and A. Dechter (1988). Belief maintenance in dynamic constraint networks. *Proceedings of the Seventh National Conference on Artificial Intelligence (AAA-88).* St. Paul, MN (distributed by Morgan Kaufmann, Los Altos, Calif.).
van Dijk, T. A. (Ed.) (1985). *Handbook of Discourse Analysis,* vols. 1 to 4. New York: Academic Press.
van Dijk, T. A., and W. Kintsch (1983). *Strategies of Discourse Comprehension.* New York: Academic Press.
Doyle, J. (1979). A truth maintenance system. *Artificial Intelligence* **12**:231–272.
Dyer, M. G. (1983a). *In-depth understanding: A Computer Model of Integrated Processing for Narrative Comprehension.* Cambridge, MA: M.I.T. Press.
Dyer, M. G. (1983b). The role of affect in narratives. *Cognitive Science* 7:211–242.
Dyer, M. G. (1989). Symbolic neuroengineering for natural language processing: A multilevel research approach. In: J. Barnden and J. Pollack (Eds.), *Advances in Connectionist and Neural Computation Theory.* Norwood, NJ: Ablex Publ. (in press).
Dyer, M. G., R. E. Cullingford, S. J. Alvarado (1987). SCRIPTS. In: S. C. Shapiro (Ed.), *Encyclopedia of Artificial Intelligence.* New York: Wiley.
Dyer, M. G., and W. G. Lehnert (1982). Question answering for narrative memory. In: J. F. Le Ny and W. Kintsch (Eds.), *Language and Comprehension.* Amsterdam: North-Holland.
Feldstein, M. and K. Feldstein (1985, Apr. 10). Judicious steps at home can lessen trade deficit (Editorial). *Los Angeles Times,* part IV, p. 5.
Flowers, M. (1982). On being contradictory. *Proceedings of the Second National Conference on Artificial Intelligence.* Pittsburgh, pp. 63–65 (distributed by Morgan Kaufmann, Los Altos, Calif.).
Flowers, M., and M. G. Dyer. Really arguing with your computer in natural language. *Proceedings of the National Computer Conference.* Las Vegas, pp. 651–659.
Flowers, M., R. McGuire, and L. Birnbaum (1982). Adversary arguments and the logic of personal attacks. In: W. G. Lehnert and M. G. Ringle (Eds.), *Strategies for Natural Language Understanding.* Hillsdale, NJ: Lawrence Erlbaum
Friedman, M. (1982, Nove. 15). Protection that hurts (Editorial). *Newsweek,* p. 90.
Greenaway, D. (1983). *Trade Policy and the New Protectionism.* New York: St. Martin's Press.
Greenaway, D., and C. Milner (1979) *Protectionism Again...? Causes and Consequences of a*

Retreat from Freer Trade to Economic Nationalism. Hobart Paper 84. London: The Institute of Economic Affairs.
Harman, G. (1986). *Change in View: Principles of Reasoning.* Cambridge, MA: M.I.T. Press.
Hintikka, J. (1962). *Knowledge and Belief.* Ithaca, NY: Cornell University Press.
Iacocca, L. A. (1986, Oct. 26). How many jobs do we save as doormat of world trade? (Editorial). *Los Angeles Times,* part IV, p. 5.
Johnson, B. C. (1981). *The Atheist Debater's Handbook.* NY: Prometheus Books.
deKleer, J. (1986). An assumption-based TMS. *Artificial Intelligence* 28(2).
Lehnert, W. G. (1978). *The Process of Question Answering: A Computer Simulation of Cognition.* Hillsdale, NJ: Lawrence Erlbaum.
Lehnert, W. G., M. G. Dyer, P. N. Johnson, P. J. Yang, and S. Harley (1983). BORIS—An experiment in in-depth understanding of narratives. *Artificial Intelligence* 20:15–62.
Los Angeles Times (1984, Feb. 16). Car quotas: Costly folly (Editorial). *Los Angeles Times,* part II, p. 6.
Los Angeles Times (1984, Dec. 9). No walls or new walls (Editorial). *Los Angeles Times,* part V, p. 4.
Minsky, M. A. (1977). Frame system theory. In: P. Johnson and O. Wason (Eds.), *Thinking: Readings in Cognitive Science.* Cambridge, MA: M.I.T. Press.
Morrow, L. (1983, Jan. 10). The protectionist temptation (Editorial). *Time,* p. 68.
Pearl, J. (1986). Fusion, propagation and structuring in belief networks. *Artificial Intelligence* 29:241–288.
Riesbeck, C. K. (1984). Knowledge reorganization and reasoning style. *International Journal of Man-Machine Studies* 20:45–61.
Samuelson, R. J. (1984, Sept. 12). Politics of trade protection: Opponents shifting balance (Editorial). *Los Angeles Times,* part II, p. 5.
Schank, R. C. (1973). Identification of conceptualizations underlying natural language. In: R. C. Schank and K. M. Colby (Eds.), *Computer Models of Thought and Language,* San Francisco: Freeman.
Schank, R. C. (Ed.) (1975). *Conceptual Information Processing.* Amsterdam: North-Holland.
Schank, R. C. (1978). What makes something "ad hoc." *Proceedings of Theoretical Issues in Natural Language Processing,* vol 2. Urbana-Champaign, Illinois, pp. 8–13.
Schank, R. C. (1982). *Dynamic Memory: A Theory of Reminding and Learning in Computers and People.* New York: Cambridge University Press.
Schank, R. C., and R. P. Abelson (1977). *Scripts, Plans, Goals, and Understanding.* Hillsdale, NJ: Lawrence Erlbaum Assoc.
Schank, R. C., and J. G. Carbonell (1979). Re: The Gettysburg Address, representing social and political acts. In: N. Findler (Ed.), *Associative Networks.* New York: Academic Press.
Spiro, R. J., B. C. Bruce, and W. F. Brewer (Eds.) (1980). *Theoretical Issues in Language Comprehension.* Hillsdale, NJ: Lawrence Erlbaum Assoc.
Stonecipher, H. W. (1979). *Editorial and Persuasive Writing.* New York: Hastings House.
Thurow, L. C. (1983, Apr. 25). The road to lemon socialism (Editorial). *Newsweek,* p. 63.
Toulmin, S. (1958). *The Uses of Argument.* New York: Cambridge University Press.
Toulmin, S., R. Reike, and A. Janik (1979). *An Introduction to Reasoning.* New York: Macmillan.
Tulving, E. (1972). Episodic and semantic memory. In: E. Tulving and W. Donaldson (Eds.), *Organization of Memory.* New York: Academic Press.
Wilensky, R. (1983). *Planning and Understanding.* Reading, MA: Addison-Wesley.
Wilks, Y., and J. Bien (1983). Beliefs, points of view, and multiple environments. *Cognitive Science* 7:95–119.
Yoffie, D. B. (1983). *Power and Protectionism.* New York: Columbia University Press.

NAME INDEX

Abelson, R. P., 196, 238, 255, 299, 306, 312, 317, 335, 342, 344
Ackerman, W., 285
Addis, T. R., 131, 154, 228, 253
Adler, J., 85, 99
Aiello, N., 199
Allen, B., 185, 190, 194
Allen, J., 187, 190, 193
Alty, J., 148–149, 154
Alvarado, S. J., 286, 287, 293, 342, 343
Amarel, S., 129, 156
Andersen, J. R., 131, 136, 154
Anderson, S. K., 156
Appelt, D., 193
Arbab, B., 244, 253
Atkinson, D. J., 193
August, S. E., 328, 343

Bachant, J., 105, 128
Bainbridge, L., 53–54, 99
Ballard, D. H., 18, 27
Bareiss, E. R., 236, 253
Barr, A., 39, 51, 159, 170, 193, 230, 253
Barrett, M., 135, 154
Barrow, H. G., 198
Bayerl, S., 10, 23, 26
Beer, R. D., 223, 224
Bell, C. E., 182, 190, 193
Benbasat, I., 75, 99
Bennet, J. S., 146, 154

Bentolila, J., 101
Berry, D. C., 142–143, 154
Bessiere, P., 182, 194
Bibel, W., 1, 11, 17, 23, 25, 26, 27
Bien, J., 344
Biggs, S. F., 156
Billault, J. P., 154
Birnbaum, L., 343
Bittinger, M. L., 285
Black, J. B., 336, 343
Bloomfield, B. P., 143, 154
Blume, S. S., 63, 99
Blythe, J., 254
Bobrow, D. G., 139, 154, 190, 198
Bollinger, T., 254
Boose, J. H., 53, 59, 68, 79, 80, 81, 92, 99, 100, 101, 106, 128, 253
Brachman, R. J., 19, 22, 26
Bradshaw, J. M., 79, 92, 100
Brailsford, J. R., 155
Bratko, I., 242, 244, 253
Brauer, W., 193
Bredeweg, B., 154
Bresina, A., 197
Bresina, J. L., 181, 193, 197
Bresnahan, T., 326, 343
Breuker, J., 131, 132, 144–146, 152–154, 156
Brewer, W. F., 344
Britton, B. K., 336, 343
Broadbent, D. E., 54, 100

Broadbent, M. H. P., 54, 100
Brown, J., 106, 118, 128
Brownston, L., 16, 26
Bruce, B. C., 344
Buchanan, B. G., 104, 105, 128, 129
Buchanan, J. R., 189, 195
Buchner, B. A., 113, 128
Bush, C. R., 287, 343

Calvieras, B., 254
Cannat, J. J., 252–253
Carbonell, J. G., 27, 52, 101, 129, 151, 154, 241, 254, 299–300, 304, 335, 343–344
Carnegie Group Inc., 191, 198
Carpenter, G. A., 223–224
Cendrowska, J., 84, 100
Chandrasekaran, B., 138, 140, 154
Chang, C. L., 285
Chapman, D., 161, 166, 168, 193
Charniak, E., 2, 12, 26, 27, 164–165, 173, 193
Cheeseman, P., 193
Chellas, B. F., 335, 343
Chen, H. H., 223–224
Chien, R. T., 193
Chilausky, R. L., 104, 112, 129, 143, 155
Chomsky, N., 145, 154
Chouraqui, E., 239, 254
Clancey, W. J., 115, 128, 137, 145–147, 149–150, 154
Clocksin, W., 190, 198
Cohen, R., 336, 343
Colby, K. M., 342, 344
Collins, H. M., 54, 100, 142–143, 154
Comb, J. B., 42, 51
Combs, D. M., 101
Coombs, M. J., 148–149, 154
Corkill, D. D., 186, 193
Corlett, R., 244, 254
Cox, P. T., 228, 254
Cross, A. D., 155
Cuddington, J. T., 327, 343
Cullingford, R. E., 343
Currie, K., 161, 178, 180, 182, 184, 187, 193, 194
Czaja, L., 193

d'Agapeyeff, A., 135, 154
Daniel, L., 185, 187, 189, 193
Davies, D. J. M., 190, 198
Davies, T. R., 239, 254
Davis, P. R., 193
Davis, R., 113, 114, 128, 136–138, 154, 193
Davoodi, M., 154
Dean, T., 161, 193, 195

Dechter, A., 335, 343
Dechter, R., 335, 343
Deering, M., 190, 198
de Hoog, R., 154
DeJong, G. F., 109, 110, 128, 227, 232, 254
de Kleer, J., 106, 118, 128, 139, 155, 198, 335, 344
Derkson, J. A., 190, 199
Dershowitz, N., 189, 193
Dhaliwal, J. S., 75, 99
Dickerson, D. J., 156
Diederich, J., 79, 81, 100
Dixon, N., 54, 100
Donaldson, W., 344
Doolen, G., 224
Doran, J. E., 186, 193
Doshi, R. S., 193
Doyle, J., 195, 198, 335, 343
Doyle, R. J., 193
Drabble, B., 187, 194
Draper, R. C., 154
Dreussi, J., 186, 188, 196
Dreyfus, H. L., 141–143, 155
Dreyfus, S. E., 141–143, 155
Drummond, M., 186–187, 194
Dubois, D., 285
Duda, R. O., 103, 128
Duffay, P., 194
Duval, B., 228, 229, 233, 254
Dyer, M. G., 286, 297, 299, 307, 309, 316, 328, 331, 335, 340, 342–344

Eason, K. D., 153, 155
Ehret, D., 100, 128
Elcock, E. W., 198
Eliot, L. B., 151, 155
Elshout, J. J., 155
Elver, E., 1, 26–27
Emde, W., 113, 115–116, 128
Enderton, H. B., 285
Erman, L. D., 186, 194
Ernst, G., 169, 173, 194
Eshelman, L. D., 79, 100, 106, 128

Fagan, L. M., 101
Fahlman, S. E., 191, 194, 199
Faletti, J., 194, 198
Farrell, R., 26
Feigenbaum, E. A., 51, 66, 100, 103, 128, 131, 135–136, 155, 159, 170, 193, 230, 252, 254
Feldman, J. A., 18, 27
Feldstein, K., 329–330, 343
Feldstein, M., 329–330, 343

NAME INDEX

Feyerabend, P., 59, 100
Fikes, R. E., 109, 128, 158, 172–173, 187–188, 194, 199
Firby, J., 161, 195
Fitzgerald, P.E., 54, 100
Flowers, M., 286, 294, 299, 307, 313, 316, 318, 335, 342–343
Foote, M. H., 156
Forbus, K., 106, 129
Forgy, C., 103, 129
Foster, J. M., 198
Fox, M. S., 185, 190–191, 194, 196, 198
Frege, G., 7
Freud, S., 80, 100
Friedman, M., 288, 292, 324–325, 327, 337, 343

Gachnig, J. G., 128
Gaines, B. R., 52–53, 55–56, 59, 63, 66, 68, 73, 75, 77, 79–81, 84, 92, 100, 101, 102, 131–132, 148, 153, 155, 253
Galantner, E., 196
Gallaire, H., 199
Gallier, J. H., 9, 27
Gammack, J. G., 81, 101, 140, 153, 155
Ganascia, J. G., 227, 254
Garber, S., 155
Garg-Janardan, C., 81, 101
Genesereth, M. R., 229, 254, 285
Genetet, S., 101
Gentner, D., 139, 155
Georgeff, M., 182, 186–187, 194
Gilbert, V. P., 26
Giles, C. L., 212–214, 222–224
Ginsberg, M. L., 256, 285
Golshani, F., 28
Gorry, G. A., 195
Gray, P. M. D., 198
Green, C. C., 170–171, 194
Green, R. H., 154
Greenaway, D., 301, 321, 327, 343
Greenberg, M., 196
Grimes, M., 101
Grossberg, S., 224

Haas, A. R., 194
Hagstrom, W. O., 62, 101
Hall, R. J., 254
Harandi, M. T., 103, 113, 129
Harber, S. D., 155
Harley, S., 344
Harman, G., 318, 344
Hart, P. E., 128, 158, 172–173, 187–188, 194
Hartoch, G. P., 224–225

Hawkins, D., 54, 60, 101, 148, 150, 155
Hayes, P. J., 14, 27, 171, 185, 187, 194, 195
Hayes-Roth, B., 182, 186, 195
Hayes-Roth, F., 42, 51, 66, 101, 104, 129, 135–136, 155, 182, 186, 195
Hayward, S., 154
Hearn, L., 129
Hendrix, G., 195, 199
Hewitt, C., 190, 195, 199
Hilbert, D., 285
Hillis, W. D., 199
Hintikka, J., 335, 344
Hinton, G. E., 18, 27, 224–225
Hirschberg, J., 156
Hobbs, R., 196
Hopfield, J. J., 221–225
Hunt, E. B., 107, 129

Iacocca, L. A., 320, 344
Inference Corporation, 191, 199
Intellicorp, 190, 199

Jackson, P., 39, 51
Janik, A., 344
Jansweiger, W. H. N., 145, 155
Jiang, Y. T., 199
Johnson, B. C., 340–341, 344
Johnson, L., 145, 146, 155
Johnson, P. E.., 144–146, 151–153, 155
Johnson, P. N., 344

Kaelbling, L. P., 187, 195
Kahn, K., 195
Kearney, F. W., 129
Kedar-Cabelli, S., 129, 232, 254–255
Keller, R., 129, 232, 255
Kelly, G. A., 80–81, 101
Kent, E., 26
Keravnou, E. T., 145–146, 155
Kerschberg, L., 49, 51
Kidd, A. L., 148–150, 155
Kintsch, W., 336–337, 343
Klinker, G., 79, 101
Knight, J. A., 156
Kodratoff, Y., 226–229, 233, 236, 238–239, 241, 244, 253–254
Kohonen, T., 223–224
Kolodner, J. L., 151, 155
Konolige, K., 186
Koomen, J., 187, 190, 193
Kornfeld, W. A., 186
Kowalski, R., 171–172, 195
Kulikowski, C., 129, 156

NAME INDEX

Laird, J. E., 236, 254
Lange, R., 105, 113, 129
Lansky, A., 160–161, 182, 187, 194–195
Larkin, J. H., 136, 155
Latombe, J. C., 194–195
Lavington, S. H., 199
Lavrac, N., 242, 244, 253
Leary, M., 138, 155
Lee, H. Y., 224
Lee, K. C., 225
Lee, R. C. T., 285
Lee, Y. C., 224–225
Lehnert, W. G., 295, 335, 343–344
Lenat, D. B., 51, 66, 101, 128, 129, 195
Le Ny, J. F., 343
Lesser, V. R., 186, 193–194
Letz, R., 26
Levesque, H. J., 19, 23, 26, 27
Lin, C. W., 225
Linster, M., 253
London, P., 195–196
Los Angeles Times, 313, 321, 323, 344
Luckham, D. C., 189, 195

McCarthy, J., 171, 195
McClelland, J. L., 27, 225
McCorduck, P., 131, 135–136, 155
McDermott, D., 2, 26–27, 164–165, 173, 193, 195
McDermott, D. V., 190, 195, 199
McDermott, J., 100–101, 105, 128–129, 137, 155, 190, 195
McGregor, J. J., 191, 198–199
McGuire, R., 343
McKinnon, R. I., 327, 343
Mahedevan, S., 129, 252, 255
Malcolm, C., 187, 197
Malone, J. R., 191, 199
Mamdani, A., 285
Manago, M., 254
Manna, Z., 9, 27
Marcus, S., 79–80, 101
Martin, N., 26
Masui, S., 190, 195
Matheson, J., 85, 99
Maxwell, T., 212, 223–224
May, M., 79, 81, 100
Mellish, C., 190, 198
Merat, F. L., 223–224
Merton, R. K., 62, 101
Mescheder, B., 117, 129
Michalski, R. S., 21, 27, 52, 101, 104, 106–108, 112, 129, 143, 227, 230, 241, 249, 253–255
Michie, D., 142–143, 155, 193, 244, 252–253

Miller, D., 161, 195
Miller, G. A., 195
Miller, P. L., 149, 156
Milner, C., 321, 327, 343
Minker, J., 199
Minsky, M., 13, 27, 223, 224, 331
Mitchell, T. M., 27, 52, 101, 109–110, 112, 129, 229, 232, 241, 252, 254–255
Mittal, S., 138, 140, 154
Mooney, R., 109–110, 128, 227, 232, 254
Moore, R. C., 196
Morik, K., 113, 115–116, 128–129
Morrow, L., 298, 322, 337–338, 344
Mostow, D. J., 196
Murray, A. M., 198
Musen, M. A., 79, 101

Neisser, U., 144, 156
Newell, A., 28, 51, 112, 129, 134–136, 144–145, 156, 169, 173, 194, 196, 254
Nguyen, H. T., 285
Nicolas, J., 229, 255
Nielsen, M., 196
Nii, H. P., 199
Nilsson, N. J., 128, 158, 172–173, 186–188, 194–196, 223, 224, 229, 254, 285
Nisbett, R. E., 54, 62, 101–102
Norman, D. A., 132, 156
Nunez, M., 244, 255

Ovchinnikov, S., 285

Pao, Y. H., 200, 205, 223–225
Papert, S., 223, 224
Partridge, D., 135–136, 142–143, 156
Pask, G., 148, 156
Pawley, G. S., 222, 224
PDP Research Group, 27
Pearl, J., 25, 27, 335, 344
Pednault, E. P. D., 196
Peterson, J. L., 196
Pietrzykowski, T., 228, 254
Pippenger, N., 196
Pollack, M. E., 148–149, 156
Pomerleau, D. A., 223, 225
Pople, H. E., 140, 156
Porter, B. W., 253
Pospesel, H., 9, 27
Post, E. L., 28, 51
Prade, H., 285
Pribram, K. H., 196
Priedetis, A., 241, 255
Puget, J.-F., 232, 255

NAME INDEX

Quillian, M. R., 12
Quinlan, J. R., 84, 101, 107–108, 129, 230, 242, 255

Rabbitt, P., 145, 156
Rappaport, A., 59, 65, 84, 101
Raven, P. F., 155
Reagan, R., 288–293, 332–333, 335, 337, 339
Reboh, R., 199
Reddy, D. R., 194
Reichgelt, H., 145, 156
Reiger, C., 196
Reisig, W., 196
Reiter, R., 199
Rich, C., 196
Rich, E., 39, 51
Rieke, R., 344
Riesbeck, C. K., 27, 151, 156, 299, 307, 344
Ringle, M. G., 343
Roach, J., 197
Robinson, A. E., 198
Rodriguez-Bachiller, A., 138, 155
Rogers, C. R., 80, 101
Rosenblatt, F., 208, 223, 225
Rosenbloom, P. L., 254
Rosenschein, S. J., 196
Ruhmann, I., 79, 81, 100
Rulifson, J. F., 190, 199
Rumelhart, D. E., 18, 27, 225
Runes, D. D., 59, 102
Russell, S. T., 239, 254

Sacerdoti, E. D., 160, 162, 170, 176–177, 182, 186, 188, 190, 196, 199
Safir, A., 129, 156
Sagalowicz, D., 199
Salvendy, G., 81, 101
Samuelson, R. J., 326, 344
Sathi, A., 196
Schank, R. C., 196, 230, 236, 238–239, 255, 299–300, 306, 335, 342, 344
Schmidt, C. G., 181, 196
Schneeberger, J., 1, 27
Schön, D. A., 62, 102
Schreiber, G., 154
Schubert, L. K., 13, 27
Schumann, J., 26
Selfridge, M., 151, 156
Sergot, M., 195
Shapiro, S. C., 343
Shaw, M. L. G., 68, 75, 77, 79, 81, 83–84, 89, 92, 100–102
Shortliffe, E. H., 101, 104–105, 128–129, 137, 156

Shriver, B., 64, 102
Shrobe, H. E., 196
Siklossy, L., 186, 188, 197
Simon, H. A., 28, 51, 135–136, 156, 159–160, 164, 169, 196–197
Siqueira, J., 232, 255
Slater, P., 84, 89, 102
Slatter, P. E., 130, 131, 136, 156
Sloman, A., 199
Smets, P., 256, 285
Smith, B., 131, 156
Smith, G., 194
Smith, R., 193
Smith, R. G., 186, 197
Smithers, T., 187, 197
Sobel, A., 190, 195
Sparck Jones, K., 153, 156
Spiro, R. J., 336, 344
Sridharan, A., 181, 197
Stachowitz, R. A., 42, 51
Stallman, R. M., 185, 197
Steele, G. L., 197
Steels, L., 140, 150, 156
Stefik, M. J., 170, 179–180, 182, 185–186, 188, 190, 197–199
Steier, D., 112, 129
Steinberg, L. I., 129, 252, 254
Stenton, S. P., 148–149, 156
Stepp, R. E., 230, 254
Stevens, A. S., 139, 155
Stich, S. P., 62, 102
Stickel, M. E., 23, 27
Stonecipher, H. W., 287, 344
Strohm, G., 185, 190, 194
Sun, G. Z., 224
Suppes, P., 285
Sussman, G. A., 169, 173, 174, 190, 199
Sussman, G. J., 185, 197, 199
Swartout, W. R., 106, 115, 129

Takefuji, Y., 221, 225
Tan, M., 100, 128
Tank, D. W., 221–225
Tate, A., 161, 167, 170, 174–175, 177–178, 182, 184–189, 193–194, 197
Tecuci, G., 236, 254
Tenenberg, J., 197
Thiagarajan, P. S., 196
Thurow, L. C., 328–329, 344
Toulmin, S., 316, 335, 340, 344
Tong, R. M., 285
Touretzky, D. S., 15, 27
Trayner, C., 186

Tulving, E., 297, 344
Turing, A., 28

Valiant, L. G., 255
van Dijk, T. A., 336–337, 343
van Harmelin, F., 145, 156
van Melle, W., 103, 129
van Someren, M., 154
Vere, S. A., 161, 178, 179, 182, 197
Vickers, J. N., 59, 101
Vilain, M. B., 197
Vrain, C., 248, 250, 251, 255

Wahl, D., 81, 102
Waldinger, R., 9, 27, 110, 129, 175–176, 186, 190, 198, 199, 232, 255
Wall, R., 285
Warren, D. H. D., 170, 176, 187, 189–190, 198
Waterman, D. A., 51, 66, 101, 104, 129
Waters, R. C., 196
Webber, B., 156
Wegner, P., 64, 102, 198
Weiss, S. C., 106, 129
Weiss, S. M., 140, 156
Welbank, M., 80, 102

Weyhrauch, R. W., 11, 27
Whiter, A. M., 189, 197
Wielemaker, J., 154
Wielinga, B., 144–146, 152, 154–156
Wier, C. C., 236, 253
Wilensky, R., 186, 198, 335, 344
Wilkins, D. E., 161, 178, 180–181, 187, 198
Wilks, Y., 344
Williams, B., 198
Williams, R. J., 223, 225
Wilson, G. V., 222, 225
Wilson, T. D., 54, 101
Winograd, T., 190, 198
Winston, P. H., 39, 51, 239, 241, 255
Woodward, J. B., 68, 75, 102
Worden, R. P., 149, 156
Wrobel, S., 113, 129

Yager, R. R., 256, 285
Yang, P. J., 344
Yoffie, D. B., 301, 344
Young, R. M., 81, 101, 139, 140, 155

Zadeh, L. A., 285
Zaulkernan, I., 155

SUBJECT INDEX

Abduction, 25, 58, 227–228, 233–234
Abstract causal domain model (ACDM), 117–127
ABSTRIPS, 186, 188
AGAPE, 238
AI planning, 157–199, 335
Airplan, 190
Analogical reasoning, 239–241, 240–242 (illus.)
AND/OR graph, 30–31
ART, 28, 191
ART1, 215
ART2, 215–216
Assertional representation, 22
Associative memory, 204, 216–220
Associative networks, 12–13
ATTENDING, 149
Authority triangles, 299–304
Autoassociative memory, 219, 224

Backward chaining, 33–36, 40
Bayesian probability, 136
Beam search, 185
Belief relationship, 335
Binary propositional logic (*see* Propositional logic)
Blackboard system, 182, 186
BLIP, 113, 115–117
BORIS, 297
Breadth-first search, 36–39, 172

CADUCEUS, 140

CASNET, 140
Causal models, 117–128, 138, 151
Certainty factor, 44–45
CMACS, 151
Cognitive psychology, 140, 142, 181
Commonsense knowledge, 24–25
Commonsense planning, 181
Conceptual dependency theory, 335
Conceptual sorting, 140
Conflict resolution, 39–40, 231, 306–307
Connectionism, 143, 340
Connectionist networks, 17
Conniver, 190
Conversation theory, 148

Data-driven strategy, 33
Deduction, 226, 271
Deep knowledge, 76, 137–138, 150
Dependency-directed search, 185
Depth-first backtracking, 184
Depth-first search, 190
Deviser I, 161, 178–180, 182, 187, 190
DISCIPLE, 236–238
Distributed planning, 186
DKAS, 113

EBG (*see* Explanation-based generalization)
EBL (*see* Explanation-based learning)
Epistemological analysis, 147
Euclidean metric, 216–217

351

SUBJECT INDEX

Evidential reasoning, 335
Expertise:
    competence models, 133, 144–147, 150, 152–153
    deep models, 133, 137–141, 152
    distributed models, 133, 147–150, 152–153
    heuristic models, 133, 135–137, 150, 152–153
    hybrid models, 150–152
    implicit models, 133, 141–144, 152–153
Explanation-based generalization, 231–233, 251
Explanation-based learning, 106, 109–112, 231–236

FACT, 191
FOCUS, 84, 89
Forbin, 161
Forward chaining, 33–36, 39, 40
Frames, 13–14, 21–22, 331
Fregean knowledge representation, 7–9, 18
Functional knowledge representation, 10–12
Functional-link net, 204, 212–214, 224

GDR (*see* Generalized delta rule)
General problem solver, 169, 173, 184
Generalized delta rule (GDR), 203, 209, 211–214, 223–224
Generalized perceptron, 209–214
GPS (*see* Generalized problem solver)

Hacker, 169, 173–174
Hadamard matrices, 220
HEARSAY II, 186
Heteroassociative memory, 219, 222
Holographic associative memory, 219, 223
Horn clause, 10

IF-THEN rules, 29
ID3, 107–109, 230, 241–244, 252
INDUCE, 241, 245–247, 252
Induction, 25, 74–75, 142–143, 227–228
Inductive learning, 241–251
Inexact reasoning, 43–45
Inference engine, 31–32
Inference mechanism, 31–32
Inheritance, 14–16
Interplan, 170, 174–175, 186
ISIS-II, 190
ISODATA, 204, 215

KEE, 28, 190
K-means algorithm, 204
Knowledge acquisition, 52, 60–66, 103–129

Knowledge acquisition (*Cont.*):
    behavior modeling, 52
    causal models, 117–128
    distributed, 113
    model-based, 103–129
    repertory grids, 52
    semantic models, 115–117
    syntactic models, 113–115, 127
    text analysis, 52
    tools, 79–85
    validation, 73–77
    virtual machine hierarchy, 71–73
Knowledge Craft, 191
Knowledge representation:
    associative networks, 12–13
    comparative evaluation, 19–23
        assertional representation, 22–23
        linear versus two-dimensional formalisms, 19–20
        multistyle representation, 23
        object-oriented descriptions, 20–21
        unary functions, 21–22
    in the form of patterns, 202–203
    frames, 13–14, 21–22
    Fregean, 7–9, 19
    functional, 10–12
    Horn-clause logic, 9–10
    modal logic, 18
    natural language, 6
    neural nets, 17–18
    object-oriented, 12, 14
    predicate logic, 7–9
    procedural, 16–17
    semantic network, 12–13, 21–22
KRL, 190
KSSO, 190

LAWALY, 186, 188
LEAP, 112
Learning:
    analogical reasoning, 239–241
    connectionist, 18
    explanation-based, 106, 109–112, 127, 231–236
    generalization, 229
    generalized perceptron, 209–214
    inductive reasoning, 241–251
    neural net, 200–225
    particularization, 229–230
    similarity-based, 103–104, 106, 109, 112, 127
    unsupervised learning, 214–216
Lerner project, 115–117

LISP, 11–12, 71, 174, 190, 231
LOOPS, 190
LOP, 10–12

Machine learning (*see* Learning)
MAXNET, 215–216
MDX, 140
Means ends analysis, 169, 173, 184, 186
Meta-argument, 340–341
Metaknowledge, 236
Metalevel planning, 186
MINPAD, 149
Modal logic, 18
Model of expertise (*see* Expertise)
MolGen, 179–180, 185–186, 188
Multidimensional scaling, 140
MYCIN, 29, 36, 45, 105, 137, 145–146, 149

Naive physics knowledge, 24
Natural language processing, 286–344
  conceptual models, 286–334
  psycholinguistic analysis, 336–337
  structural models, 336
Neo-CRIB, 146
NEOMYCIN, 146, 149
NETL, 191
Neural nets, 17–18, 200–225, 340
NEXPERT, 84, 90
NLP (*see* Natural language processing)
NOAH, 160–161, 170, 176–178, 180, 184, 186, 188
NonLin, 161, 164, 166, 177–178, 180, 182, 185–189
Nonmonotonic reasoning, 16, 25

OAV, 29
Object-attribute-value, 29
Object-oriented knowledge representation, 12, 14, 20–21, 64, 81
One-then-best backtracking, 185
OpEd, 286–342
Open planning architecture (*see* O-Plan)
O-Plan, 161, 178, 180, 182–184, 190
OPM, 186
Opportunistic search, 185–186
OPS5, 28, 40

Parallel processing, 203
Parser, 297–298
Pattern recognition, 205–208
PEARL, 190
Perceptron, 203–204, 208–214, 223, 225

Plan representation:
  action ordering, 160–162, 178, 181
  least commitment, 162–163, 179–180, 182, 184
  state space, 159–160, 163, 180–182
PLANNER, 190
Planning (*see* AI planning)
PLANX10-D, 181
POPLER, 190
Predicate logic, 7–10
  first-order, 7–10
  higher-order, 8–9
PRINCESS, 25–26
Procedural attachment, 11
Procedural reasoning system (PRS), 182
Procedural representation, 16–17
Production systems, 28–51
Prolog, 10–12, 17, 43, 149, 171, 185, 190, 231–233
Propositional logic, 256–285
  basic deduction procedure, 271–275
  direct method of inference, 276–281
  equivalence of propositions, 262–266
  equivalent formulas, 267
  normal forms, 266–271
  representing, 256–258
  resolution approach to inference, 281–285
  rules of reasoning, 275–276
  truths of compound propositions, 259–262
Protocol analysis, 143, 146
Protos, 236–237
Prototype theory, 142
PRS, 182

QA3, 171, 173
QA4, 190
QLISP, 190
Qualitative reasoning, 139

R1, 137
Reasoning by analogy, 328–329
RETE algorithm, 16
ROGET, 146
Rule-based systems:
  advantages, 48
  components of, 31–32, 46–47
  conflict resolution, 39–40
  control strategies, 33–39
  development of, 45–46
  disadvantages, 49–50, 200–202, 340, 342
  integration with databases, 49–50
  knowledge representation, 16, 29–30
  uncertainty, 43–45

Rule-based systems (*Cont.*):
   validation, 40–43
Rule models, 113–115

Schema, 178
Script, 238–239, 335
Semantic networks, 12–13, 21–22
Shallow knowledge, 137–138
Similarity-based learning, 103–104, 106–109, 112
SIPE, 161, 178, 180–181, 187–189
Situation calculus, 171
SRL, 191
STRIPS, 158, 172–173, 181
Structural matching, 241, 247–252

Teiresias, 113–115
Teleology, 173, 178, 187
Theorem proving, 9–10, 16, 23, 25, 35, 170–171, 173, 250

Traveling salesperson problem, 221–223
Truth maintenance, 335
Turing machine, 28
TWEAK, 161

Unary functions, 21
Uncertainty, 43–45
UNITS, 190
Unsupervised learning, 214–216

Validation of knowledge base, 40–43
VLSI circuit design, 112

Walsh functions, 219–220
Walsh holographic memory, 222–224
Warplan, 170, 176, 187, 189–190
Working memory, 16, 32

XCON, 105, 137, 143